The Designer's Guide to the Cortex-M Processor Family

The Designer's Guide to the Cortex-M Processor Family

Third Edition

Trevor Martin

Hitex (UK) Ltd., Coventry, England, United Kingdom

Newnes is an imprint of Elsevier
The Boulevard, Langford Lane, Kidlington, Oxford OX5 1GB, United Kingdom
50 Hampshire Street, 5th Floor, Cambridge, MA 02139, United States

ISBN: 978-0-323-85494-8

For Information on all Newnes publications
visit our website at https://www.elsevier.com/books-and-journals

Publisher: Mara E. Conner
Acquisitions Editor: Tim Pitts
Editorial Project Manager: Zsereena Rose Mampusti
Production Project Manager: Erragounta Saibabu Rao
Cover Designer: Christian J. Bilbow

Typeset by MPS Limited, Chennai, India

Contents

About the author

Trevor Martin is a senior technical specialist in Hitex UK, where he has worked over 25 years. Trevor has worked with a wide range of microcontrollers and associated development tools. Since the launch of the Cortex-M3 processor in 2004, Trevor has contributed numerous articles and application notes for many of the leading Cortex-M-based microcontrollers. Having an extensive knowledge of the Cortex-M processor family, Trevor is also familiar with many of the development techniques, application software, and communication protocols required for today's embedded applications.

Foreword

This is the third edition of a book that has become a dear friend to many software developers using Arm Cortex-M-based microcontrollers. When it was first published in 2013, the dominance of the Arm architecture in all things embedded was not in sight and the term "IoT" was not an everyday term.

Since the beginnings in the mid-2000s, Arm Cortex-M-based microcontrollers have evolved from low-power 16-bit replacements to high-performance processors running up to 1 GHz. The list of supported devices on Arm's device database is close to 10,000 entries—all applications can be covered with a specific MCU from one of Arm's silicon partners. But not only the computing performance was increased, also many interfaces, especially for Internet connectivity, have been added even to the tiniest device. Adding safety and security features while maintaining the low-power heritage and low-cost advantage is a big challenge for software developers.

Arm's "Common Microcontroller Software Interface Standard" or in short CMSIS helps to simplify software reuse, reduce the learning curve for microcontroller developers, speed up project build and debug, and thus reduce the time to market for new applications. It plays an important role in every Cortex-M software developer's day-to-day work. From its humble beginnings in 2009 up to now, it grew from a vendor-independent hardware abstraction layer to a set of tools, APIs, frameworks, and flows.

CMSIS is supported throughout Arm's wide ecosystem. It is an integral part of many SDKs and IDEs that are available to developers. Lots of third-party software vendors (including the major cloud service providers) deliver their software stacks as Open-CMSIS-Packs. New development flows built on these packs help developers to maintain their software and enable CI/CD DevOps flows. Using Arm Virtual Hardware, this enables automated build and test in the cloud. These new flows ensure high software quality throughout the whole development process.

While it has become an impressive toolbox software developers can choose from, it is not always easy to understand which CMSIS component is used and when. This book's tutorial-based approach helps developers to understand the various components and how to use them effectively.

Trevor is an experienced professional and has held hundreds of workshops on embedded development. His book is a condensed version of his expertise and takes you through all steps that are required to create embedded applications successfully. Being very hands-on, both new and experienced developers can learn new approaches to embedded software development. I hope you enjoy reading it as much as I did.

Christopher Seidl

Preface

ARM first introduced the Cortex-M processor family in 2004. Since then, the Cortex-M processor family has gained wide acceptance as general-purpose processors for small microcontrollers. At the time of writing, there are several thousands of standard microcontrollers that feature the Cortex-M processor and the pace of development shows no sign of slowing down. The Cortex-M processor is now well established as an industry standard architecture for embedded systems. As such, the knowledge of how to use it is becoming a requisite skill for professional developers. This book is intended as both an introduction to the Cortex-M processor family and a guide to the techniques used to develop an application software to run on them. The book is written as a tutorial, and the chapters are intended to be worked through in order. Each chapter contains a set of examples that present the key principles outlined in this book using a minimal amount of code. Each example is designed to be built with the community edition of the Keil MDK. These examples run in a simulator so that you can use this book without any additional hardware. That said the examples can also be run on low-cost hardware modules that are widely available through the Internet.

This book can be divided into two sections. In this first section, we will examine the Cortex-M processors, and the software support and tools used to develop the application code.

Chapter 1 provides an introduction and feature overview of each processor in the Cortex-M family.

Chapter 2 introduces the basics of building a C project for a Cortex-M processor.

Chapter 3 provides an architectural description of the Cortex-M3 and its differences from the other Cortex-M processors.

Chapter 4 introduces the CMSIS programming standard for Cortex-M processors.

Chapter 5 extends Chapter 3 by introducing the more advanced features of the Cortex-M architecture.

Chapter 6 covers the Cortex-M7 and introduces its new architectural features and extended memory model.

Chapter 7 discusses on the next generation of Cortex-M processors introduced by the Armv8-M architecture and the new extensions to the existing programmer's model.

Chapter 8 provides a description of the CoreSight debug system and its real-time features.

Chapter 9 discusses the math and DSP support available on the Cortex-M4 and how to design real-time DSP applications.

In the second half of this book, we will look more closely at how to construct a software program for a Cortex-M processor. This section emphasizes the use of an RTOS-based layered software architecture that promotes software testing and code reuse.

Chapter 10 introduces the use of an RTOS on a Cortex-M processor. This is a key framework that helps to design and manage more complex codebases.

Chapter 11 examines some real-world techniques that can be used when developing an RTOS-based project.

Chapter 12 discusses on the CMSIS Driver specification. It starts with installing, followed by validating and using a standard driver. We will next look at how a CMSIS driver is constructed and how to extend the range of drivers with a custom driver profile.

Chapter 13 provides an introduction to software testing on a Cortex-M microcontroller and introduces an important design technique called Test-Driven Development (TDD).

Chapter 14 looks at designing reusable software components within a TDD framework. The CMSIS Pack specification is used to create a device-independent component that can be installed into the toolchain.

Chapter 15 defines a layered software architecture based on the techniques learned in the preceding chapters.

With Chapter 16 (the final chapter), we will look just over the horizon at near-future developments in silicon and software.

This book is useful for students, beginners, and advanced and experienced developers alike. However, it is assumed that you have a basic knowledge of how to use microcontrollers and that you are familiar with the the 'C' programming language. In addition, it is helpful to have basic knowledge of how to use the μVision debugger and IDE.

Trevor Martin

Acknowledgments

I would like to thank Tim Pitts, Zsereena Rose Mampusti, and Saibabu Rao Erragounta of Elsevier and Joseph Yui of ARM along with Christopher Seidl and all the team at Keil.

CHAPTER 1

Introduction to the Cortex-M Processor Family

Introduction

The objective of this book is to provide you with a fundamental understanding of the Cortex-M processor family and an overview of the essential software libraries, standards, development techniques, and tools required to design effective and efficient application code.

Book Structure

This book is arranged as a tutorial and it is best to work through it chapter by chapter. Each chapter contains a number of hands-on examples that use the community edition of the Keil MDK toolchain. The Keil MDK is the industry-leading reference toolchain for Cortex-M microcontrollers and the community edition license allows unrestricted use for noncommercial projects. The MDK debugger also provides a sophisticated simulator that allows you to run most of the examples in this book without the need for additional hardware. In the initial chapters, we will look at the general Cortex-M programmer's model and the features of each Cortex-M processor. Alongside the exploration of the processor hardware, we will see how to access these features from our development software. We will also review a set of specifications called the "Common Microcontroller Software Interface Standard", which provides software and tools that enable software development over the whole range of Cortex-M processors. In the second half of the book, we will look at more sophisticated software development techniques and then in the last few chapters bring these together to create a modular component-based general-purpose software architecture that can be used to develop a wide range of applications. So let's make a start with an overview of the current range of Cortex-M processors.

External URL

Throughout the book, there are a lot of external links and references. For the paper copy of this book, there is a pdf that contains a set of clickable links for each chapter. This pdf can be downloaded from the GitHub repository which contains the book example set (see Chapter 2: Developing Software for the Cortex-M Family). If any of the URL links stop working, I will update the pdf with an alternative.

The Designer's Guide to the Cortex-M Processor Family.
DOI: https://doi.org/10.1016/B978-0-323-85494-8.00016-4

Cortex-M Processor Family

In the late 1990s and early noughties, several Silicon vendors adopted the now "classic" ARM7 and ARM9 processors as the CPU for a new generation of high-performance microcontrollers. While this represented a huge step forward in performance, both the ARM7 and ARM9 were only CPUs each silicon vendor had to provide their own bus structure, interrupt handling, and power management. So while the CPU was the same across different families, every implementation was different and the overall market remained very fragmented. With the introduction of the next-generation Cortex processor, this all changed. In addition to all its technical benefits, the Cortex-M processor provides a standard core that is the same to use across all the different manufacturer's devices. This creates a range of benefits which are summarized below.

- Learn to use the Cortex-M processors once then reuse that knowledge many times.
- Create industry-wide standards to accelerate code development.
- A standard core means that silicon vendors must innovate in order to differentiate their devices from competitors.
- A standard processor creates a wide and deep ecosystem for supporting software, tools, and hardware.

Cortex Profiles

In 2004 Arm introduced its new Cortex family of processors. The Cortex processor family is subdivided into three different profiles (Fig. 1.1). Each profile is optimized for different segments of embedded systems applications.

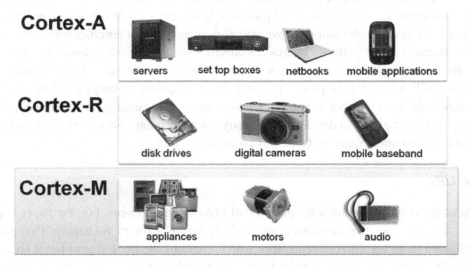

Figure 1.1
The Cortex processor family has three profiles: Application, Real Time, and Microcontroller.

The Cortex-A profile has been designed as a high-end application processor. Cortex-A processors are capable of running feature-rich operating systems (OSs) such as embedded versions of Windows and Linux. The key applications for Cortex-A are consumer electronics such as smartphones, tablet computers, and set-top boxes. The second Cortex profile is Cortex-R. This is the real-time profile that delivers a high-performance processor which is the heart of an application-specific device. Very often, a Cortex-R processor forms part of a "system-on-chip" design that is focused on a specific task such as hard disk drive control, automotive engine management, and medical devices. The final profile is Cortex-M or the microcontroller profile. Unlike earlier Arm CPUs, the Cortex-M processor family has been designed specifically for use within a small microcontroller to provide high-performance processing and real-time deterministic interrupt handling coupled with low-power consumption.

Cortex-M Hardware Architectures

The Cortex-M processor family is spread across four architectural revisions that provide a range of features and extensions to the core processor. The four revisions are Armv6-M, Armv7-M, Armv8-M, and Armv8.1-M. Each of these architectures has the same core programmer's model and is upward compatible with the next architectural revision. Fig. 1.2 shows the key processor features available to each architectural revision.

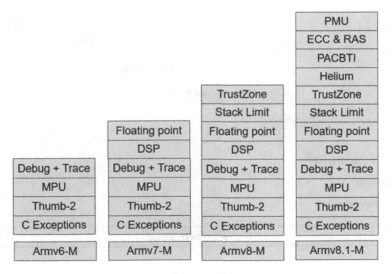

Figure 1.2
The Cortex-M profile has five different variants with a common programmer's model.

The full range of Cortex-M processors is shown in Fig. 1.3. Starting with the simplest devices the Armv6-M architecture contains three processors: Cortex-M0, Cortex-M1, and Cortex-M0 + . The Cortex-M0 and Cortex-M0 + are the smallest processors in the family. They allow silicon manufacturers to design low-cost, low-power devices that can replace existing 8-bit microcontrollers while still offering 32-bit performance. The Cortex-M1 has much of the same features as the Cortex-M0 but has been designed as a "soft core" to run inside a Field Programmable Gate Array device.

The next group of Cortex-M processors exists within the Armv7-M architecture. These are the Cortex-M3, Cortex-M4, and Cortex-M7. The Cortex-M3 is the mainstay of the Cortex-M family and was the first Cortex-M variant to be launched. It has enabled a new generation of high-performance 32-bit microcontrollers that can be manufactured at a very low cost. Today, there are many Cortex-M3-based microcontrollers available from a wide variety of silicon manufacturers. This represents a seismic shift where Cortex-M-based microcontrollers have essentially replaced traditional 8/16-bit microcontrollers and even other 32-bit microcontrollers. The next highest performing member of the Cortex-M family is the Cortex-M4. This has all the features of the Cortex-M3 and adds support for digital signal processing (DSP). The Cortex-M4 also includes hardware floating-point support for single-precision calculations. The Corex-M7 is the Armv7-M architecture processor with the highest level of performance while still maintaining the Cortex-M programmer's model. The Cortex-M7 has also been designed for use in high reliability and safety-critical systems.

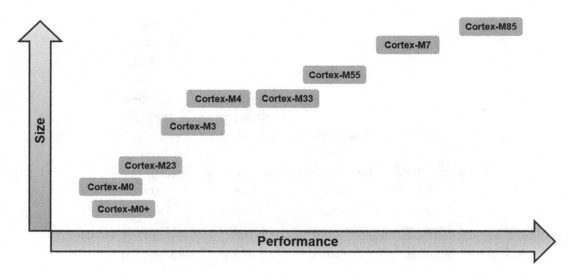

Figure 1.3
The Cortex-M profile has nine different variants with a common programmer's model.

The next generation of Cortex-M processors was introduced with the Armv8-M Architecture. The Armv8-M architecture introduces a mainline (Full) processor in the form of Cortex-M33 and a baseline (minimal) processor with Cortex-M23. The Armv8-M architecture also introduced Stack limit checking and the Trust Zone security peripheral. Stack limit checking is mainly used as a security extension and provides hardware monitoring of the processor stack overflows. The TrustZone security peripheral is used to create IoT devices that are secure against common software hacking techniques without the need to add a separate secure element processor.

The latest architectural revision is Armv8.1-M. The Armv8.1-M architecture introduced the Helium Vector Processing extension that supports single-cycle arithmetic operations for multiple integer or floating-point vectors. The Armv8.1-M architecture also provides additional important safety features called Error-Correcting Codes (ECC) and Reliability Availability and Serviceability (RAS), to monitor the condition of the processor during run time. The Armv8.1-M architecture processors and Caches are monitored using ECC and failures are reported through a standardized RAS interface. The Armv8.1-M also contains a powerful Performance Monitoring Unit (PMU) that can be configured to send detailed diagnostics to an external monitoring tool. Finally, the Armv8.1 includes an important security extension. The "Pointer Authentication Code Branch Target Identification" (PACBTI) extension is designed to detect manipulation of pointer addresses through common hacking techniques such as Stack Smashing and Return Orientated Programming. The Armv8.1-M architecture currently contains two processors: Cortex-M55 and Cortex-M85. These are both very new and at the time of writing and Cortex-M55 microcontrollers are just being released. The Cortex-M85 has also just been announced by Arm so real silicon is a couple of years away, but Cortex-M85-based devices will be the highest performing Cortex-M microcontrollers.

We can now have a closer look at each processor. The next section will introduce the Armv7-M processors first and in particular, the Cortex-M3 as this processor serves as a good benchmark for the rest of the Cortex-M processors. Once you understand the Cortex-M3, it is easy to scale up or down to the other processors.

Armv7-M

Cortex-M3

Today, the Cortex-M3 is the most widely used of all the Cortex-M processors. This is partly because it has been available not only for the longest period of time but also because it meets the requirements for a general-purpose microcontroller. This typically means it has a good balance between high performance, low-power consumption, and low cost (Fig. 1.4).

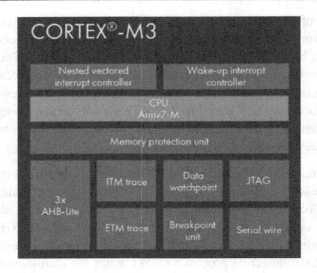

Figure 1.4
The Cortex-M3 was the first Cortex-M device available. It is a complete processor for a general-purpose microcontroller.

The heart of the Cortex-M3 is a high-performance 32-bit CPU. The Cortex-M3 is a reduced instruction set computer where most instructions will execute in a single cycle.

Figure 1.5
The Cortex-M3 CPU has a three-stage pipeline with branch prediction.

This is partly made possible by a three-stage pipeline with separate fetch decode and execute units (Fig. 1.5).

Figure 1.6
The Cortex-M3 CPU can execute most instructions in a single cycle. This is achieved by the pipeline executing one instruction, decoding the next, and fetching a third.

So while one instruction is being executed, a second is being decoded, and a third is being fetched (Fig. 1.6). This is great when the code is going in a straight line, however, when the program branches, the pipeline must be flushed and refilled with new instructions before execution can continue. This could potentially make branches quite expensive in terms of processing power. However, the Cortex-M3 and Cortex-M4 include an instruction fetch unit that can handle speculative branch target fetches which can reduce the bench penalty. The Cortex-M7 includes a full Branch Target Address Cache unit which is even more efficient. This helps the Cortex-M3 to have a sustained processing power of 1.24 DMIPS/MHz, while the Cortex-M7 reaches 2.31 DMIPS/MHz. In addition, the Cortex-M3 processor has a hardware integer math unit with hardware divide and single-cycle multiply. The Cortex-M3 processor also includes a Nested Vector Interrupt Controller (NVIC) that can service up to 240 interrupt sources. The NVIC is common across all Cortex-M processors and provides fast deterministic interrupt handling. From an interrupt being raised to reaching the first line of "C" in the interrupt service routine takes just 12 cycles every time. The NVIC also contains a standard timer called the SysTick timer. This is a 24-bit countdown timer with auto-reload. This timer is present on all of the different Cortex-M processors and is used to provide regular periodic interrupts. A typical use of this timer is to provide a timer tick for small footprint Real-Time Operating Systems (RTOS). We will have a look at such an RTOS in Chapter 10, Using a Real-Time Operating System. Also, next to the NVIC is the Wake-up Interrupt Controller (WIC); this is a small area of the Cortex-M processor that is kept alive when the processor is in low-power mode. The WIC can use the interrupt signals from the microcontroller peripherals to wake up the Cortex-M processor from a low-power mode. The WIC can be implemented in various ways and in most cases does not require a clock to function; also, it can be in a separate power region from the main Cortex-M processor. This allows 99% of the Cortex-M processor to be placed in a low-power mode with just minimal current being used by the WIC.

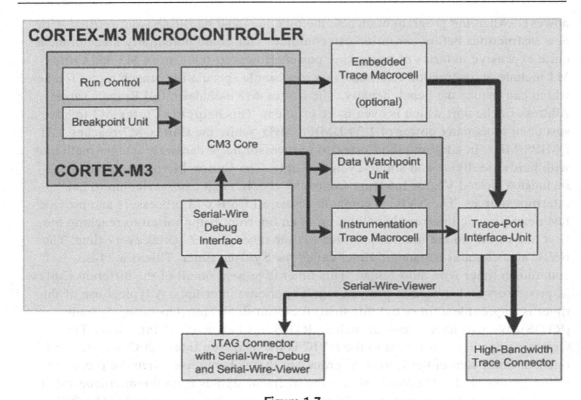

Figure 1.7
The Cortex-M debug architecture is independent of the CPU and contains up to three real-time
trace units in addition to the run control unit.

One thing that is commonly missed when looking at a manufacturer's data sheet is that the
Cortex-M family also has a very advanced debug architecture called CoreSight (Fig. 1.7).
This should not be confused with the earlier Joint Test Action Group (JTAG) debug
interface found on the "classic" ARM7/9 processors. The JTAG interface provided a means
to download the application code into the on-chip flash memory and then exercise the code
with basic run/stop debugging. While a JTAG debugger provided a low-cost way of
debugging, it had two major problems. The first was a limited number of breakpoints,
generally, two with one being required for single-stepping code, and second, when the CPU
was executing code, the microcontroller became a black box with the debugger having no
visibility to the CPU, memory, or peripherals until the microcontroller was halted. The
CoreSight debug architecture within the Cortex-M processors is much more sophisticated
than the old ARM7 or ARM9 processors. It allows up to eight hardware breakpoints to be
placed in code or data regions. CoreSight also provides three separate trace units that
support advanced debug features without intruding on the execution of the Cortex CPU.
The Cortex-M3 and Cortex-M4 are always fitted with a Data Watchpoint and Trace (DWT)

unit and an Instrumentation Trace Macrocell (ITM) unit. The debug interface allows a low-cost debugger to view the contents of memory and peripheral registers "on the fly" without halting the CPU, and the DWT can stream the contents of program variables in real time without using any processor resources. The ITM is a second trace unit that provides a debug communication method between the running code and the debugger user interface. During development, the standard IO channel can be redirected to a console window in the debugger. This allows you to instrument your code with printf() debug messages that can then be read in the debugger while the code is running. This can be useful for trapping complex runtime problems. The ITM is also very useful during software testing as it provides a way for a test harness to dump data to the PC without needing any specific hardware on the target. The ITM is actually more complex than a simple UART, as it provides 32 communication channels that can be used by different resources within the application code. For example, we can provide extended debug information about the performance of an RTOS by placing the code in the RTOS kernel that uses an ITM channel to communicate with the debugger. The final trace unit is called the Embedded Trace Macrocell (ETM). The ETM is an optional fit and is not present on all Cortex-M devices. Generally, a manufacturer will fit the ETM on their high-end microcontrollers to provide extended debug capabilities. The ETM provides instruction trace information that allows the debugger to build an assembler and High-Level Language trace listing of the code executed. The ETM also enables more advanced tools such as code coverage monitoring and timing performance analysis. These debug features are often a requirement for safety-critical and high-integrity code development.

Cortex-M4

The Cortex-M4 is an enhanced version of the Cortex-M3. The additional features on the Cortex-M4 are focused on supporting DSP algorithms. Typical algorithms are transforms such as Fast Fourier Transform, digital filters such as Finite Impulse Response filters, and control algorithms such as a Proportional Internal Differential control loop. With its DSP features, the Cortex-M4 has created a new generation of Arm-based devices that can be characterized as Digital Signal Controllers. These devices allow you to design applications that combine microcontroller-type functions with real-time signal processing. In Chapter 9, Practical DSP for Cortex-M Microcontrollers, we will look at the Cortex-M4 DSP extensions in more detail and also how to construct software that combines real-time signal processing with typical event-driven microcontroller code (Fig. 1.8).

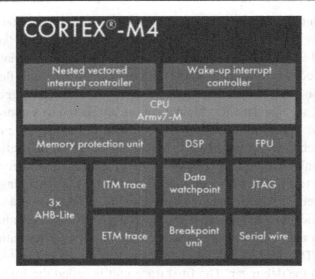

Figure 1.8
The Cortex-M4 is fully compatible with the Cortex-M3 but introduces a hardware floating-point unit and additional DSP instructions.

The Cortex-M4 has the same basic structure as the Cortex-M3 with the same CPU programmer's modes, NVIC, CoreSight debug architecture, MPU, and bus interface. The enhancements over the Cortex-M3 are partly to the instruction set where the Cortex-M4 has additional DSP instructions in the form of SIMD instructions. The hardware multiply accumulate has also been improved so that many of the 32×32 arithmetic instructions are single cycle (Table 1.1).

Table 1.1: MAC arithmetic operations and instructions

Operation	Instructions
$16 \times 16 = 32$	SMULBB, SMULBT, SMULTB, SMULTT
$16 \times 16 + 32 = 32$	SMLABB, SMLABT, SMLATB, SMLATT
$16 \times 16 + 64 = 64$	SMLALBB, SMLALBT, SMLALTB, SMLALTT
$16 \times 32 = 32$	SMULWB, SMULWT
$(16 \times 32) + 32 = 32$	SMLAWB, SMLAWT
$(16 \times 16) \pm (16 \times 16) = 32$	SMUAD, SMUADX, SMUSD, SMUSDX
$(16 \times 16) \pm (16 \times 16) + 32 = 32$	SMLAD, SMLADX, SMLSD, SMLSDX
$(16 \times 16) \pm (16 \times 16) + 64 = 64$	SMLALD, SMLALDX, SMLSLD, SMLSLDX
$32 \times 32 = 32$	MUL
$32 \pm (32 \times 32) = 32$	MLA, MLS
$32 \times 32 = 64$	SMULL, UMULL
$(32 \times 32) + 64 = 64$	SMLAL, UMLAL
$(32 \times 32) + 32 + 32 = 64$	UMAAL
$32 \pm (32 \times 32) = 32$ (upper)	SMMUL, SMMULR
$(32 \times 32) = 32$ (upper)	SMMLA, SMMLAR, SMMLS, SMMLSR

The Cortex-M4 has a set of SIMD instructions aimed at supporting DSP algorithms. These instructions allow a number of parallel arithmetic operations in a single processor cycle (Fig. 1.9).

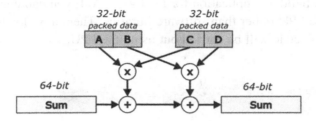

Figure 1.9
The SIMD instructions can perform multiple calculations in a single cycle.

The SIMD instructions work with 16-bit or 8-bit data which has been packed into 32-bit word quantities. So, for example, we can perform two 16-bit multiplies and sum the result into a 64-bit word. It is also possible to pack the 32-bit works with 8-bit data and perform a quad 8-bit addition or subtraction. As we will see in Chapter 9, Practical DSP for Cortex-M Microcontrollers, the SIMD instructions can be used to vastly enhance the performance of DSP algorithms such as digital filters that are basically performing lots of multiply and sum calculations on a pipeline of data. The Cortex-M4 processor may also be fitted with a hardware FPU. This choice is made at the design stage by the microcontroller vendor, so like the ETM and MPU, you will need to check the microcontroller datasheet to see if it is present. The FPU within the Cortex-M4 supports single-precision floating-point arithmetic calculations using the IEEE 754 standard. Cycle times for each operation are shown in Table 1.2.

Table 1.2: Cycle times for FPU arithmetic operations

Operation	Cycle Count
Add/Subtract	1
Divide	14
Multiply	1
Multiply Accumulate	3
Fused MAC	3
Square Root	4

On small microcontrollers, floating-point math has always been performed by software libraries provided by the compiler tool. Typically, such libraries can take hundreds of instructions to perform a floating-point multiply. So, the addition of floating-point hardware that can do the same calculation in a single cycle gives an unprecedented performance boost. The FPU can be thought of as a coprocessor that sits alongside the Cortex-M4 CPU. When a calculation is performed, the floating-point values are transferred directly from the FPU registers to and from the SRAM memory store, without the need to use the CPU registers. While this may sound involved the entire FPU transaction is managed by the compiler. When you build an application for the Cortex-M4, you can compile code to automatically use the FPU rather than software libraries. Then, any floating-point calculations in your C code will be carried out using the FPU.

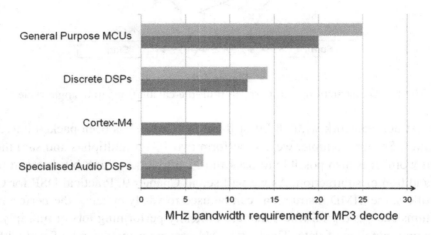

Figure 1.10
MP3 decode benchmark.

With optimized code, the Cortex-M4 can run DSP algorithms far faster than standard microcontrollers and even some dedicated DSP devices (Fig. 1.10). Of course, the weasel word here is "optimized," this means having a good knowledge of the processor and the DSP algorithm you are implementing and then hand-coding the algorithm making use of compiler intrinsics to get the best level of performance. Fortunately, Arm provides a fully open-source DSP library that implements many commonly required DSP Algorithms as easy-to-use library functions. We will look at using this library in Chapter 9, Practical DSP for Cortex-M Microcontrollers.

Cortex-M7

The Cortex-M7 is the highest performance Armv7-M architecture processor currently available. This is actually a bit of an understatement. The benchmark figures for the Cortex-M7 versus the Cortex-M4 are shown below in Table 1.3. Please note that these figures are shown per MHz of the CPU frequency. As well as significantly outperforming the Cortex-M4, the Cortex-M7 can run at much higher frequencies currently in excess of 1 GHz. In short, it leaves the Cortex-M4 in the dust. However, the Cortex-M7 still maintains the

Cortex-M programmers model so if you have used an earlier Cortex-M processor moving to the Cortex-M7 is not a major challenge.

Table 1.3: Cortex-M4 versus Cortex-M7 benchmark

Benchmark	Cortex-M4	Cortex-M7
CoreMark/MHz	3.54	2.31
DMIPS/MHz	1.26	5.29

The Cortex-M7 achieves this boost in performance levels with some architectural enhancements and a more sophisticated memory system. The Cortex-M7 CPU has a superscalar architecture; this means that it has two parallel three-stage pipelines which can dual-issue instructions. The CPU is also capable of processing different groups of instructions in parallel. The CPU also has a "Branch Target Address Cache" that improves the performance of high-level statements such as conditional branches and more importantly loops. When the same source code is compiled for a Cortex-M4 and a Cortex-M7, it will take significantly fewer cycles to run on the Cortex-M7 (Fig. 1.11).

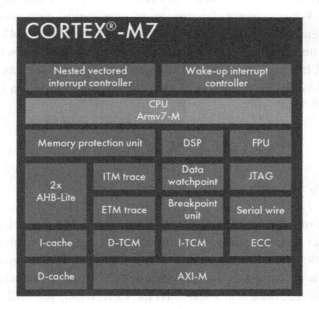

Figure 1.11
CortexM7 Processor.

The Cortex-M7 also introduces a more complex bus structure. The first Arm-based microcontrollers used an ARM7 processor that used a bus called the Advanced

High-Performance bus (AHB). As the complexity of microcontrollers grew, this bus became a bottleneck when there were several bus masters (CPU and DMA units) within the microcontroller. With the introduction of the Cortex-M family, the AHB was upgraded to the AHB matrix lite. This in effect is a set of busses that allow multiple bus masters to access memory in parallel. Internal bus arbitration only occurs if two bus masters try to access the same group of peripherals or block of memory. The Cortex-M7 introduces the Advanced Extensible Interface (AXI-M). The AXI-M is a high-performance 64-bit interface that supports multiple outstanding memory transactions. It also opens up a lot of possibilities for silicon developers to design multicore systems and is a step toward network-on-chip designs.

For a Developer, the most significant difference between a Cortex-M4 and a Cortex-M7 is a more complex memory system. In order for the Cortex-M7 processor to achieve very high levels of performance, it has a memory hierarchy. The CPU has two regions of memory called the Instruction and Data Tightly Coupled Memories (TCM). The I-TCM and D-TCM are blocks of zero wait-state memory which can be up to 64 MB in size. This ensures that any critical routines or data can be accessed by the processor without any delays. The processor can also include two caches. These units provide up to 64 kB of Instruction and Data cache for system memory which is located on the AXI-M bus. When you are developing an application, you need to understand and manage this memory system. Like the Cortex-M4 the Cortex-M7 has additional DSP capability in the form of SIMD instructions and can be fitted with a single or double precession Floating-Point Unit. The Cortex-M7 processor can be synthesized with additional safety features such as "Error-Correcting Codes" on its bus interfaces and a "Built-In Self Test" unit.

Armv6-M

Cortex-M0

The Cortex-M0 was introduced a few years after the Cortex-M3 was released. The Cortex-M0 is a much smaller processor than the Cortex-M3 and can be as small as 12k gates in its minimum configuration. The Cortex-M0 is typically designed into microcontrollers that are intended to be very low-cost devices and\or intended for low-power operation. However, the important thing is that once you understand the Cortex-M3, you will have no problem using the Cortex-M0; the differences are mainly transparent to high-level languages (Fig. 1.12).

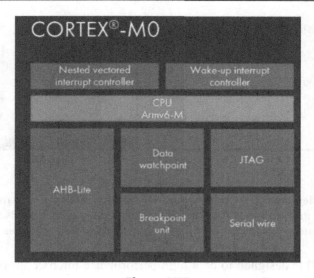

Figure 1.12
The Cortex-M0 is a reduced version of the Cortex-M3 while still keeping the same programmer's model.

The Cortex-M0 processor has a CPU that can execute a subset of the Thumb-2 instruction set. Like the Cortex-M3, it has a three-stage pipeline but no branch speculation fetch, therefore branches and jumps within the code will cause the pipeline to flush and refill before execution can resume. The Cortex-M0 also has a Von Neumann bus architecture, so there is a single path for code and data. While this makes for a simple design, it can become a bottleneck and reduce performance. Compared to the Cortex-M3, the Cortex-M0 achieves 0.96 DMIPS/MHz, which while less than the Cortex-M3 is still about the same as an ARM7 which has three times the gate count. So, while the Cortex-M0 is at the bottom end of the Cortex-M family, it still packs a lot of processing power. The Cortex-M0 processor has the same NVIC as the Cortex-M3, but it is limited to a maximum of 32 interrupt lines from the microcontroller peripherals. The NVIC also contains the SysTick timer that is fully compatible with the Cortex-M3. Most RTOS that run on the Cortex-M3 and Cortex-M4 will also run on the Cortex-M0, though the vendor will need to do a dedicated port and recompile the RTOS code. As a developer, the biggest difference you will find between using the Cortex-M0 and the Cortex-M3 is within its debug capabilities. While on the Cortex-M3 and Cortex-M4 there is extensive real-time debug support, the Cortex-M0 has a more modest debug architecture. On the Cortex-M0, the DWT unit does not support data-trace and the ITM is not fitted, so we are left with basic run control (i.e., run, halt, single stepping and breakpoints, and watchpoints) and on-the-fly memory/ peripheral accesses. This is still an enhancement from the JTAG support provided on ARM7 and ARM9 CPUs.

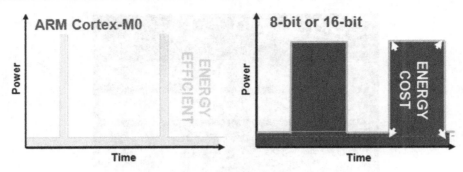

Figure 1.13
The Cortex-M0 is designed to support low-power standby modes. Compared to 8- or 16-bit MCU, it can stay in sleep mode for a much longer time because it needs to execute fewer instructions than an 8/16-bit device to achieve the same result.

While the Cortex-M0 is designed to be a high-performance microcontroller processor, it has a relatively low gate count. This makes it ideal for both low-cost and low-power devices. The power consumption of the Cortex-M0 in a typical 90 nM fabrication technology is 12.5 μW/MHz when running and almost zero when in its low-power sleep mode. While other 8-bit and 16-bit architectures can also achieve similar low-power figures, they need to execute far more instructions than the Cortex-M0 to achieve the same end result (Fig. 1.13). This means extra cycles, and extra cycles would mean more power consumption. If we pick a good example for Cortex-M0 such as a 16 × 16 multiply, then the Cortex-M0 can perform this calculation in one cycle. In comparison, a typical 8-bit architecture like the 8051 will need at least 48 cycles and a 16-bit architecture will need 8 cycles. This is not only a performance advantage but also an energy efficiency advantage as well (Table 1.4).

Table 1.4: Number of cycles taken for a 16 × 16 multiply against typical 8- and 16-bit architectures

8-bit Example (8051)		16-bit Example	ARM Cortex-M

MOV A, XL ; 2 bytes
MOV B, YL ; 3 bytes
MUL AB; 1 byte
MOV R0, A; 1 byte
MOV R1, B; 3 bytes
MOV A, XL ; 2 bytes
MOV B, YH ; 3 bytes
MUL AB; 1 byte
ADD A, R1; 1 byte
MOV R1, A; 1 byte
MOV A, B ; 2 bytes
ADDC A, #0 ; 2 bytes
MOV R2, A; 1 byte
MOV A, XH ; 2 bytes
MOV B, YL ; 3 bytes
	MUL AB; 1 byte
	ADD A, R1; 1 byte
	MOV R1, A; 1 byte
	MOV A, B ; 2 bytes
	ADDC A, R2 ; 1 bytes
	MOV R2, A; 1 byte
	MOV A, XH ; 2 bytes
	MOV B, YH ; 3 bytes
	MUL AB; 1 byte
	ADD A, R2; 1 byte
	MOV R2, A; 1 byte
	MOV A, B ; 2 bytes
	ADDC A, #0 ; 2 bytes
	MOV R3, A; 1 byte
		MOV R1,&MulOp1
		MOV R2,&MulOp2
		MOV SumLo,R3
		MOV SumHi,R4
		
		(Memory mapped
		 multiply unit)
			MULS r0,r1,r0

Time: 48 clock cycles* **Code size:** 48 bytes		**Time:** 8 clock cycles **Code size:** 8 bytes	**Time:** 1 clock cycle **Code size:** 2 bytes

*1 clock cycle = oscillator/12.

Like the Cortex-M3, the Cortex-M0 also has the WIC feature. While the WIC is coupled to the Cortex-M0 processor, it can be placed in a different power domain within the microcontroller (Fig. 1.14).

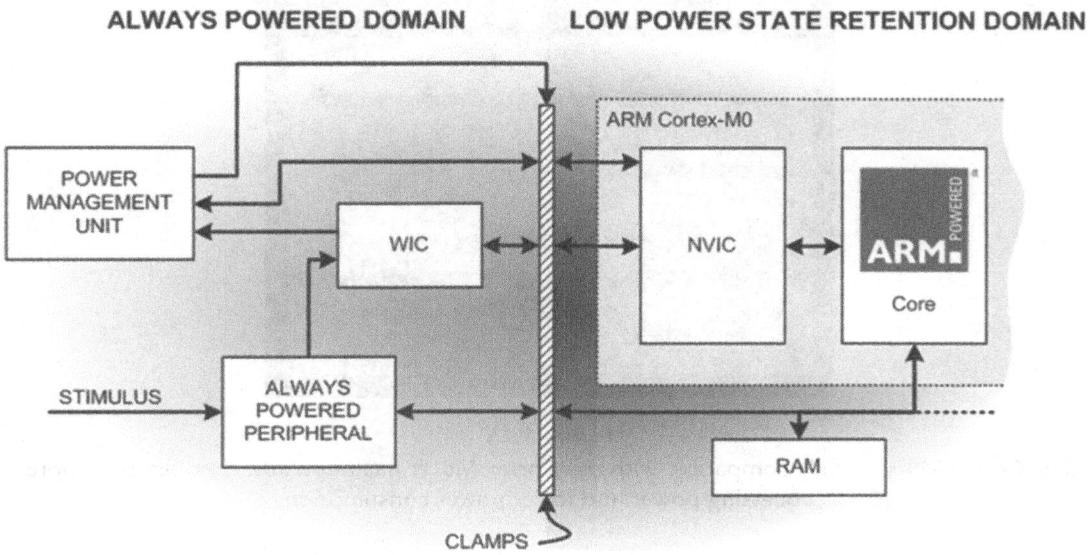

ALWAYS POWERED DOMAIN   **LOW POWER STATE RETENTION DOMAIN**

CLAMPS

Figure 1.14
The Cortex-M processor is designed to enter low-power modes. The WIC can be placed in a separate power domain.

This allows the microcontroller manufacturer to use their expertise to design very low-power devices where the bulk of the Cortex-M0 processor is placed in a dedicated low-power domain which is isolated from the microcontroller peripheral power domain. These kinds of architected sleep states are critical for designs that are intended to run from batteries.

Cortex-M0 +

The Cortex-M0 + processor is the second generation ultra-low-power Cortex-M core. It has complete instruction set compatibility with the Cortex-M0 allowing you to use the same compiler and debug tools. As you might expect, the Cortex-M0 + has some important enhancements over the Cortex-M0 (Fig. 1.15).

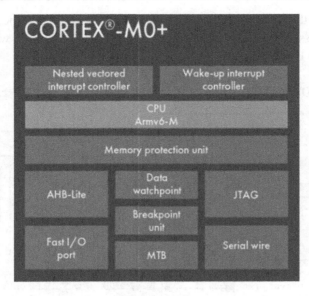

Figure 1.15

The Cortex-M0 + is fully compatible with the Cortex M0. It has more advanced features more processing power and lower power consumption.

The defining feature of the Cortex-M0 + is its power consumption, which when fabricated in a 90-nM technology, is just 9.37 μW/MHz compared to 12.5 μW/MHz for the Cortex-M0 and 31 μW/MHz for the Cortex-M3. One of the Cortex-M0 + key architectural changes is a move to a two-stage pipeline. When the Cortex-M0 and Cortex-M0 + execute a conditional branch, the instructions in the pipeline are no longer valid. This means that the pipeline must be flushed every time there is a branch. Once the branch has been taken the pipeline must be refilled to resume execution. While this impacts performance it also means accessing the flash memory and each access costs energy as well as time. By moving to a two-stage pipeline, the number of flash memory accesses and hence the runtime energy consumption is also reduced (Fig. 1.16).

Figure 1.16

The Cortex M0 + has a two-stage pipeline compared to the three-stage pipeline used in other Cortex-M processors.

Another important feature added to the Cortex-M0 + is a new peripheral I/O interface that supports single-cycle access to peripheral registers. The single-cycle I/O interface is a standard part of the Cortex-M0 + memory map and uses no special instructions or paged addressing. Registers located within the I/O interface can be accessed by normal C pointers from within your application code. The I/O interface allows faster access to peripheral registers with less energy use while being transparent to the application code. The single-cycle I/O interface is separate from the advanced high-speed bus (AHB) lite external bus interface, so it is possible for the processor to fetch instructions via the AHB lite interface while making data accessible to the peripheral registers located within the I/O interface (Fig. 1.17).

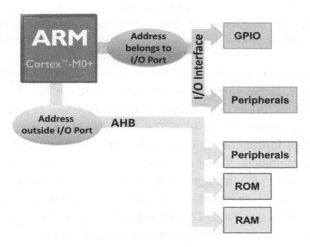

Figure 1.17
The I/O port allows single-cycle access to GPIO and peripheral registers.

The Cortex-M0 + is designed to support fetching instructions from 16-bit flash memories. Since most of the Cortex-M0 + instructions are 16-bit, this does not have a major impact on performance but does make the resulting microcontroller design simpler, smaller, and consequently lower cost. The Cortex-M0 + has some Cortex-M3 features missing from the original Cortex-M0. This includes the MPU, which we will look at in Chapter 5, Advanced Architecture Features, and the ability to relocate the vector table to a different position in memory. These two features provide an improved OS support and support for more sophisticated software designs with multiple application tasks on a single device.

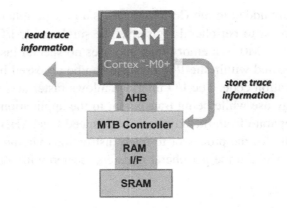

Figure 1.18
The Micro Trace Buffer can be configured to record executed instructions into a section of user SRAM. This can be read and displayed as an instruction trace in the PC debugger.

The Cortex-M0 + also has an improved debug architecture compared to the Cortex-M0. As we will see in Chapter 8, Debugging With CoreSight, it supports the same real-time access to peripheral registers and SRAM as the Cortex-M3 and Cortex-M4. In addition, the Cortex-M0 + has a new debug feature called Micro Trace Buffer (MTB) (Fig. 1.18). The MTB allows executed program instructions to be recorded into a region of SRAM set up by the programmer during development. When the code is halted this instruction trace can be downloaded and displayed in the debugger. This provides a snapshot of code execution immediately before the code was halted. While this is a limited trace buffer, it is extremely useful for tracking down elusive bugs. The MTB can be accessed by standard JTAG/Serial wire debug adaptor hardware, for which you do not need an expensive trace tool.

Armv8-M

The introduction of Armv8-M architectures represents the next generation of Cortex-M processors. Each of the new processors inherits the same programming model as the Armv6/7 architecture processors but introduces significant new extensions to the original processors. Many of the new microcontrollers that use Armv8-M and Armv8.1-M processors are designed to use a 40-nM silicon wafer process that squares the gate count over the earlier Armv6 and Armv7 devices which typically use a 90-nM process. So, we get an improved processor with lower power operation, plus more flash, ram, and many more complex peripherals.

Cortex-M33/M23

The first processors introduced in Armv8-M architecture were the Cortex-M33 which is an Armv8-M Mainline processor roughly equivalent to the Cortex-M4 and the Cortex-M23 which is a reduced baseline processor roughly equivalent to a Cortex-M0 + (Fig. 1.19).

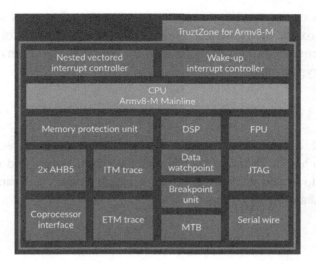

Figure 1.19
CortexM33 Processor.

In both these processors, the most important extension is the TrustZone security peripheral. TrustZone is used to provide a defense against a software or network attack and is essential for any device that will be connected to the internet, specifically IoT devices. TrustZone is used to partition the microcontroller memory resources between a Non-Secure world and a Secure world. The Secure world is used as an isolated security island to protect sensitive data and code (Fig. 1.20).

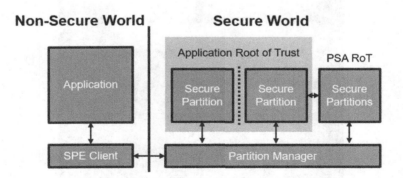

Figure 1.20
TrustZone creates a secure world partition that is used to host a range of security services.

Alongside, the Armv8-M processor's Arm has developed an open-source "The Platform Security Architecture" (PSA). The PSA provides a methodology, security model, Trusted Firmware, and tools to create secure devices.

The Armv8.1-M processors also have a range of additional refinements in the form of new CPU registers that provide hardware stack limit checking and support for tightly integrated co-processors that provide hardware acceleration for common algorithms such as those found in DSP and cryptographic libraries.

Armv8.1-M

The Armv8.1-M architecture adds a number of important processor extensions to create devices with a much higher arithmetic performance for both integer and floating-point calculations. This allows you to create low-power secure devices that can also run advanced DSP and Machine Learning algorithms.

Cortex-M55

The first Cortex processor released in the Armv8.1-M architecture was the Cortex-M55. While broadly similar in features to the Cortex-M4 the Cortex-M55 has a higher CPU benchmark and a supporting memory and bus infrastructure similar to the Cortex-M7 (Fig. 1.21).

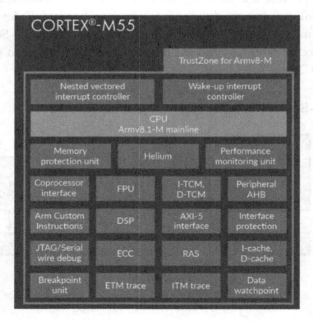

Figure 1.21
Cortex-M55 Processor.

The Cortex-M55 adds a number of CPU extensions most importantly the Helium Cortex-M Vector Extension MVE. While the Cortex-M4 and M7 can perform multiple integer calculations in a single cycle using the SIMD instructions. The MVE configures the Floating-point registers as eight 128-bit vectors that can be packed with integer or floating-point values and processed with a range of arithmetic and complex maths calculations in a single cycle. Like the standard floating-point unit the MVE is supported directly by the compiler and also optimized versions of the Arm DSP library.

On the earlier Cortex-M processors, some silicon manufacturers have implemented hardware protection and monitoring using memory Error Correction codes. Within the Armv8.1-M architecture, Arm has added ECC support to their processors with a standardized reporting interface. The ECC protection is used to monitor cache and TCM memory regions. The condition of the processor is then reported through a standard Reliability, Availability, and Serviceability RAS interface. In addition, there is also a sophisticated PMU that can be configured to report a wide range of different run time processor activity to an external tool.

Cortex-M85

Cortex-M85 is the latest Cortex-M processor to be released by Arm. At the time of writing, it has just been announced and will take a couple of years to find its way into real silicon. While Arm has not released any performance benchmarks, it is clear that it will be the highest performing Cortex-M processor to date (Fig. 1.22).

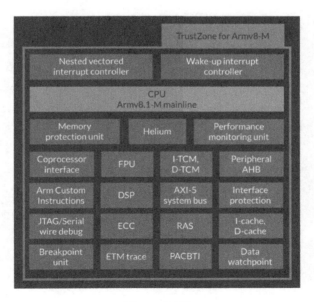

Figure 1.22
CortexM85 Processor.

The Cortex-M85 inherits all the features of the Armv8-M and Armv8.1-M architecture and adds an additional security feature called "PACBTI." This hardware feature is used to defend against common software hacking attacks that aim to corrupt a pointer address or a function return address. Since this type of software attack is the most common form of hacking attempted and also one of the most devastating, this is a very important feature.

Conclusion

This book is covering two topics. An introduction to the Cortex-M processor hardware and also an introduction to developing software for Cortex-M-based microcontrollers. With the introduction of the Cortex-M processor, we now have a low-cost hardware platform that is capable of supporting more sophisticated software design. The last decade has seen the adoption of Real Time Operating Systems (RTOS) and middleware libraries needed to support the more complex peripherals found on Cortex-M devices. So, alongside understanding the low-level features of the Cortex-M processors, we also need to use more sophisticated design techniques and more extensive software testing. Additionally, as the size of the code base for a typical application is ever increasing, we also need to pay more attention to the code structure and overall software architecture.

CHAPTER 2

Developing Software for the Cortex-M Family

Introduction

One of the big advantages of using a Cortex-M processor is that it has a wide and growing range of development tool support. There are toolchains available at zero cost up to several thousand dollars, depending on the depth of your pockets and the type of application you are developing. Today there are five main toolchains that are used for Cortex-M development (Table 2.1).

Table 2.1: Cortex-M processor toolchains

Development Tool
GNU GCC with free and commercial IDE
Greenhills
IAR Embedded Workbench for Arm
Keil Microcontroller Development Kit for Arm (MDK)
Tasking VX Toolset for Arm

Strictly speaking, the GNU GCC is a compiler linker toolchain and does not include an integrated development environment or a debugger. A number of companies have created a toolchain around the GCC compiler by adding their own IDE and debugger to provide a complete development system. Some of these are listed in the appendix. There are quite a few, so this is not intended to be a comprehensive list.

Keil Microcontroller Development Kit

In this tutorial, we are going to use the Keil Microcontroller development Kit (MDK) (Fig. 2.1). The Keil MDK provides a complete development environment for all Cortex-M-based microcontrollers.

The Designer's Guide to the Cortex-M Processor Family.
DOI: https://doi.org/10.1016/B978-0-323-85494-8.00006-1

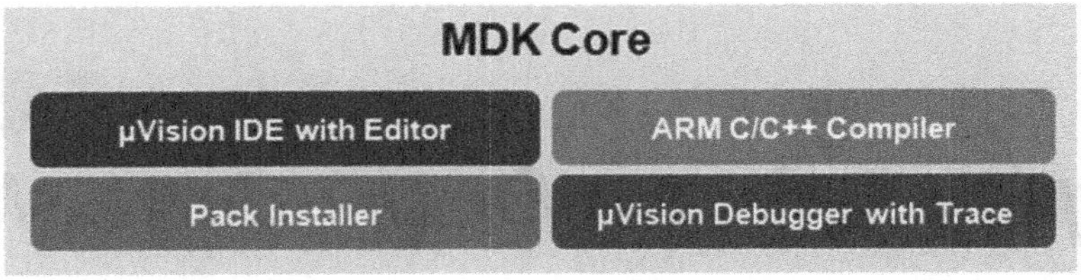

MDK Core

µVision IDE with Editor	ARM C/C++ Compiler
Pack Installer	µVision Debugger with Trace

Figure 2.1
The MDK Core installation contains an IDE, Compiler, and Debugger. Device and middleware support is added through software packs.

The MDK includes its own development environment called Microvision, which acts as an editor, project manager, and debugger. One of the great strengths of the MDK is that it uses the Arm "C" Compiler. This is a very widely used "C/C++" Compiler that has been continuously developed by Arm since their first CPUs were created. The MDK is also a reference toolchain for the "Common Microcontroller Software Interface Standard," which we will meet in Chapter 4, Common Microcontroller Software Interface Standard. The MDK also includes an integrated Real-Time Operating System called RTX. All of the Cortex-M processors are capable of running an operating system, and we will look at using an RTOS in Chapter 9, Practical DSP for Cortex-M Microcontrollers. As well as including an RTOS the MDK includes a DSP library that can be used on the Cortex-M4 and the Cortex-M3. We will look at this library in Chapter 8, Debugging With CoreSight.

Community Edition

While the MDK is a commercial toolchain, there are two free-to-use versions, as shown in Table 2.2.

Table 2.2: MDK free-to-use versions

Version	Supported Code Size	Permitted Uses
MDK Lite	32K Image size	Commercial and noncommercial
MDK Community	Unlimited	Noncommercial only

The exercises that accompany this book are designed to work with the MDK lite version but will also work with the community edition. Instructions for installing the community edition license are provided in the Github example repository described in the MDK Installation section below.

Software Packs

The initial MDK installation is only for the core toolchain components. This consists of the µVision IDE, the Compiler, and debugger, plus a utility called the pack installer. The core toolchain does not contain any support for specific Cortex-M microcontrollers. Support for a specific family of Cortex-M microcontrollers is installed through a software pack system. The pack installer allows you to select and install support for a family of microcontrollers. Once selected, a "Device Family Pack" will be downloaded from a Pack Repository website and installed into the toolchain. This software pack system can also be used to distribute software libraries and other software components. We will look at how to make a software component for code reuse in Chapter 14, Software Components.

The Tutorial Exercises

There are a couple of key reasons for using the MDK as the development environment for this book. First, it includes the Arm "C" Compiler, which is the industry reference compiler for Arm processors. Second, it includes a software simulator that models each of the Cortex-M processors and the peripherals for a range of Cortex-M-based microcontrollers. This allows you to run most of the tutorial examples in this book without the need for a hardware debugger or evaluation board. The simulator is a very good way to learn how each Cortex-M processor works, as you can get as much detailed debug information from the simulator as from a hardware debugger. The MDK also includes a Real-Time Operating System (RTOS) that supports the CMSIS-RTOSv2 specification. We will see more of this in Chapter 10, Using a Real-Time Operating System, but it is basically a universal API for Cortex-M RTOS. While we can use the simulator to experiment with the different Cortex-M processors, there comes a point when you will want to run your code on some real hardware. In the first exercise, we will also look at how to connect the debugger to a typical evaluation board and run our project on real hardware.

Installation

For the practical exercises in this book, we need to install the MDK lite toolchain and the "Device Family Support" packs for the microcontrollers we will use, and the tutorial example pack.

First, Download the Keil MDK lite from http://www.keil.com.

The MKD-Arm is available as a community edition of the full toolchain, which allows you to build projects with an unlimited code size for noncommercial use. It includes the fully working Compiler, RTOS, and Simulator and works with compatible hardware debuggers.

Run the downloaded executable to install the MDK onto your PC.

Open a browser and go to the repository shown below.

The repository holds the examples pack and instructions to install the community edition license

https://github.com/DesignersGuide/Cortex-M_Processor_Examples

The examples are held in the file:

Elsevier.Cortex-M_designers_Guide. < version > .pack

Download the latest version of the pack and save it to your hard disk.

Make sure it has a pack extension; some browsers may rename the file by giving it a zip extension.

Once you have downloaded the Tutorial Examples Pack double click on the file to install the contents into the MDK toolchain.

Once the examples are installed, we need to add the device family support packs for the Cortex-M microcontrollers we are planning to use. This is done through the pack installer utility.

If the pack installer is not open start the uVision IDE (Fig. 2.2).

Figure 2.2
Microvision IDE Icon.

Open the pack installer using its toolbar icon (Fig. 2.3).

Figure 2.3
Pack installer toolbar Icon.

In the pack installer, click on the devices tab and navigate through to select the ST Microelectronics STM32F1 Series (Fig. 2.4).

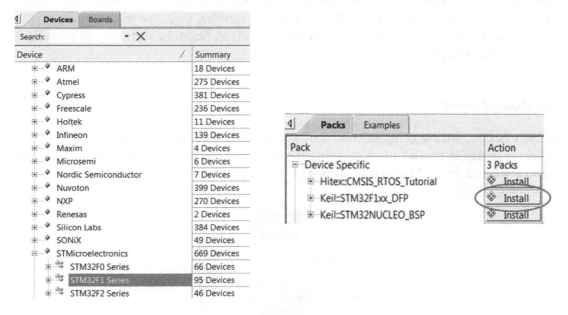

Figure 2.4
Use the pack installer to download support files for the STM32F1 family of microcontrollers.

In the packs tab select the Keil::STM32F1xx_DFP and press the install button.

Repeat this process to install support for the following device families (Table 2.3).

Table 2.3: Required Device family support packs

Silicon Vendor	Device	Device Pack
ST Microelectronics	STM32F7 Series	Keil::STM32F7xx_DFP
NXP	LPC1700 Series	Keil::LPC1700_DFP

In the Generic section of the Packs window select and install or if necessary, update the following packs shown in Table 2.4.

Table 2.4: Required generic software packs

Generic Pack
Arm::CMSIS
Arm::CMIS-Driver_Validation
Arm-Packs::Unity
Keil::Arm_Compiler

Exercise 2.1: Building a First Program

Now that the toolchain support packs and examples are installed, we can look at setting up a project for a typical small Cortex-M-based microcontroller. Once the project is configured, we can get familiar with the Microvision IDE, build the code and take our first steps with the debugger.

The Blinky Project

In this example, we are going to build a simple project called Blinky. The code in this project is designed to read a voltage using the microcontrollers ADC and send its value to a console window in the Microvision debugger (Fig. 2.5).

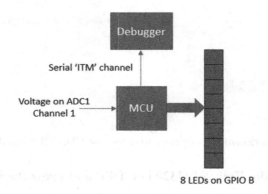

Figure 2.5
The blinky project hardware consists of an analog voltage source and a bank of LEDs.

The code also flashes a group of LEDs in sequence. There are eight LEDs attached to port pins on GPIO port B.

To view, the examples, open the pack installer from within the microvision IDE.

Start the uVision IDE.

Open the pack installer using its toolbar icon

Select the boards tab and select the "Cortex-M Designers Guide" (Fig. 2.6).

Figure 2.6
Pack installer using the Boards and Examples tabs.

Now click on the examples tab to see the tutorial examples.

To show the exercises in numerical order, click on the gray block at the head of the column.

Select the examples tab and press the copy button for "Ex 2.1 First Project."

Set the install directory to a suitable place on your hard drive. In the following instructions, this location will be called <path>\First Project.

This is a multiproject workspace (Fig. 2.7) that contains a minimal Exercise project plus a completed Solution for reference.

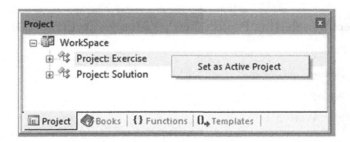

Figure 2.7
Multiproject workspace within the Microvision IDE.

Ensure the Exercise project is active by placing the cursor on it, right-clicking and selecting "Set as active project."

The active project is then highlighted by a dark bar.

The next few steps show how to start a project from scratch to reach this point. I suggest you first use the example project, then come back and redo everything from scratch. This will give you a bit of practice in setting up the initial project.

For now, review the following section and resume the instructions at "Configuring the Run-Time Environment."

Configuring a Project from Scratch

In Microvision close any currently open project by selecting "project\close project" from the main menu bar (Fig. 2.8).

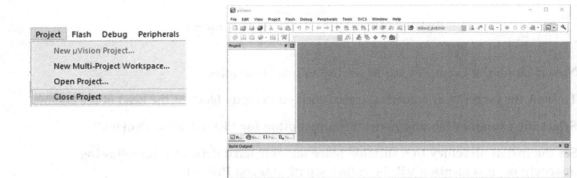

Figure 2.8
Close any open project.

Start a new project by selecting Project\New μVision project (Fig. 2.9).

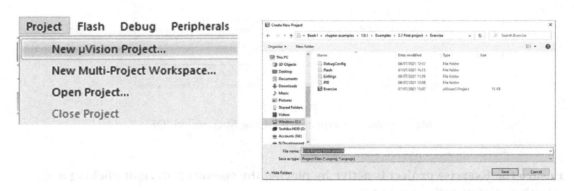

Figure 2.9
Create a new project and save it in the First Project directory.

This will open a menu asking for a project name and directory.

You can give the project any name you want but make sure you select the "<path> \First Project\exercise" Directory.

This directory contains the "C" source code files we will use in our project.

Enter your project name and click Save.

Next, select the microcontroller to use in the project.

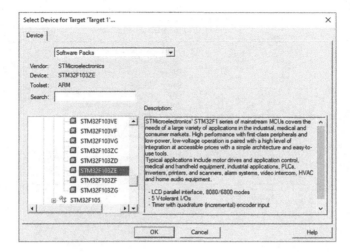

Figure 2.10
Select the STM32F103ZE from the device database.

Once you have selected the project directory and saved the project name, a new dialog with a device database will be launched (Fig. 2.10). Here we must select the microcontroller we are going to use for this project. Navigate the device database and select ST Microelectronics, the "STM32F103 Series," and then the STM32F103ZE, then click OK. This will configure the project for this device; this includes setting up the correct compiler options, linker script file, simulation model, debugger connection, and Flash programming algorithms.

When you have selected the STM32F103ZE click OK.

Configuring the Run Time Environment

The software components used in a project are managed by three icons on the microvision toolbar (Fig. 2.11).

RTE   Pack   Pack
Manager Installer

Figure 2.11
Toolbar Icons for managing the "Run Time Environment."

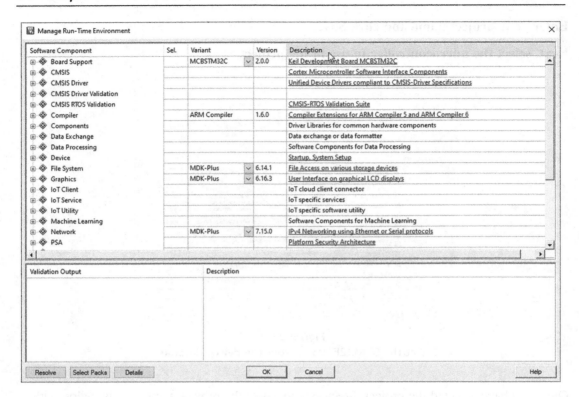

Figure 2.12
The Run Time environment manager allows you to add software components to your project to rapidly build a development "platform."

The pack installer can be relaunched from here. The second icon is the Pack Manager this opens a menu that allows you to control which installed version of the pack to use. This project is configured to use the latest pack versions. In a real project, you should fix the version to ensure compatibility between different workstations. On the toolbar press the RTE Icon to open the Run Time Environment Manager.

The Run-Time Environment Manager RTE (Fig. 2.12) allows you to select software components that have been installed through the pack system and add them to our project. This helps you to build and manage complex software platforms very quickly. For help on any of the software components, click on the blue link in the description column to access the documentation.

For now, we need to add the minimum support for our initial project.

In the RTE tick the "CMSIS::Core" box and the "Device::Startup box" (Fig. 2.13).

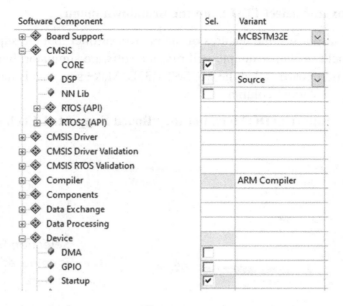

Figure 2.13
Select the "CMSIS::Core" and "Device::Startup" components.

This adds the initial Startup code and the support functions for the Cortex-M processor that we will meet in Chapter 4, Common Microcontroller Software Interface Standard.

The project also uses the STDIO printf() function to display the ADC conversion values. We can use the RTE to configure the STDIO channel. We will look at these options in Chapter 7, Armv8-M. For now, we will connect the printf() output to a console window in the debugger using a debug serial channel called the Instrumentation Trace ITM.

In the RTE select the Compiler:: IO:STDOUT option (Fig. 2.14).

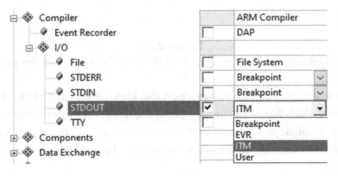

Figure 2.14
Directing the STDOUT channel to the debugger.

Check the tick box and select ITM from the dropdown menu.

This example will also use some features of the microcontroller that are supported through a board support package. This example will run in a software simulator, but it is also targeted an evaluation board called the MCBSTM32E. Make sure this is the board selected in the Board Support\Variant column.

Tick the "Board Support::ADC" box and the "Board Support::LED box" (Fig. 2.15).

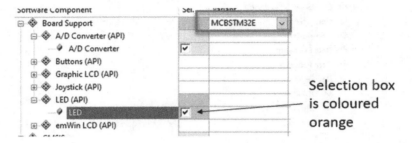

Figure 2.15
The board support components require a subcomponents to work. Until this is resolved the "sel." Column will be colored Orange.

When you enable the board support components, the selection column box will turn orange. This indicates that additional support files are required to use the selected component.

The validation output window will show the required additional subcomponents (Fig. 2.16).

Validation Output	Description
⊟ ⚠ Keil.MCBSTM32C::Board Support:LED	Additional software components required
⊟ require Device:GPIO	Select component from list
◈ Keil::Device:GPIO	GPIO driver used by RTE Drivers for STM32F1 Series

Figure 2.16
The LED functions require the GPIO driver to be added to the project.

To add in the GPIO driver, you can either open the Device section of the RTE and manually add in the GPIO driver or simply press the RTE "Resolve" button (Fig. 2.17), and all of the component dependencies will be resolved automatically. If there are multiple options that cannot be resolved, they will be shown in the validation box, and you will need to set them manually.

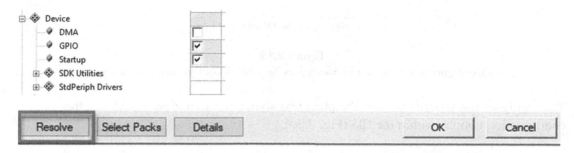

Figure 2.17
Select the GPIO support file manually or press the resolve button to add the required components automatically.

Now that we have selected all of the components required by our project, press the OK button, and the support files will be added to the project window (Fig. 2.18).

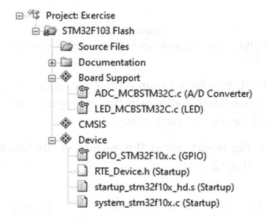

Figure 2.18
Initial project with selected software components.

The RTE components are added to folders shown as a green diamond. Our project source code will be placed in the manilla folder called "Source Group 1."

There are two main types of component files. The files with a yellow key are read-only and do not normally need to be edited. They are not stored in the project directory but are held within a pack repository within the MDK installation. The remaining files are configuration header files or source code you may need to edit.

Double click on the startup_stm32F10x.hd.s file to open it in the editor.

Click on the configuration wizard tab at the bottom of the editor window (Fig. 2.19).

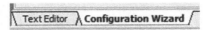

Figure 2.19
Configuration files can be viewed as Text or as configuration wizards.

The configuration wizard converts the plain text source file to a view that shows the configuration options within the file (Fig. 2.20).

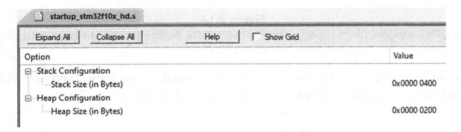

Figure 2.20
The configuration wizard allows you to view and modify #defines within a header or source file.

This view is created by XML tags in the source file comments. Changing the values in the configuration wizard modifies #define values in the underlying source code. In this case, we can set the size of the stack space and heap space.

In the project view, click the books tab at the bottom of the window. If this is not visible, select view\books (Fig. 2.21).

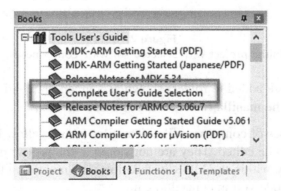

Figure 2.21
Tool chain help is located in the books window.

In the books window, the "Complete Users Guide" opens the help system for the uVision and compiler manuals.

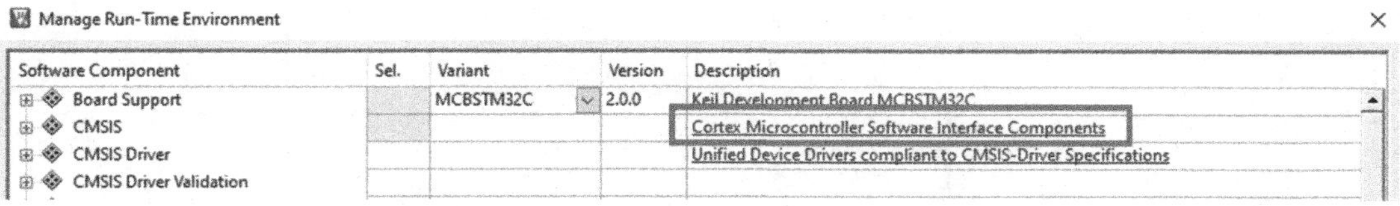

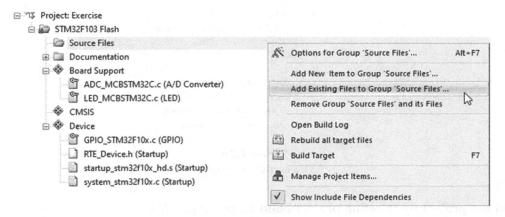

Figure 2.22
Additional help files for Software components are located in the RTE dialog.

In the RTE, the blue hyperlink in the Description column opens the help files for a specific software component (Fig. 2.22).

Switch back to the project view and add the project "C" source files.

Highlight the "Source Group" folder in the Project window, right-click and select "Add Existing Files to Group Source Group 1" (Fig. 2.23).

Figure 2.23
Add an existing file to the project source group.

This will open an "Add files to Group" dialog box. In the dialog box, add the Blinky.c project file (Fig. 2.24).

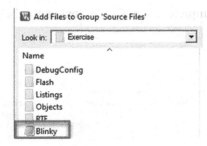

Figure 2.24
Add Blinky.c located in the project directory.

The project should now contain Blinky.c and the RTE components (Fig. 2.25).

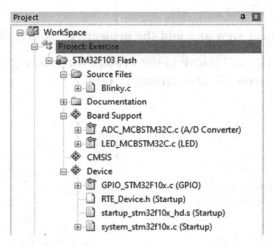

Figure 2.25
The complete project.

Build the project by selecting project\build target (Fig. 2.26).

Figure 2.26
Build the project using the menu or the icons on the toolbar.

The build system will compile each of the ".c" modules in turn and then link them together to make a final application program. The output window shows the result of the build process and reports any errors or warnings (Fig. 2.27).

```
Build Output
compiling system_stm32f10x.c...
linking...
Program Size: Code=2450 RO-data=418 RW-data=12 ZI-data=1052
".\Flash\Exercise.axf" - 0 Error(s), 0 Warning(s).
Build Time Elapsed:  00:00:01
```

Figure 2.27
The final program size is reported in the Build Output window.

The program size is also reported as a set of program memory regions, as shown in Table 2.5.

Table 2.5: Linker section memory types

Section	Description
Code	Size of the executable image
RO-data	Size of the code constants in the Flash memory
RW-data	Size of the initialized variable in SRAM
ZI-data	Size on uninitialized variables in the SRAM

If errors or warnings are reported in the build window clicking on them will take you to the line of code in the editor window.

Open the Options for Target dialog (Fig. 2.28).

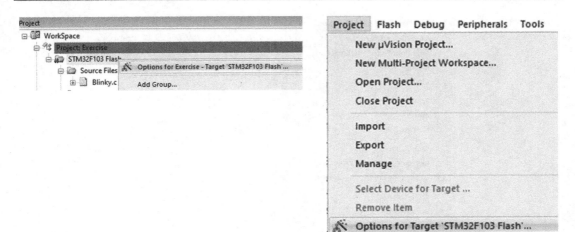

Figure 2.28
Open the project global options.

This can be done in the project menu by right-clicking the project name and selecting "Options for Target" or by selecting the same option in the Project menu from the main toolbar.

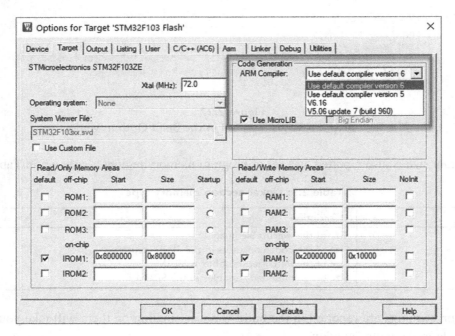

Figure 2.29
The target menu defines the project memory map.

The "Options for Target" dialog holds all of the global project settings. We will look at these in detail later, but for now, the MDK contains two versions of the Arm compiler (Fig. 2.29). The version 6.xx compiler is an entirely new codebase that uses the latest compiler technology and produces the most efficient code.

The code generation box should be set to "default compiler version 6."

Now select the debug tab (Fig. 2.30).

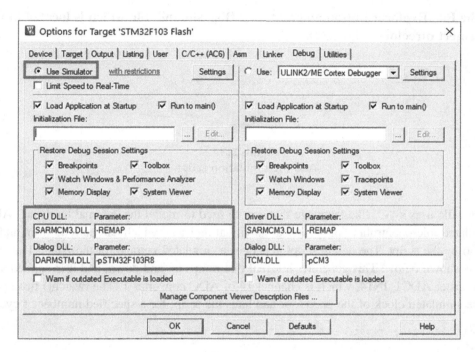

Figure 2.30
Select the simulator and the STM32F103RB simulation model.

The debug menu is split into two halves. The simulator options are on the left, and the hardware debugger is on the right. During project development, you will normally use the hardware debug options. The simulator is useful for learning and also software testing, as we will see in Chapter 14, Software Components.

Click the Use Simulator radio box.

Check that the Simulator Dialog DLL is set to DARMRM.DLL, and the parameter is −pSTM32F103R8 (Fig. 2.30).

Now we can add a simulation script to provide some input values to the ADC (Fig. 2.31).

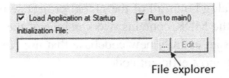

File explorer

Figure 2.31
Open the file explorer to select the simulation script.

Press the File Explorer button and add the "Dbg_sim.ini" file which is located in the First project directory (Fig. 2.32).

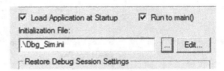

Figure 2.32
Add the simulation script.

The script file uses a "C" like language that can be used to model the external hardware. All of the simulated microcontroller "pins" appear as virtual registers, which can be read from and written to by the script. The debug script generates a simulated voltage for the ADC. The script for this is shown below. This generates a signal that ramps up and down, and it is applied to the virtual register ADC1_IN14, which is channel 14 of ADC converter 1. The twatch() function reads the simulated clock of the processor and halts the script for a specified number of cycles.

```
Signal void Analog (float limit) {
 float volts;
 printf ("Analog (%f) entered.\n", limit);
 while (1) {          /* forever */
  volts = 0;
  while (volts <= limit) {
   ADC1_IN1 = volts;   /* analog input-2 */
   twatch (250000);    /* 250000 Cycles Time-Break */
   volts += 0.1;       /* increase voltage */
  }
  volts = limit;
  while (volts >= 0.0) {
   ADC1_IN1 = volts;
   twatch (250000);    /* 250000 Cycles Time-Break */
   volts -= 0.1;       /* decrease voltage */
  }
 }
}
```

Click OK to close the options for target dialog.

Now start the debugger and run the code (Fig. 2.33).

Figure 2.33
Start the debugger using the menu or icon.

This will connect μVision to the Simulation model and download the Project image into the simulated memory of the microcontroller. Once the program image has been loaded, the microcontroller is reset, and the code is run until it reaches main() ready to start debugging (Fig. 2.34).

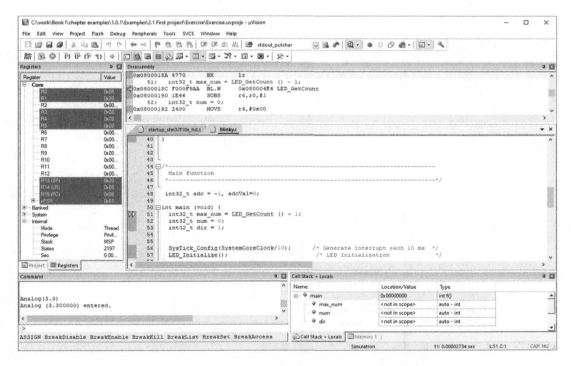

Figure 2.34
The debug view.

The μVision debugger is divided into a number of windows that allow you to examine and control the execution of your code. The key windows are shown in Fig. 2.35.

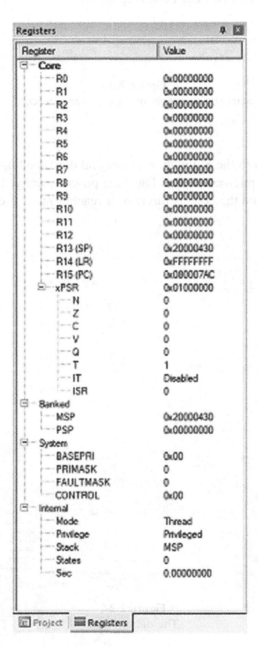

Figure 2.35
Register window.

The register window (Fig. 2.35) displays the current contents of the CPU register file (R0-R15), the program Status register (xPSR), and also the Main Stack pointer and the Process stack Pointer. We will look at all of these registers in the next chapter.

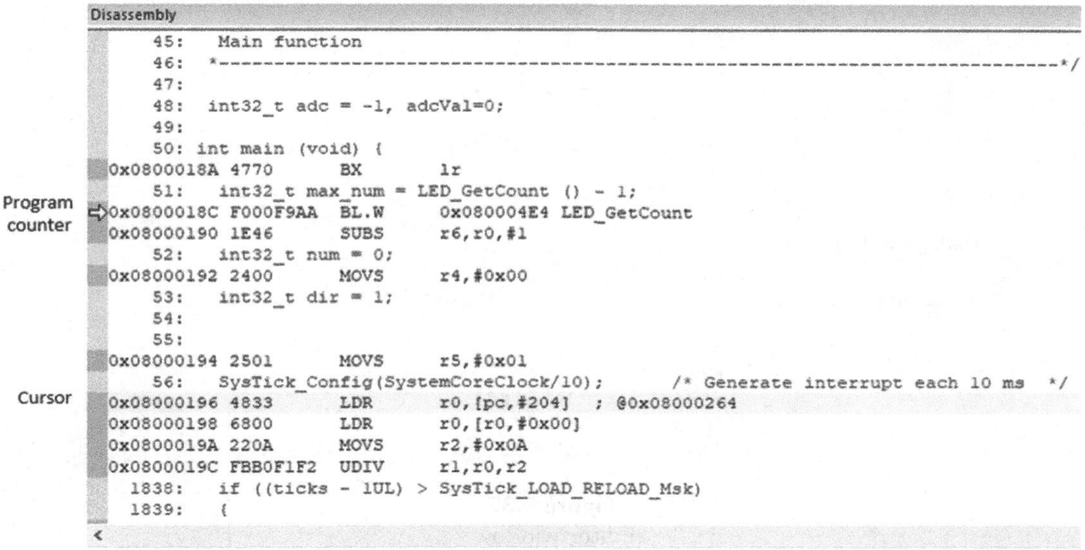

```
Disassembly
        45:    Main function
        46:   *----------------------------------------------------------------------*/
        47:
        48:   int32_t adc = -1, adcVal=0;
        49:
        50: int main (void) {
0x0800018A 4770         BX        lr
        51:    int32_t max_num = LED_GetCount () - 1;
0x0800018C F000F9AA     BL.W         0x080004E4 LED_GetCount
0x08000190 1E46         SUBS      r6,r0,#1
        52:    int32_t num = 0;
0x08000192 2400         MOVS      r4,#0x00
        53:    int32_t dir = 1;
        54:
        55:
0x08000194 2501         MOVS      r5,#0x01
        56:    SysTick_Config(SystemCoreClock/10);       /* Generate interrupt each 10 ms   */
0x08000196 4833         LDR       r0,[pc,#204]   ; @0x08000264
0x08000198 6800         LDR       r0,[r0,#0x00]
0x0800019A 220A         MOVS      r2,#0x0A
0x0800019C FBB0F1F2     UDIV      r1,r0,r2
      1838:    if ((ticks - 1UL) > SysTick_LOAD_RELOAD_Msk)
      1839:    {
<
```

Program counter

Cursor

Figure 2.36
Disassembly window.

As its name implies, the disassembly window (Fig. 2.36) will show you the low-level assembler listing interleaved with the high-level "C" code listing. One of the great attractions of the Cortex-M family is that all of your project code can be written in a high-level language such as "C\C + +." You never or very rarely need to write low-level assembly routines. However, it is useful to be able to "read" the low-level assembly code to see what the Compiler is doing. The disassembly window shows the absolute address of the current instruction. Next is shown the OP Code. This is either a 16-bit instruction or a 32-bit instruction. The raw OP Code is then displayed as an assembler mnemonic. The current location of the program counter is shown by the yellow arrow in the left-hand margin. The dark gray blocks indicate the location of executable lines of code.

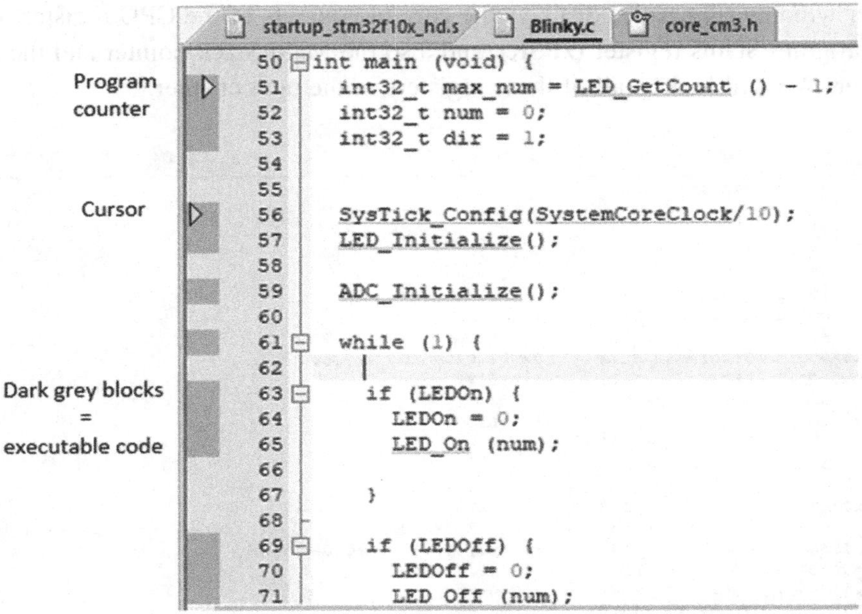

Figure 2.37
Editor window.

The source code editor window (Fig. 2.37) has a similar layout to the disassembly window. This window displays the high-level "C" source code. The current location of the program counter is shown by the yellow arrow in the left-hand margin. The blue arrow shows the location of the cursor. Like the disassembly window, the dark gray blocks indicate the location of executable lines of code. The source window allows you to have a number of project modules open. Each source module can be reached by clicking the tab at the top of the window.

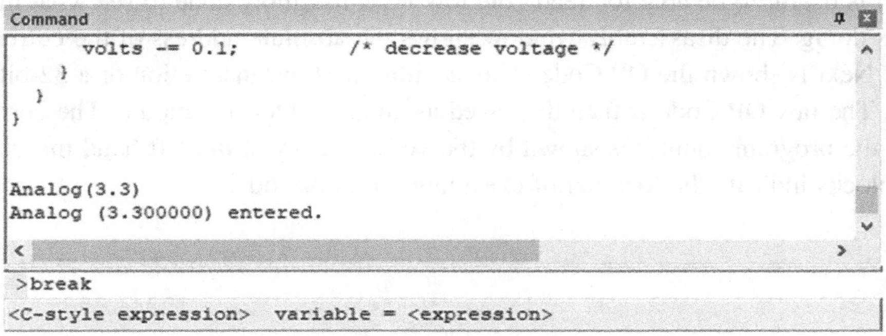

Figure 2.38
Debugger command line.

The command window (Fig. 2.38) allows you to enter debugger commands to directly configure and control the debugger features. These commands can also be stored in a text file and executed as a script when the debugger starts.

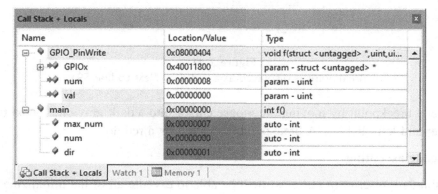

Figure 2.39
Variable watch window.

Next to the command window is a group of watch windows (Fig. 2.39). These windows allow you to view local variables, global variables, and the raw memory.

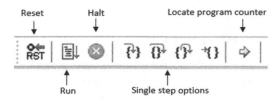

Figure 2.40
Run control toolbar.

You can control the execution of the code through icons on the toolbar (Fig. 2.40). The code can be single stepped a "C" or assembler line at a time or run at full speed and halted. The same commands are available through the debug menu, which also shows the function key shortcuts that you may prefer.

Set a breakpoint on the main while loop in Blinky.c (Fig. 2.41).

```
startup_stm32f10x_hd.s   Blinky.c
54
55
56    SysTick_Config(SystemCoreClock/10);
57    LED_Initialize();
58
59    ADC_Initialize();
60
```

Figure 2.41
A Breakpoint is displayed as a red dot (Next to line 59).

You can set a breakpoint by moving the mouse cursor into a dark gray block next to the line number and left-clicking. A breakpoint is marked by a red dot.

Start the code executing.

With the simulation script in place, we will be able to execute all of the initializing code and reach the breakpoint.

Now spend a few minutes exploring the debugger run control (Fig. 2.42).

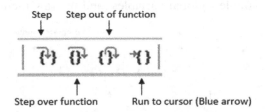

Figure 2.42
Toolbar single-step options.

Use the single-step commands, set a breakpoint and start the simulator running at full speed. If you lose what's going on, exit the debugger by selecting debug/start/stop and then restart again. This ensures that the initial script is restarted and you are back to square one.

Add a variable to the watch window.

Once you have finished familiarizing yourself with the run control commands within the debugger locate the main() function within Blinky.c. Just above main() is the declaration for a variable called adcVal. We can view the contents of the variable in a number of ways.

Hover the cursor over the variable, and a tooltip will display its current value (Fig. 2.43).

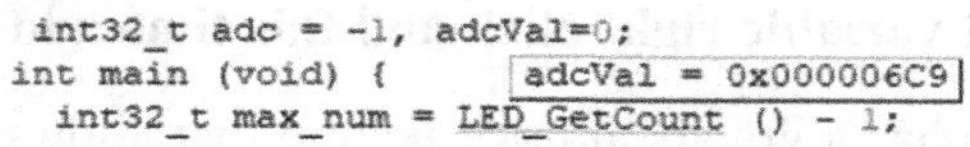

```
int32_t adc = -1, adcVal=0;
int main (void) {        adcVal = 0x000006C9
    int32_t max_num = LED_GetCount () - 1;
```

Figure 2.43
a Tooltip shows the current value of a global variable

Highlight this variable right-click and select "Add adcVal to" (Fig. 2.44).

This will display a range of visualization options, watch, memory, and Analyzer.

Select watch 1 to add the variable to a permanent viewing window.

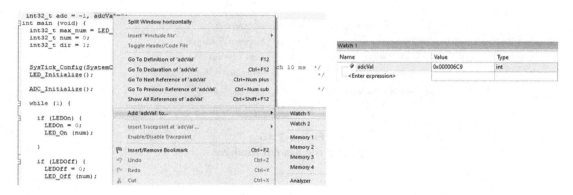

Figure 2.44
Add the adcVal to Watch 1.

Now start the code running, and you will be able to see the adcVal variable updating in the watch window.

The simulation script feeds a voltage to the simulated microcontroller, which in turn provides converted results to the application code.

If the watch window is not updating, click in the view menu and check that "Periodic Window Update" is enabled (Fig. 2.45).

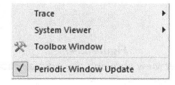

Figure 2.45
The watch windows will refresh in real time if the Periodic Update is enabled.

Now highlight the adcVal variable right click and this time add it to the Analyzer.

The μVision debugger also has a graphical trace feature that allows you to visualize the historical values of a given global variable.

This may cause a legacy logic analyzer window to open. While this still works it has been superseded by a more sophisticated System analyzer that integrates a wide range of real-time data into a single view.

Open the system analyzer by selecting the view menu\analysis windows and system analyzer (Fig. 2.46).

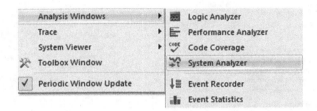

Figure 2.46
Opening the System Analyzer.

You may need to adjust the time scale to get a clear view of the adcVal data. You can add any other global variables or peripheral registers to the system analyzer window. The system analyzer window (Fig. 2.47) can also display the activity of interrupts, RTOS threads, and user-defined tags that have been added to the code.

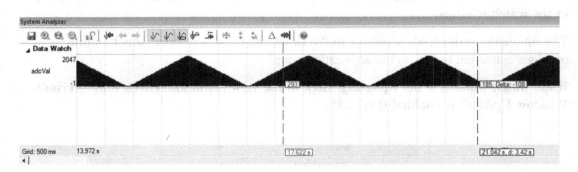

Figure 2.47
System Analyzer view showing the adcVal data over time.

Now view the state of the user peripherals.

The simulator has a model of the whole microcontroller, not just the Cortex-M processor, so it is possible to examine the state of the microcontroller peripherals directly.

Select peripherals\General purpose IO\GPIOB (Fig. 2.48).

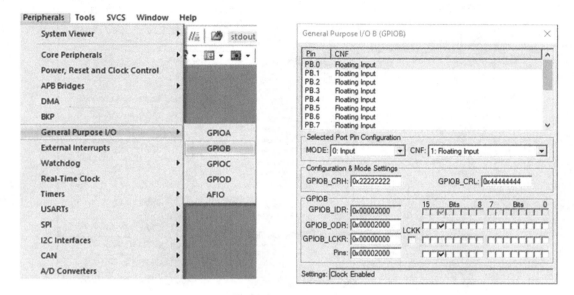

Figure 2.48
Under peripherals we can see the state of all the user peripherals on the chip.

This will open a window that displays the current state of the microcontroller GPIO Port B. As the simulation runs, we can see the state of the port pins. If the pins are configured as inputs, we can manually set and clear them by clicking the individual "Pins" boxes.

You can do the same for the ADC by selecting ADC1.

When the code is running, it is possible to see the current configuration of the ADC and the conversion results. You can also manually set the input voltage by entering a new value in the Analog Inputs boxes (Fig. 2.49).

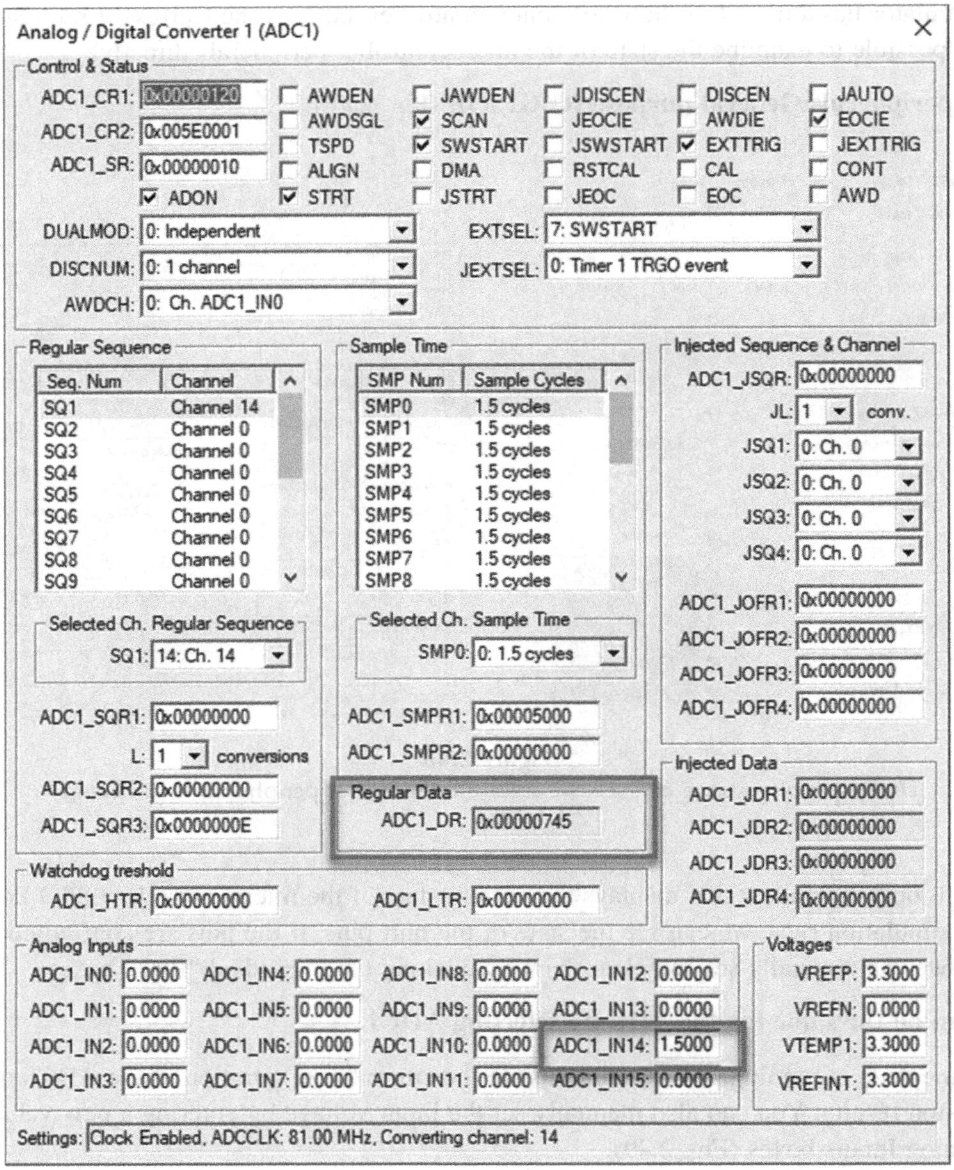

Figure 2.49
ADC peripheral window.

The simulator also includes a terminal that provides an I\O channel for the microcontrollers UARTS and the ITM debug channel.

Select the view\serial windows\Debug(printf) window (Fig. 2.50).

This opens a console-type window that displays a debug console that is connected to the microcontroller through its internal debug hardware. As we will see later, this is very useful for software testing within the simulator and on real hardware.

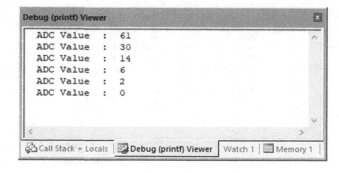

Figure 2.50
ITM console window.

The simulator also boasts some advanced analysis tools, including trace, code coverage, and performance analysis.

Open the View\trace menu and select trace data and enable trace recording (Fig. 2.51).

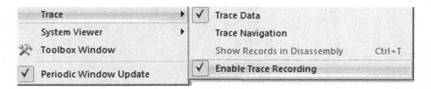

Figure 2.51
Enabling Instruction the trace.

This will open the instruction trace window(Fig. 2.52). The trace records a history of each instruction executed.

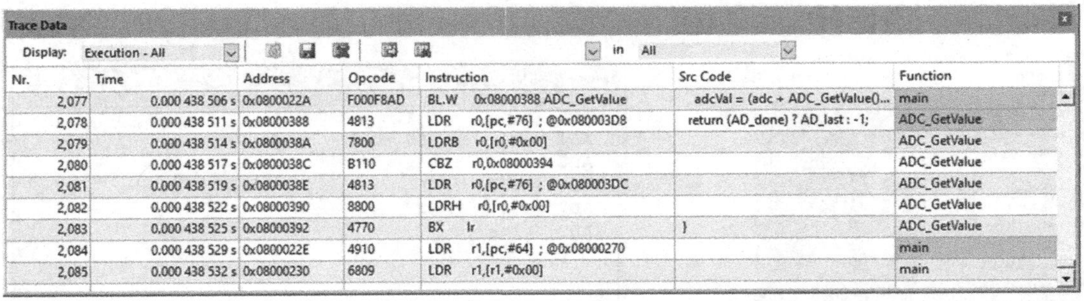
Trace Data
Display: Execution - All
in All
Nr. Time Address Opcode Instruction Src Code Function
2,077 0.000 438 506 s 0x0800022A F000F8AD BL.W 0x08000388 ADC_GetValue adcVal = (adc + ADC_GetValue()... main
2,078 0.000 438 511 s 0x08000388 4813 LDR r0,[pc,#76] ; @0x080003D8 return (AD_done) ? AD_last : -1; ADC_GetValue
2,079 0.000 438 514 s 0x0800038A 7800 LDRB r0,[r0,#0x00] ADC_GetValue
2,080 0.000 438 517 s 0x0800038C B110 CBZ r0,0x08000394 ADC_GetValue
2,081 0.000 438 519 s 0x0800038E 4813 LDR r0,[pc,#76] ; @0x080003DC ADC_GetValue
2,082 0.000 438 522 s 0x08000390 8800 LDRH r0,[r0,#0x00] ADC_GetValue
2,083 0.000 438 525 s 0x08000392 4770 BX lr } ADC_GetValue
2,084 0.000 438 529 s 0x0800022E 4910 LDR r1,[pc,#64] ; @0x08000270 main
2,085 0.000 438 532 s 0x08000230 6809 LDR r1,[r1,#0x00] main

Figure 2.52
Instruction Trace.

Now open the View\Analysis\Code coverage and View\Analysis\Performance Analyzer windows (Fig. 2.53).

The "Performance Analysis" window shows the number of calls to a function and its cumulative run time.

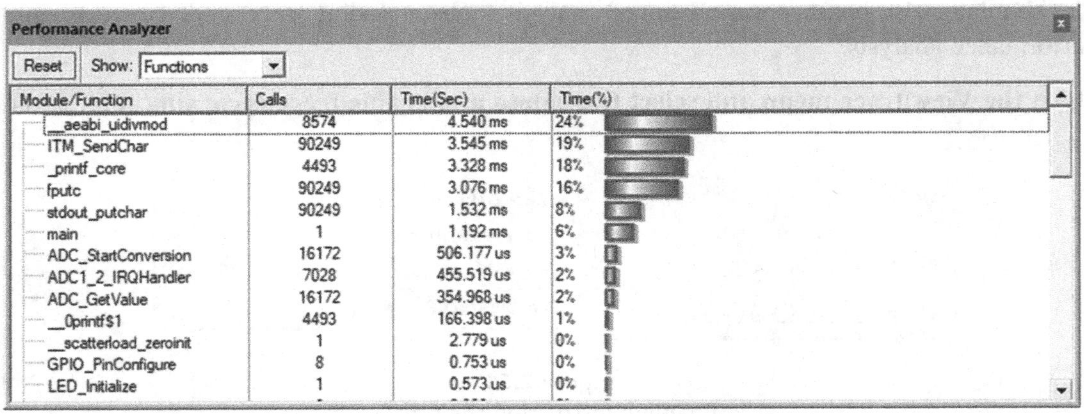
Performance Analyzer
Reset Show: Functions
Module/Function Calls Time(Sec) Time(%)
__aeabi_uidivmod 8574 4.540 ms 24%
ITM_SendChar 90249 3.545 ms 19%
_printf_core 4493 3.328 ms 18%
fputc 90249 3.076 ms 16%
stdout_putchar 90249 1.532 ms 8%
main 1 1.192 ms 6%
ADC_StartConversion 16172 506.177 us 3%
ADC1_2_IRQHandler 7028 455.519 us 2%
ADC_GetValue 16172 354.968 us 2%
__0printf$1 4493 166.398 us 1%
__scatterload_zeroinit 1 2.779 us 0%
GPIO_PinConfigure 8 0.753 us 0%
LED_Initialize 1 0.573 us 0%

Figure 2.53
Performance Analyzer window.

The code coverage window gives a digest of the number of executed and partially executed lines in each function (Fig. 2.54).

Both "code coverage" and "performance analysis" are essential for validating and testing software. In Chapter 8, Debugging With CoreSight, we will see how this information can be obtained from a real microcontroller.

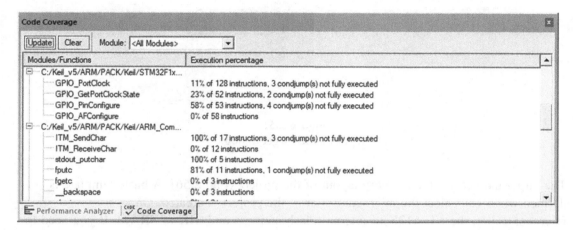

Figure 2.54
Code Coverage window.

Project Configuration

Now that you are familiar with the basic features of the debugger, we can look in more detail at how the project code is constructed.

First, quit the debugger by selecting debug\start\stop debug session.

Open the "options for target" dialog (Fig. 2.55).

All of the key global project settings can be found in the "Options for Target" dialog box.

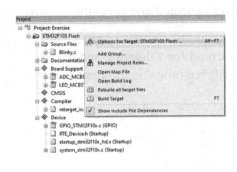

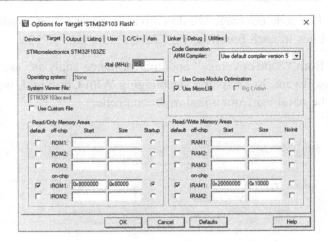

Figure 2.55
Options for Target dialog.

The target tab defines the memory layout of the project (Fig. 2.56). A basic template is defined when you select the microcontroller as the project is created. On this microcontroller, there is 512K of internal Flash memory and 64K of SRAM. If you need to define a more complex memory layout, it is possible to create additional memory regions to subdivide the volatile and nonvolatile memories.

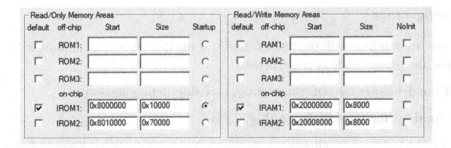

Figure 2.56
The Target menu defines the project memory map.

The more complex memory map above has split the internal FLASH into two blocks and defined the lower FLASH block as the default region for code and constant data. As the IROM2 default box is unchecked, the region is declared to the linker, but nothing will be placed into the upper block (IROM2) unless you explicitly tell the linker to do this. Similarly, the SRAM has been split into two regions, and the upper region (IRAM2) is unused unless you explicitly tell the linker to use it. When the linker builds the project, it looks for the "RESET" code label. The linker then places the reset code at the base of the code region designated as the Startup region. The initial Startup code will write all of the internal SRAM to zero unless you tick the NoInit box for a given SRAM region. Then, the SRAM will be left with its startup garbage values. This may be useful if you want to allow for a soft reset where some system data is maintained.

If you want to place objects (code or data) into an unused memory region, select a project module right-click and open its local options (Fig. 2.57).

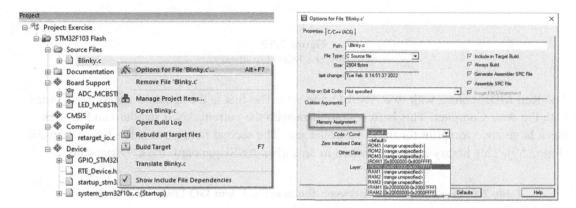

Figure 2.57
Opening the local project options.

In its local options, the memory assignment boxes allow you to force the different memory objects in a module into a specific code region.

Back in the main "Options for Target" menu, there is an option to set the External Crystal frequency used by the microcontroller (Fig. 2.58).

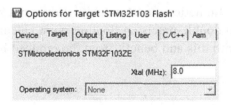

Figure 2.58
The Crystal (Xtal) frequency is only used by the simulator.

This will be a standard crystal value that can be multiplied by an internal phase-locked loop to generate a high-frequency clock for the CPU and microcontroller peripherals. This option is only used to provide the input frequency for the simulation model and nothing else within the IDE.

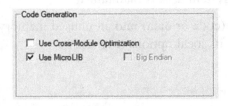

Figure 2.59
Selecting the MicroLib library.

The Keil MDK comes with two ANSI library sets. The first is the standard library that comes with the Arm Compiler. This is fully compliant with the current ANSI standard and as such, has a large code footprint for microcontroller use. The second library set is the Keil MicroLIB (Fig. 2.59). This library has been written to an earlier ANSI standard, the C99 standard.

Table 2.6: Size comparison between the Standard Arm ISO Libraries and the Keil MicroLib library

Processor	Object		Standard	MicroLib	% saving
Cortex-M0(+)	Thumb	Library Total	16,452	5996	64
		RO Total	19,472	9016	54
Cortex-M3\M4	Thumb-2	Library Total	15,018	5796	63
		RO Total	18,616	8976	54

By selecting MicroLIB you will save at least 50% of the ANSI library code footprint versus the Arm compiler libraries (Table 2.6). So try to use microLIB were ever possible. However, it does have some limitations most notably, it does not support all of the functions in standard lib and double-precision floating-point calculations. In a lot of applications, you can live with this and benefit from the reduced library footprint.

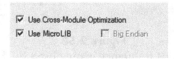

Figure 2.60
Cross Module optimization enables a multipass compile process for best code generation.

The "Use cross-module optimization" tick box enables a multipass linking process that fully optimizes your code (Fig. 2.60). When you use this option, the code generation is changed, and the code execution may no longer map directly to the "C" source code. So do not use this option when you are testing and debugging code as you will not be able to accurately follow it within the debugger source code window. We will look at the other options, system viewer file, and the operating system in the Chapters 10. Using a Real-Time Operating System and 14 Software Components.

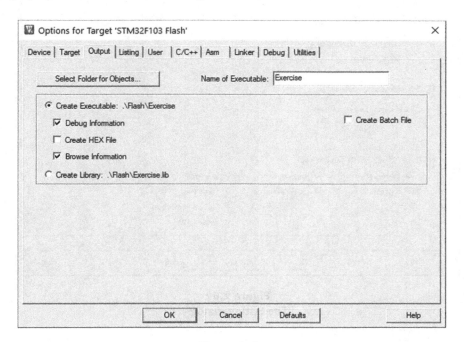

Figure 2.61
The output dialog.

The output menu (Fig. 2.61) allows you to control the final image of your project. Here we can choose between generating a standalone program by selecting "create executable," or we can create a library that can be added to another project. The default is to create a standalone project with debug information. When you get toward the end of a project, you will need to select the "create hex file" option to generate a HEX32 file that can be used with a production programmer. If you want to build the project outside of μVision, select "create batch file" and this will produce a <Project name>.bat DOS batch file that can be run from another program to rebuild the project outside of the IDE. By default, the name of the final image is always the same as your project name. If you want to change this, simply change the "Name of executable" field. You can also select a directory to store all of the project, Compiler, and linker-generated files. This ensures that the original project directory only contains your original source code. This can make life easier when you are archiving projects (Fig. 2.62).

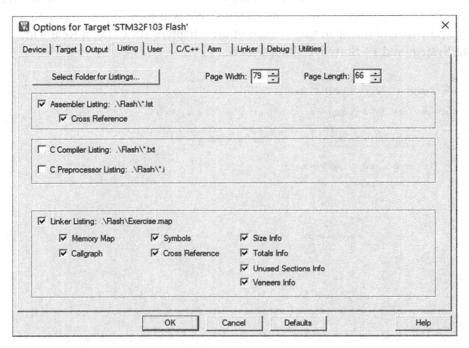

Figure 2.62
The Listing dialog.

The listing tab allows you to enable compiler listings and linker map files. By default, the linker map file is enabled. A quick way to open the map file is to select the project window and double click on the project root. The linker map file contains a lot of information that can seem incomprehensible at first, but there are a couple of important sections that you should learn to read and keep an eye on when developing a real project. The first is the "memory map of the image." This shows you a detailed memory layout of your project. Each memory segment is shown against its absolute location in memory. Here you can track which objects have been located to which memory region. You can also see the total amount of memory resources allocated, the location of the stack and also if it is enabled the location of the heap memory (Fig. 2.63).

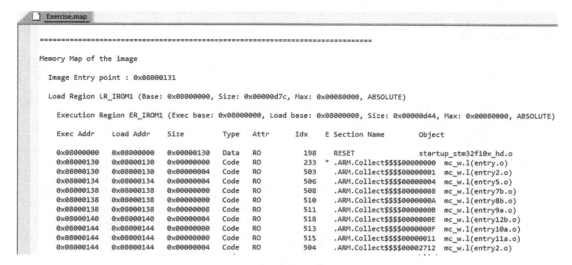

Figure 2.63
Linker map file symbols listing.

The second section gives you a digest of the memory resources required by each module and library in the project together with details of the overall memory requirement. The image memory usage is broken down into the code size, the code data size. This is the amount of nonvolatile memory used to store the initializing values to be loaded into SRAM variables on startup. In simple projects, this initializing data is held as a simple ROM table which is written into the correct RAM locations by the startup code. However, in projects with large amounts of initialized data, the Compiler will switch strategies and use a compression algorithm to minimize the size of the initializing data. On startup, this table is decompressed before the data is written to the variable locations in memory. The RO Data entry lists the amount of nonvolatile memory used to store code literals. The SRAM usage is split into initialized RW data and uninitialized ZI Data (Fig. 2.64).

```
=============================================================================
Image component sizes

    Code (inc. data)   RO Data   RW Data   ZI Data    Debug   Object Name

    336          20        0         3         0     230349   adc_mcbstm32e.o
    328          56        0        20         0     233832   blinky.o
    704          36        0         0         0       5171   gpio_stm32f10x.o
    284           4       64         0         0       5747   led_mcbstm32e.o
    296          60        0         4         0       4175   retarget_io.o
     36           8      304         0      1024        836   startup_stm32f10x_hd.o
    480          38        0        20         0       1947   system_stm32f10x.o

    ----------------------------------------------------------------------
   2464         222      400        48      1024     482057   Object Totals
      0           0       32         0         0          0   (incl. Generated)
      0           0        0         1         0          0   (incl. Padding)

    ----------------------------------------------------------------------
```

Figure 2.64
Linker map file sections listing.

The next tab is the user tab. This allows you to add external utilities to the build process.
The menu allows you to run a utility program to pre or postprocess files in the project. A
utility program can also be run before each module is compiled. Optionally you can also
start the debugger once the build process has finished (Fig. 2.65).

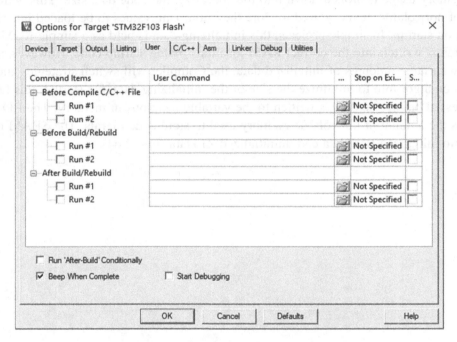

Figure 2.65
The User dialog.

The code generated by the Compiler is controlled by the C/C++ tab. This controls the code generation for the whole project. However, the same menu is available in the local options for each source module. This allows you to have global build options and then different build options for selected modules. In the local options menu, the option tick boxes are a bit unusual in that they have three states (Fig. 2.66).

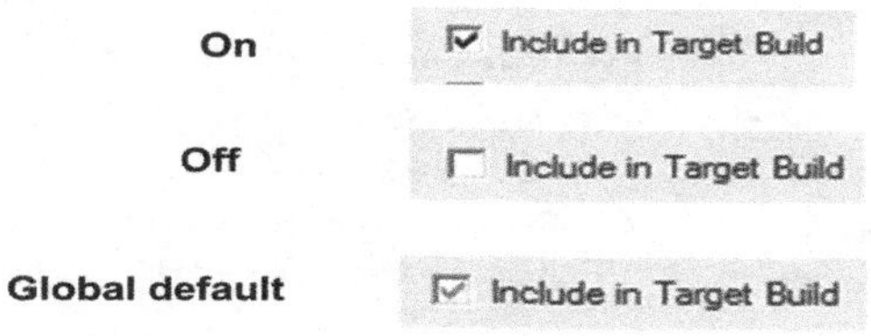

On   ☑ Include in Target Build

Off   ☐ Include in Target Build

Global default   ☑ Include in Target Build

Figure 2.66
The dialog tick boxes have three states, On, Off, and Inherit default.

They can be unchecked, checked with a solid black tick or checked with a gray tick. Here the gray tick means "inherit the global options," and this is the default state.

The most important option in this menu is optimization control. During development and debugging, you should leave the optimization level at O1, which provides a good debug view of the code. As you increase the optimization level, the Compiler will use more and more aggressive techniques to optimize the code. At the high optimization level, the generated code no longer maps closely to the original source code, which then makes using the debugger very difficult. For example, when you single step the code, its execution will no longer follow the expected path through the source code. Setting a breakpoint can also be hit, and miss as the generated code may not exist on the same line as the source code. There are also optimization options to generate the smallest code image or the fastest code image (Fig. 2.67).

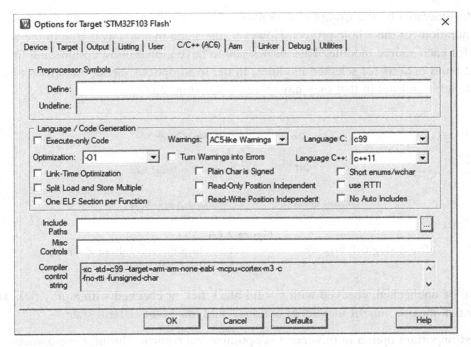

Figure 2.67
The C/C++ dialog.

The compiler menu also allows you to enter any #defines that you want to pass to the source module when it is being compiled. If you have structured your project over several directories, you may also add local include paths to directories with project header files. The misc controls text box allows you to add any compiler switches that are not directly supported in the dialog menu. Finally, the full compiler control string is displayed. This includes the CPU options, project include paths and library paths and the make dependency files (Fig. 2.68).

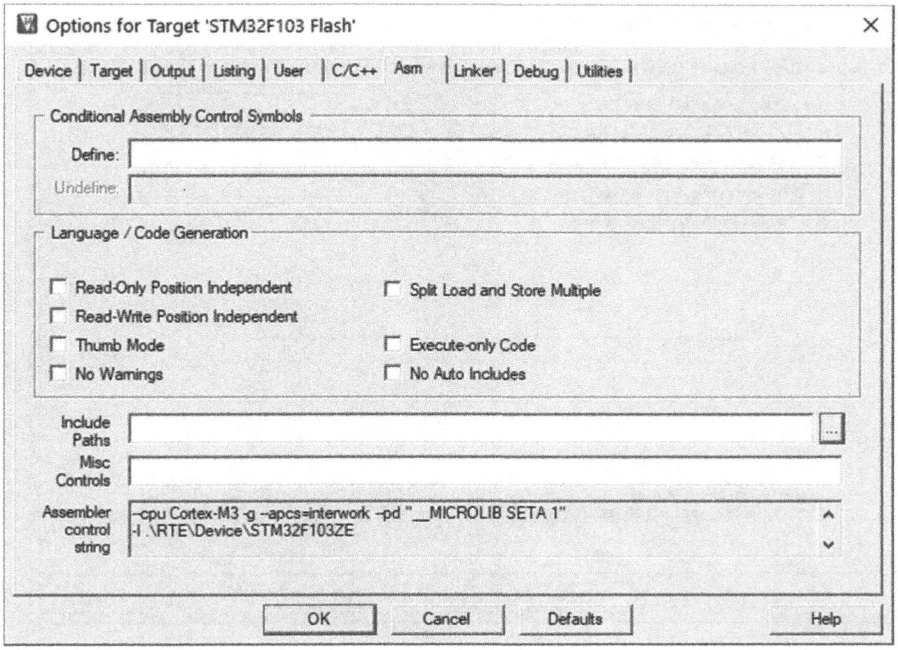

Figure 2.68
The Assembler dialog.

There is also an assembler options window that includes many of the same options as the "C/C++" menu. However, most Cortex-M projects are written completely in "C/C++," so with luck, you will never have to use this menu! (Fig. 2.69).

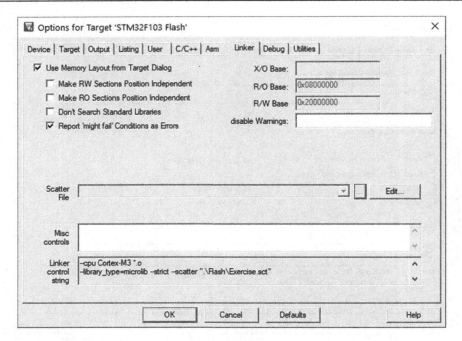

Figure 2.69
The Linker dialog.

By default, the linker menu imports the memory layout from the target menu. This memory layout is converted into a linker command file called a "scatter" file. The scatter file provides a text description of the memory layout to the linker so it can create a final image. An example of a scatter file is shown below.

```
LR_IROM1 0x08000000 0x00080000 {   ; load region size_region
 ER_IROM1 0x08000000 0x00080000 { ; load address = execution address
  *.o (RESET, +First)
  *(InRoot$$Sections)
  .ANY (+RO)
  .ANY (+XO)
 }
RW_IRAM1 0x20000000 0x00008000 {   ; RW data
 Blinky.o (+ZI)
 .ANY (+RW +ZI)
 }
RW_IRAM2 0x20008000 UNINIT 0x00080000 {
 Blinky.o (+RW)
 }
}
```

The scatter file defines the ROM and RAM regions and the program segments that need to be placed in each segment. The example scatter file first defines a ROM region of 512K. All of this memory is allocated in one bank. The scatter file also tells the linker to place the reset segment containing the vector table at the start of this section. Next, the scatter file tells the linker to place all the remaining nonvolatile segments in this region. The scatter file also defines two banks of RAM of 32K. The linker is then allowed to use both pages of RAM for initialized and uninitialized variables. This is a simple memory layout that maps directly onto the memory of the microcontroller. If you need to use a more sophisticated memory layout, you can add extra memory regions in the target menu, and this will be reflected in the scatter file. If, however, you need a complex memory map that cannot be defined through the target menu, then you will need to write your own scatter file. The trick here is to get as close as you can with the target menu and then hand edit the scatter file (Fig. 2.70).

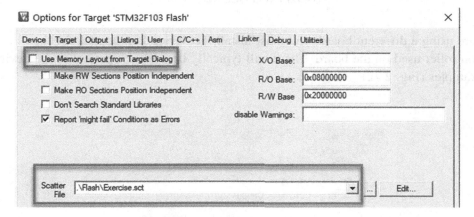

Figure 2.70
Using a custom scatter file.

If you are using your own scatter file, you must then uncheck the "use Memory Layout from Target" Dialog and then manually select the new scatter file using the scatter file text box. Always change the name of a custom scatter file from the default so that it is not accidentally overwritten by the IDE.

Exercise 2.2: Hardware Debug

In this section, we will look at configuring the debugger to use a hardware debug interface rather than the internal simulator. In this exercise, we will use the STM32F7 Discovery board but the instructions below can be applied to any other hardware debug setup (Fig. 2.71).

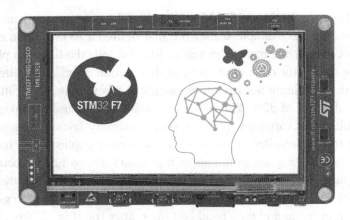

Figure 2.71
STM32F746G Discovery Board.

If you are using a different board, you must install the device family pack for the microcontroller used on the board. There will typically be a Blinky project in the device pack examples (Fig. 2.72).

Figure 2.72
Make Hardware Debug the active project.

Connect the Discovery board to the PC using the USB mini connector.

Build the project.

The project is created and built in exactly the same way as the simulator version (Fig. 2.73).

Open the options for target dialog and the debug tab.

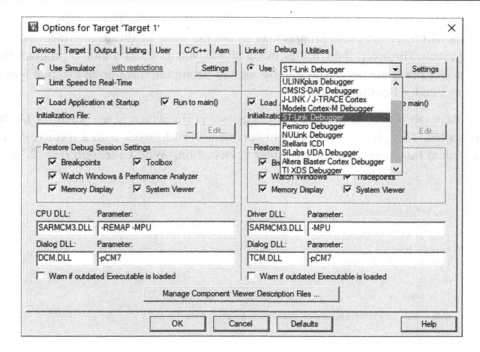

Figure 2.73
Selecting the hardware debug interface.

In the debug menu, the "Use" option has been switched to select a hardware debugger rather than the simulator. We can also select the debug hardware interface. For the Discovery board, the ST-Link debugger is used (Fig. 2.74).

Now open the utilities menu.

Figure 2.74
Setting the Flash programming algorithm from the Utilities Settings dialog.

When the project is created, the Utilities menu will have the "Use Debug Driver" box ticked. This will force the Flash programmer to use the same hardware interface selected in the Debug menu. However, the Utilities menu allows you to select a tool to program the microcontroller Flash memory. This can be another debug interface or an external tool such as a Silicon Vendor Bootloader tool.

The final Pack tab is used to provide a debugger script file which is run when the hardware debug is started. This provides any custom options needed to work with a specific device. We will look at this more closely in Chapter 8, Debugging With CoreSight (Fig. 2.75).

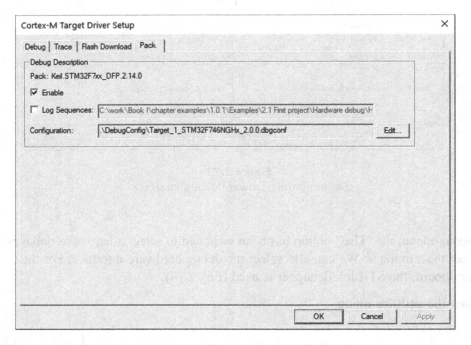

Figure 2.75
The Packs tab provides dedicated script file to support custom device features.

The most common Flash programming problems are listed below.

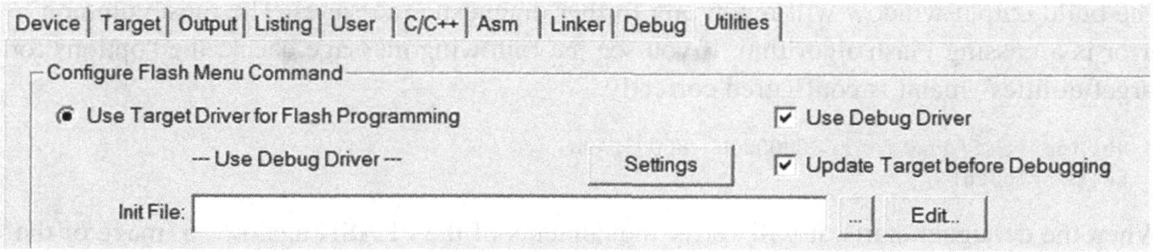

Device | Target | Output | Listing | User | C/C++ | Asm | Linker | Debug | Utilities |

Configure Flash Menu Command

⦿ Use Target Driver for Flash Programming   ☑ Use Debug Driver

--- Use Debug Driver ---   Settings   ☑ Update Target before Debugging

Init File: [   ]   ... | Edit.. |

Figure 2.76
Update Target must be selected to automatically program the FLASH when the debugger is started.

One point worth noting in the Utilities menu is the "Update Target before Debugging" tick box (Fig. 2.76).

When this option is ticked, the FLASH memory will be reprogrammed when the debugger starts. If it is not checked, then you must manually reprogram the FLASH by selecting Flash\Download from the main toolbar (Fig. 2.77).

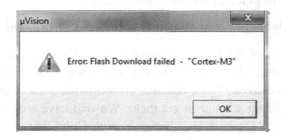

Flash   Debug   Peripherals   Tools

⬇ Download   F8

Erase

Configure Flash Tools...

Figure 2.77
Manually downloading the program image to Flash.

If there is a problem programming the FLASH memory, you will get the following error window pop up (Fig. 2.78).

μVision   ✕

⚠   Error: Flash Download failed - "Cortex-M3"

OK

Figure 2.78
Flash programming Error.

The build output window will report any further diagnostic messages. The most common error is a missing Flash algorithm. If you see the following message check, the "options for target\utilities" menu is configured correctly.

```
No Algorithm found for: 00000000H - 000032A3H
Erase skipped!
```

When the debugger starts, it will verify the contents of the FLASH against an image of the program. If the FLASH does not match the current image, you will get a memory mismatch error, and the debugger will not start. This means that the FLASH image is out of date, and the current version needs to be downloaded into the FLASH memory (Fig. 2.79).

Figure 2.79
Flash validation error.

Select cancel to close both of these dialogs without making any changes.

Start the debugger.

When the debugger starts, it is now connected to the hardware and will download the code into the FLASH memory of the microcontroller. We can now use the debugger to control the real microcontroller in place of the simulation model.

Experiment with the debugger now that it is connected to the real hardware.

You will notice that some of the features available in the simulator are not present when using the hardware module. These are the Instruction Trace, Code Coverage, and Performance Analysis windows. These features are available with hardware debug, but you need a more advanced debug adapter to get them. We will have a deeper look at the CoreSight debug system in Chapter 8, Debugging With CoreSight.

How to Get Out of Jail Free

It is possible to program the FLASH memory and accidentally lock yourself out of the microcontroller. This can happen because the device executes a sleep instruction before the debug adapter can gain control of the processor or the initializing code misconfigures the clock tree, and the onboard debug hardware does not communicate correctly with the debug adapter. Fortunately, most Cortex-M-based microcontrollers have several different boot modes (User FLASH, RAM, System Memory) that are controlled by the state of GPIO pins immediately after reset. To recover the microcontroller, simply change the boot state so that the user FLASH code is not executed. You can then reconnect the debug adapter and erase the user FLASH.

Startup Barrier

If you are experimenting with something that may upset the CPU (Power, Clock Tree Watchdog) you can place a startup barrier at the very beginning of your code. This can be as simple as a delay loop that allows the debugger to connect before the contentious code executes. This way we can guarantee to connect the debugger and always be able to erase the FLASH, if the code tips the processor into an unstable state.

Third-Party Configuration Tools

Many silicon vendors now offer a free configuration tool as part of their software support packages. These are extremely useful tools that can be used to generate the initial configuration code for system peripherals such as the MCU clock tree and device peripherals. The Microvision IDE is designed to integrate with these tools, but it is important to use them in the correct order. First, you must create an initial minimal project in microvision. Then in the RTE Device branch, if the third-party tool is supported, there will be an option to create the device configuration code. For example, the STM32 range of microcontrollers is supported by the CubeMX configuration tool.

Download and install CubeMX.

In the hardware debug project open the RTE.

Uncheck the STM32Cube Framework::Classic box.

Select the device::STM32Cube Framework and check the STM32CubeMX box.

Now press the Triangle button (Fig. 2.80).

Figure 2.80
Launch the CubeMX tool from within Microvision.

This will launch the CubeMX tool and allow you to configure the project. When you generate the project code, it will be added to the Microvision project using the pack system and an intermediate "meta" description. If you do it the other way round, that is, use CubeMX to create the project, the final generated project does not use the pack system, and the resulting project is more difficult to maintain over time.

Conclusion

By the end of this chapter, you should be able to set up a basic Cortex-M project, build the code and be able to debug it in the simulator or on a suitable hardware module. In the next chapter, we will start to look at the Cortex-M family of processors in more detail and then look at some practical issues involved in developing software to run on them.

CHAPTER 3

Cortex-M Architecture

Introduction

In this chapter, we will have a closer look at the Cortex-M processor Architecture. The bulk of this chapter will concentrate on the Cortex-M3 processor. Once we have a firm understanding of the Cortex-M3, we will look at the key differences in the Cortex-M0, M0 + , and M4. There are some significant additions in the Cortex-M7 processor, and we will look at these in Chapter 6, Cortex-M7 Processor. Once you are familiar with the classic processors based on the Armv6-M and Armv7-M architecture, we will look at the next generation of Cortex-M processors based on the Armv8-M architecture in Chapter 7, Armv8-M. Throughout the chapter, there are a number of exercises that demonstrate a particular feature of the processor. These are intended to give you a deeper understanding of each topic and can be used as a reference when developing your own code.

Cortex-M Instruction Set

As we have seen in Chapter 1, Introduction to the Cortex-M Processor Family, the Cortex-M processors are RISC-based processors and, as such, have a small instructions set. The Cortex-M0 has just 56 instructions, Cortex-M3 has 74, and Cortex-M4 has 137 with an optional additional 32 for the floating-point unit (FPU). The Arm CPU's ARM7 and ARM9, originally used in microcontrollers, have two instruction sets: the Arm (32 bit) and the THUMB (16 bit) instruction set (Fig. 3.1). The ARM instruction set was designed to get maximum performance from the CPU, while the THUMB instruction set gave an excellent code density that allowed programs to fit into the limited memory resources of a small microcontroller. The developer had to decide which function was compiled with the Arm instruction set and which was compiled with the THUMB instruction set. Then the two groups of functions could be "interworked" together to build the final program. The Cortex-M instruction set is based on the earlier 16-bit THUMB instruction set found in the Arm processors but extends the THUMB instruction to create a combined instruction set with a blend of 16- and 32-bit instructions.

The Designer's Guide to the Cortex-M Processor Family.
DOI: https://doi.org/10.1016/B978-0-323-85494-8.00008-5

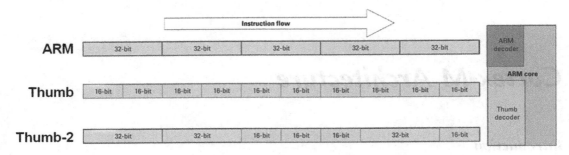

Figure 3.1
The ARM7 and ARM9 CPU had separate 32- and 16-bit instruction sets. The Cortex-M processor has a single instruction set that is a blend of 16- and 32-bit instructions.

The Cortex-M instruction set, called THUMB-2, is designed to make writing code in high-level languages much simpler and more efficient. The good news is that your whole Cortex-M project can be coded in a high-level language such as "C/C++" without the need for any hand-coded assembler. It is useful to be able to "read" THUMB-2 assembly code via a debugger disassembly window to check what the compiler is up to, but in most projects, you will never need to write an assembly routine. There are some useful THUMB-2 instructions that are not reachable using the "C" language but most compiler tool chains provide intrinsic macros which can be used to access these instructions from within your "C" code. We will look at a set of standardized compiler intrinsic function in Chapter 4, Common Microcontroller Software Interface Standard.

Programmer's Model and CPU Registers

The Cortex-M processors inherit the Arm RISC, load, and store method of operation. This means that to do any kind of data processing instruction such as ADD or SUBTRACT, the data must first be loaded into the CPU registers, the data processing instruction is then executed, and the result is stored back in the main memory. This means that code executing on a Cortex-M processor revolves around the central CPU registers (Fig. 3.2).

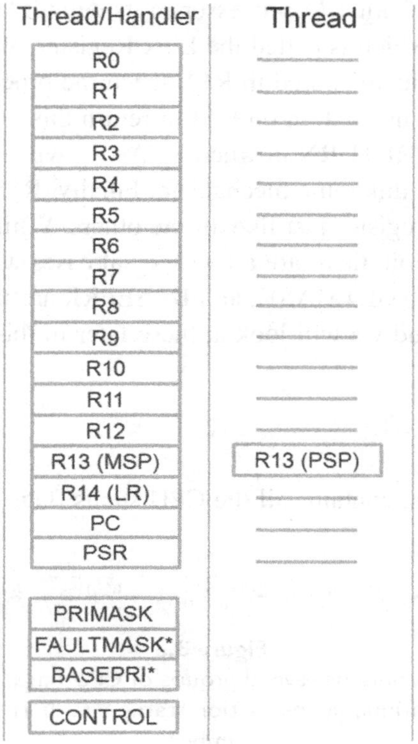

Thread/Handler	Thread
R0	————
R1	————
R2	————
R3	————
R4	————
R5	————
R6	————
R7	————
R8	————
R9	————
R10	————
R11	————
R12	————
R13 (MSP)	R13 (PSP)
R14 (LR)	————
PC	————
PSR	————

PRIMASK	
FAULTMASK*	
BASEPRI*	
CONTROL	

Figure 3.2

The Cortex-M CPU registers consist of 16 data registers, a program status register, and 4 special function registers. R13—R15 are used as the stack pointer, link register, and program counter. R13 is a banked register that allows the Cortex-M CPU to operate with dual stacks.

Within all of the Cortex-M processors, the CPU register file consists of 16 data registers followed by the Program Status Register (PSR) and a group of configuration registers. All of the data registers (R0—R15) are 32 bits wide and may be accessed by all of the THUMB-2 load and store instructions. The remaining CPU registers may only be accessed by two dedicated instructions, Move Register to Special Register (MRS) and Move Special Registers to Register (MSR). The Registers R0—R12 are general user registers that are used by the compiler as it sees fit. The registers R13—R15 have special functions. The compiler uses R13 as the stack pointer. This is actually a banked register with two R13 registers. When the Cortex-M processor comes out of reset, the second R13 register is not enabled, and the processor runs in a "simple" mode with one stack pointer referred to as the Main Stack Pointer (MSP). It is possible to enable the second R13 register by writing to the Cortex-M CPU CONTROL register, then the processor will be configured to run with two stacks. We will look at this in more detail in Chapter 5, Advanced Architecture Features, but for now, we will use the Cortex-M processor in its default "simple" mode. After the stack pointer, we have R14 which is called the Link Register. When a procedure is called, the return address is automatically stored in R14, when the processor reaches the end of a procedure, it uses the branch instruction on R14 to return this means that the instruction set does not contain a dedicated RETURN instruction. As we will see later, a return from interrupt is managed through this same mechanism. Finally, R15 is the program counter, and you can operate on this register just like all the others. While we don't need to access the PC during normal operation, there are a few special cases where this can be useful. The CPU registers, PRIMASK, FAULTMASK, and BASEPRI, are used to temporarily disable or limit interrupt handling, and we will look at these later in this chapter.

Program Status Register

The PSR, as its name implies, contains all the CPU status flags (Fig. 3.3).

Figure 3.3
The Program status register contains several groups of CPU flags. These include the condition codes (NZCVQ), Interrupt continuable instruction status bits, If Then flag, and current exception number.

The PSR has a number of alias fields that are masked versions of the full register. The three alias registers are the Application Program Status Register, Interrupt Program Status Register, and the Execution Program Status Register. Each of these alias registers contains a subset of the full register flags and can be used as a shortcut if you need to access part of the PSR. The PSR is generally referred to as the xPSR to indicate the full register rather than any of the alias subsets (Fig. 3.4).

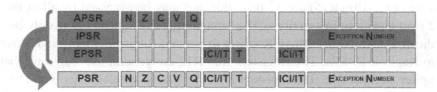

Figure 3.4
The Program status register has three alias registers that provide access to specific subregions or the Program status register. Hence, the generic name for the program status register is xPSR.

In a normal application program, your code will not make explicit access to the xPSR or any of its alias registers. Any use of the xPSR will be made by compiler-generated code. However, as a programmer, you need to have an awareness of the xPSR and the flags contained in it.

The most significant four bits of the xPSR are the condition code bits, **N**egative, **Z**ero, **C**arry, and o**V**erflow. These will be set and cleared depending on the results of a data processing instruction. The result of THUMB-2 data processing instructions can set or clear these flags. However, updating these flags is optional.

```
SUB R8, R6, #240        Perform a subtraction and do not update the condition code flags
SUBS R8, R6, #240       Perform a subtraction and update the condition code flags
```

This allows the compiler to perform an instruction that updates the condition code flags, then perform some additional instructions that do not modify the flags, and then perform a conditional branch on the state of the xPSR condition codes.

Q Bit and Saturated Math's Instructions

The Q bit follows the xPSR condition code flags and is called the saturation flag. The Cortex-M3/M4 and Cortex-M7 processors have a special set of instructions called the saturated maths instructions. There is a potential problem with standard "C" variables in that they are vulnerable to wraparound. If we increment such a variable and it reaches its maximum value, any further increment will cause it to roll round to zero (Fig. 3.5). Similarly, if a variable reaches its minimum value and is then decremented, it will roll round to the maximum value.

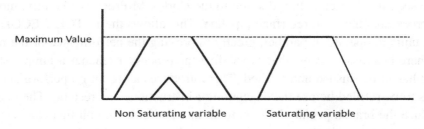

Figure 3.5

A standard variable will rollover to zero when it hits its maximum value. This is very dangerous in a control algorithm. The Cortex-M CPU supports saturated maths instructions which stick at their maximum and minimum values.

While this is a problem for most applications, it is especially serious for applications such as motor control and safety-critical applications. The Cortex-M3/M4 and M7 saturated maths instructions prevent this kind of wraparound. When you use the saturated maths instructions, if the variable reaches its maximum or minimum value, it will stick (saturate) at that value. Once the variable has saturated, the Q bit will be set. The Q bit is a "sticky" bit and must be cleared by the application code. The saturated maths instructions are not used by the "C" compiler by default. If you want to use the saturated maths instructions, you have to access them by using compiler intrinsics or by using the CMSIS-Core functions shown below.

```
uint32_t__SSAT(uint32_t value, uint32_t sat)
uint32_t__USAT(uint32_t value, uint32_t sat)
```

Interrupts and Multicycle Instructions

The next field in the PSR is the "Interrupt Continuable Instruction" (ICI) and "If Then" (IT) instruction flags. Most of the Cortex-M processor instructions are executed in a single cycle. However, some instructions such as Load Store Multiple, Multiply, and Divide take multiple cycles. If an interrupt occurs while these instructions are executing, they have to be suspended while the interrupt is served. Once the interrupt has been served, we have to resume the multicycle instruction. The ICI field is managed by the Cortex-M processor, so you do not need to do anything special in your application code. It does mean that when an exception is raised, reaching the start of your interrupt routine will always take the same amount of cycles regardless of the instruction currently being executed by the CPU.

Conditional Execution—If Then blocks

As we have seen in Chapter 1, Introduction to the Cortex-M Processor Family, most of the Cortex-M processors have a three-stage pipeline. This allows the FETCH DECODE and EXECUTE units to operate in parallel, greatly improving the performance of the processor. However, there is a disadvantage that every time the processor reaches a jump instruction, the pipeline has to be flushed and refilled. This introduces a big hit on performance as the pipeline has to be refilled before the execution of instructions can resume. The Cortex-M3 and M4 reduce the branch penalty by having an instruction fetch unit that can carry out speculative branch target fetches. However, for small conditional branches, the Cortex-M processor has another trick up its sleeve. For a small conditional branch, for example

```
if(Y = = 0x12C){
   I + + ;
   }else{
   I−; }
```

which compiles to less than four instructions, the Cortex-M processor can compile the code as an IF THEN condition block. The instructions inside the If THEN block are extended with a condition code (Table 3.1). This condition code is compared to the state of the Condition Code Flags in the PSR. If the condition matches the state of the flags, then the instruction will be executed. If it does not, then the instruction will still enter the pipeline but will be executed as a NOP. This technique eliminates the branch and hence avoids the need to flush and refill the pipeline. Even though we are inserting NOP instructions, we still get better performance levels than by using a more standard compare and branch approach.

Table 3.1: Instruction condition codes

Condition Code	xPSR Flags Tested	Meaning
EQ	Z = 1	Equal
NE	Z = 0	Not Equal
CS or HS	C = 1	Higher or Same (Unsigned)
CC or LO	C = 0	Lower (Unsigned)
MI	N = 1	Negative
PL	N = 0	Positive or Zero
VS	V = 1	Overflow
VC	V = 0	No Overflow
HI	C = 1 and Z = 0	Higher (Unsigned)
LS	C = 0 or Z = 1	Lower or same (Unsigned)
GE	N = V	Greater than or equal (Signed)
LT	N ! = V	Less than (Signed)
GT	Z = 0 and N = V	Greater than (Signed)
LE	Z = 1 and N ! = V	Less than or equal (Signed)
AL	None	Always execute

To trigger an IF THEN block, we use the data processing instructions to update the PSR condition codes. By default, most instructions do not update the condition codes unless they have an S suffix added to the assembler opcode. This gives the compiler a great deal of flexibility in applying the IF THEN condition.

```
ADDS R1,R2,R3      //perform and add and set the xPSR flags
ADD R2,R4,R5       //Do some other instructions but do not modify the xPSR
ADD R5,R6,R7
IT VS              //IF THEN block conditional on the first ADD instruction
SUBVS R3,R2,R4
```

So our 'C' IF THEN ELSE statement can be compiled into four instructions.

```
CMP      r6,#0x12C
ITE      EQ
STREQ    r4,[r0,#0x08]
STRNE    r5,[r0,#0x04]
```

The CMP compare instruction is used to perform the test and will set or clear the Zero Z flag in the PSR. The IF THEN block is created by the If THEN IT instruction. The IT instruction is always followed by one conditionally executable instruction and optionally up to four conditionally executable instructions. The format of the IT instruction is as follows

```
IT x y z cond
```

The x, y, and z parameters enable the second, third, and fourth instructions to be part of the conditional block. There can be further THEN instruction or ELSE instructions. The cond parameter is the condition applied to the first instruction. So

ITTTE NE   A four instruction If Then block with three THEN instructions which execute when the Z = 1 followed by an ELSE instruction which executes when Z = 1

ITE GE   A two-instruction IF THEN block with one THEN instruction which executes when N = V and one ELSE instruction which executes when N! = V

The use of conditional executable IF THEN blocks is left up to the compiler. Generally, at low levels of optimization, IF THEN blocks are not used, this gives a good debug view. However, at high levels of optimization, the compiler will make use of IF THEN blocks. So normally, there will not be any strange side effects introduced by the conditional execution technique, but there are a few rules to bear in mind.

First conditional code blocks cannot be nested. Generally, the compiler will take care of this rule. Second, you cannot use a GOTO statement to jump into a conditional code block. If you do make this mistake, the compiler will warn you and not generate such illegal code.

Third, the only time you will really notice execution of an IF THEN condition block is during debugging. If you single step the debugger through the conditional statement, the conditional code will appear to execute even if the condition is false. If you are not aware of the Cortex-M condition code blocks, this can be a great cause of confusion!

The next bit in the PSR is the T or Thumb bit. This is a legacy bit from the earlier Arm CPUs and is set to one, if you clear this bit, it will cause a fault exception. In previous CPUs the T bit was used to indicate that the THUMB 16-bit instruction set was running. It is included in the Cortex-M PSR to maintain compatibility with earlier Arm CPUs and allow legacy 16-bit THUMB code to be executed on the Cortex-M processors. The final field in the PSR is the Exception Number field. When an interrupt or exception is being processed, the exception or interrupt channel number is stored in the Exception Number field. As we will see later this field is not used by the application code when handling an interrupt, though it can be a useful reference when debugging.

Exercise 3.1: Saturated Maths and Conditional Execution

In this exercise, we will use a simple program to examine the CPU registers and make use of the saturated math's instructions. We will also rebuild the project to use conditional execution.

Open the Pack Installer.

Select the Boards tab and "The Designers Guide Tutorial Examples."

Select the example tab and Copy "Exercise 3.1: Saturation and Conditional Execution."

This program increments an integer variable "a" from 0 to 300 and copies it to an unsigned char variable "c."

```
int a,range = 300;
unsigned char c;
int main (void){
while (1){
for(a = 0;a<range;a + +){
            c = a;
}}
```

Build the program and start the debugger.

Add the two variables to the system analyzer and run the program.

In Fig. 3.6 the system analyzer window shows that while the integer variable performs as expected the "standard_var" which is an unsigned char saturates when it reaches 255 and "rolls over" to zero then begins incrementing again.

Stop the debugger and modify the code as shown below to use saturated maths.

```
int a,range = 300;
unsigned char c;
int main (void){
while (1){
for(a = 0;a<range;a + +){
    c = __USAT (a, 9);
    }
}}
```

This code replaces the equate with a saturated intrinsic function that saturates on the ninth bit of the integer value. This allows values 0–255 to be written to the byte value, any other values will saturate at the maximum allowable value of 255.

Build the project and start the debugger.

Run the code and view the contents of the variables in the system analyzer (Fig. 3.6).

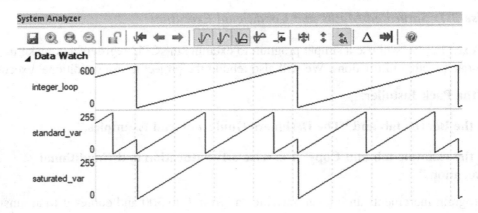

Figure 3.6
Using the saturation instructions a variable cal "saturate" at a selected bit boundary.

Now the unsigned char variable saturates rather than "rolling over." This is still wrong but not as potentially catastrophically wrong as the rollover case. If you change the bit boundary value "c" will saturate at lower values.

In the registers window click on the xPSR register to view the flags (Fig. 3.7).

⊟ xPSR	0x69000000
N	0
Z	1
C	1
V	0
Q	1
T	1
IT	Disabled
ISR	0

Figure 3.7
When a variable saturates the Q bit in the xPSR will be set.

In addition to the normal NVCZ condition code flags the saturation Q bit is set.

Exit the debugger and modify the code as shown below.

```
#define Q_FLAG 0x08000000
int a,range = 300;
unsigned char c;
unsigned int APSR;
register unsigned int apsr__asm("apsr");
int main (void){
```

```
while (1){
    for(a = 0;a<range;a+ +){
        c = __SSAT(a, 9);
    }
APSR = __get_APSR ();
if(APSR&Q_FLAG){
    range−;
}
apsr = apsr&~Q_FLAG;
}}
```

Once we have written to the unsigned char variable, it is possible to read the APSR (Application alias of the xPSR) and check if the Q bit is set. If the variable has saturated, we can take some corrective action and then clear the Q bit for the next iteration.

Build the project and start the debugger.

Run the code and observe the variables in the watch window.

Now when the data is over range, the unsigned char variable will saturate and gradually the code will adjust the range variable until the output data fits into unsigned char variable.

Set breakpoints on lines 19 and 22 to enclose the Q bit test (Fig. 3.8).

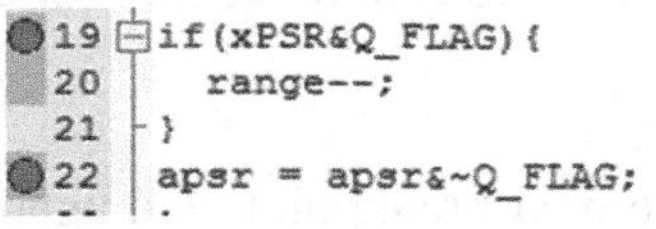

Figure 3.8
Breakpoints on lines 19 and 22.

Reset the program and then run the code until the first breakpoint is reached.

Open the disassembly window and examine the code generated for the Q bit test.

```
if(((xPSR & Q_FLAG)! = 0))
    MOV   r0,r1
    LDR   r0,[r0,#0x00]
    AND   r0,r0,#0x8000000
    CMP   r0,#0x00
    BEQ   0x080001E2
    {
        range−;
    }
    LDR   r0,[pc,#44];
    LDRH  r0,[r0,#0x00]
    SUB   r0,r0,#0x01
```

```
LDR   r1,[pc,#36];
STRH  r0,[r1,#0x00]
B   0x080001EA
else
{
  locked = 1;
}
MOV   r0,#0x01
LDR   r1,[pc,#36];
STRH  r0,[r1,#0x00]
set_apsr(0;
```

Also, make a note of the value in the state counter. This is the number of cycles used since reset to reach this point (Fig. 3.9).

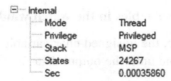

Figure 3.9
Value of the states counter at the first breakpoint.

Now run the code until it hits the next breakpoint, and again make a note of the state counter (Fig. 3.10).

Figure 3.10
Value of the states counter at the second breakpoint.

Stop the debugger and open the "options for target/C tab" (Fig. 3.11).

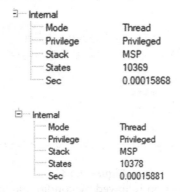

Figure 3.11
Change the optimization level.

Change the optimization level from Level 0 to Level 3 (Fig. 3.3).

Close the options for target and rebuild the project.

Now repeat the cycle count measurement by running to the two breakpoints (Fig. 3.12)

Internal	
Mode	Thread
Privilege	Privileged
Stack	MSP
States	10369
Sec	0.00015868

Internal	
Mode	Thread
Privilege	Privileged
Stack	MSP
States	10378
Sec	0.00015881

Figure 3.12
Cycle counts for the first and second breakpoint.

Now the Q bit test takes nine cycles as opposed to the original 18.

Examine the disassembly code for the Q bit test.

```
if(((xPSR & Q_FLAG)! = 0))
    STR    r2,[r0,#0x08]
    AND    r2,r2,#0x8000000
    CMP    r2,#0x00
    ITTE   NE
    SUBNE  r1,r1,#1
    {
        range-;
    }
        else
    {
        locked = 1;
    }
    STRHNE   r1,[r0,#0x04]
    STRHEQ   r4,[r0,#0x06]
    apsr = apsr&~Q_FLAG;
```

At higher levels of optimization, the compiler has switched from test and branch instructions to conditional execution instructions. Here, the assembler is performing a bitwise AND test on R1 which holds the current value of the xPSR. This will set or clear the Z flag in the xPSR. The ITT instruction sets up a two-instruction conditional block. The instructions in this block perform a subtract and store if the Z flag is zero, otherwise they pass through the pipeline as NOP instructions.

Remove the breakpoints. Run the code for a few seconds then halt it.

Set a breakpoint on one of the conditional instructions (Fig. 3.13).

```
0x080001BC BF1A      ITTE    NE
0x080001BE 1E49      SUBNE   r1,r1,#1
    31:                      range--;
    32:             }
    33:       else{
0x080001C0 8081      STRHNE  r1,[r0,#0x04]
    34:         locked = 1;
    35:       }
0x080001C2 80C4      STRHEQ  r4,[r0,#0x06]
```

Figure 3.13
If then instruction followed by conditional instructions.

Start the code running again.

The code will hit the breakpoint even though the breakpoint is within an IF statement that should no longer be executed. This is simply because the conditional instructions are always executed.

Cortex-M Memory Map and Busses

While each Cortex-M microcontroller family will have a different range of peripherals and memory sizes, Arm has defined a basic memory template that all devices must adhere to. This provides a standard layout, so all the vendor-provided memory and peripherals are located in common memory regions.

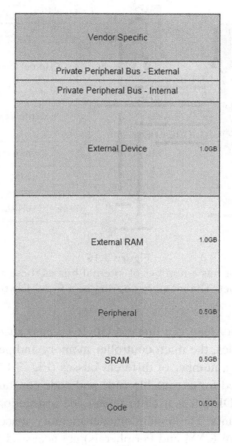

Figure 3.14
The Cortex-M memory map has a standard template that splits the 4 Gbyte address range into specific memory regions. This memory template is common to all Cortex-M devices.

The Cortex-M memory template defines eight regions that cover the 4 Gbyte address space of the Cortex-M processor (Fig. 3.14). The first three regions are each 0.5 Gbyte in size and are dedicated to the executable code space, internal RAM, and internal peripherals. The next two regions are dedicated to external memory and memory-mapped devices, both of these regions are 1 Gbyte in size. The final three regions make up the Cortex-M processor memory space and contain the configuration and Control registers for the Cortex-M processor called the system control block (SCB) plus the debugger registers which are located in a region called the System Control Space. An address space is also provided at the top of the address map for vendor-specific registers.

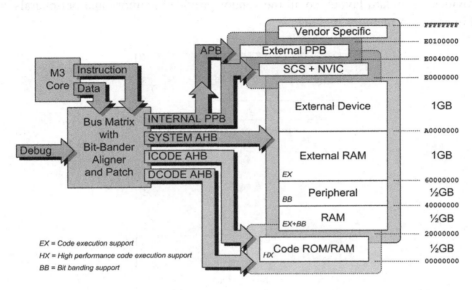

Figure 3.15
While the Cortex-M processor has a number of internal busses these are essentially invisible to the software developer. The memory appears as a flat 4 Gbyte address space.

While the Cortex-M memory map is a linear 4 Gbyte address space with no paged regions or complex addressing modes, the microcontroller memory and peripherals are connected to the Cortex-M processor by a number of different busses (Fig. 3.15). The first 0.5 Gbyte of the address space is reserved for executable code and code constants. This region has two dedicated busses. The ICODE bus is used to fetch code instructions, while the DCODE bus is used to fetch code constants. The remaining user memory spaces (Internal RAM and Peripherals plus the External RAM and Peripherals) are accessed by a separate system bus. The Cortex-M processor memory space has an additional private peripheral bus that the CPU uses to access its own peripherals and configuration registers. While this may look complicated as far as your application code is concerned, you have one seamless linear memory space. The Cortex-M processor will use the separate busses to optimize its access to different memory regions.

As mentioned earlier, most of the instructions in the THUMB-2 instruction set are executed in a single cycle. The Cortex-M3 can run up to 200 MHz, and in fact, some system-on-chip designs manage to get the processor running even faster. However, the current FLASH memory used to store the program has an access time of around 50 MHz. So there is a basic problem of pulling instructions out of the FLASH memory fast enough to feed the Cortex-M processor. When you are selecting a Cortex-M microcontroller, it is important to study the data sheet to see how the Silicon Vendor has solved this problem. Typically, the FLASH memory will be arranged as 64- or 128-bit-wide memory, so one read from the FLASH memory can load multiple instructions. These instructions are then held in a "memory accelerator" unit which then feeds the instructions to the Cortex-M processor as required. The memory accelerator is a form of simple cache unit which is designed by the Silicon Vendor. Normally this unit is disabled after reset and enabled by the supplied startup code. As part of the initial bring-up tests, you will need to make sure it is switched on, or the Cortex-M processor will be running directly from the FLASH memory. The overall performance of the Cortex-M processor will depend on how successfully this unit has been implemented by the Designers of the Microcontroller.

Write Buffer

The Cortex-M3 and M4 contain a single-entry data write buffer. This allows the CPU to make an entry into the write buffer and continue on to the next instruction while the write buffer completes the write to the real RAM or peripheral register. This avoids stalling the processor while it waits for the write to complete. If the write buffer is full, the CPU is forced to wait until it has finished its current write. While this is normally a transparent process to the application code, there are some cases where it is necessary to wait until the write to the RAM or peripheral register has finished before continuing program execution. For example, if we are enabling an external bus on the microcontroller, it is necessary to wait until the write buffer has finished writing to the peripheral register and the bus is enabled before trying to access memory located on the external bus. The Cortex-M processor provides some memory barrier instructions to deal with these situations.

Memory Barrier Instructions

The memory barrier instructions (Table 3.2) halt the execution of the application code until the write stage of an instruction has finished executing. They are used to ensure a critical section of code has been completed before continuing the execution of the application code.

Table 3.2: Memory barrier istructions

Instruction	Description
DMB	Ensures all memory accesses are finished before a fresh memory access is made
DSB	Ensures all memory accesses are finished before the next instruction is executed
ISB	Ensures that all previous instructions are completed before the next instruction is executed. This also flushes the CPU pipeline

System Control Block

In addition to the CPU registers, the Cortex-M processors have a group of memory-mapped configuration and status registers located near the top of the memory map starting at 0xE000 E008.

We will look at the key features supported by these registers thought the rest of this book, but a summary is given in Table 3.3.

Table 3.3: The Cortex processor has memory-mapped configuration and status registers located in the system control block

Register	Size in Words	Description
Auxiliary Control	1	Allows you to customize how some processor features are executed
CPU ID	1	Hardwired ID and revision numbers from Arm and the Silicon Manufacturer
Interrupt Control and State	1	Provides Pend bits for the SysTick and NMI interrupts and extended interrupt Pending\active information
Vector Table Offset	1	Programmable address offset to move the Vector table to a new location in Flash or SRM memory
Application Interrupt and Reset Control	1	Allows you to configure the PRIGROUP and generate CPU and microcontroller resets
System Control	1	Controls configuration of the processor sleep modes
Configuration and control	1	Configures CPU operating mode and some fault exceptions
System Handler Priority	3	These registers hold the 8-bit priority fields for the configurable Processor exceptions
System Handler Control and State	1	Shows the cause of a bus, memory management, or usage fault
Configurable Fault Status	1	Shows the cause of a bus, memory management, or usage fault
Hard Fault Status	1	Shows what event caused a hard fault
Memory Manager Fault Address	1	Holds the address of the memory location that generated the memory fault
Bus Fault Address	1	Holds the address of the memory location that generated the memory fault

Memory Access

The Cortex-M instruction set has addressing instructions that allow you to load and store 8-, 16-, and 32-bit quantities. Unlike the ARM7 and ARM9, the 16- and 32-bit quantities do not need to be aligned on a word or half-word boundary. This gives the compiler and linker maximum flexibility to fully pack the RAM memory (Fig. 3.16). However, there is a penalty to be paid for this flexibility because unaligned transfers take longer to carry out. There are also some restrictions on unaligned accesses. The Cortex-M instruction set contains Load and Store Multiple instructions, which can transfer multiple registers to and from memory in one instruction. This takes multiple processor cycles but uses only one 2- or 4-byte instruction. This allows for very efficient stack manipulation and block memory copy. The load and store multiple instructions only work for word-aligned data. Also, unaligned data is for user data only, you must ensure that the stacks are word-aligned. The MSP initial value is determined by the linker but the second stack pointer the Process Stack Pointer (PSP) is enabled and initialized by the user, so it is up to you to get it right. We will look at using the PSP in Chapter 5, Advanced Architecture Features.

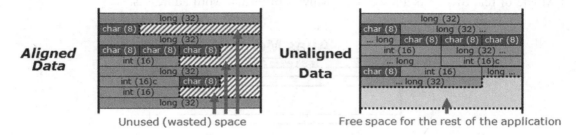

Figure 3.16
Unlike the earlier ARM7 and ARM9 CPUs the Cortex processor can make unaligned memory accesses. This allows the Compiler and linker to make the best use of the device SRAM.

Bit Manipulation

In a small embedded system, it is often necessary to set and clear individual bits within the RAM and peripheral registers. By using the standard addressing instructions, we can set and clear individual bits by using the "C" language bitwise AND and OR commands. While this works ok, the Cortex-M processors provide a more efficient bit manipulation method.

The Cortex-M processor provides a method called "bit banding," which allows individual SRAM and Peripheral register bits to be set and cleared in a very efficient manner.

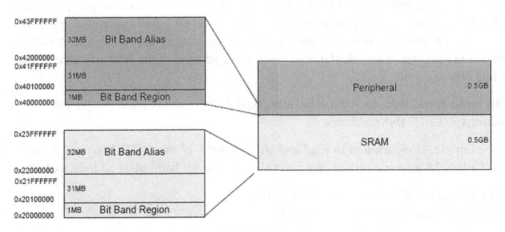

Figure 3.17
Bit banding is a technique that allows the first 1 Mbyte of the SRAM region and the first 1 MByte of the peripheral region to be bit addressed.

The first 1 Mbyte or the SRAM region and the first 1 MByte of the Peripheral region are defined as Bit Band regions (Fig. 3.17). This means that every memory location in these regions is bit addressable. In practice for today's microcontrollers, all of their internal SRAM and peripheral registers are bit addressable. Bit Banding works by creating an alias word address for each bit of real memory or peripheral register bit. This means that the 1 Mbyte of real SRAM is aliased to 32 Mbytes of virtual word addresses.

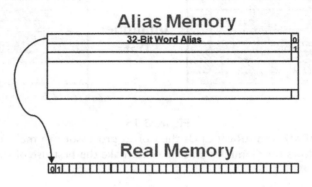

Figure 3.18
Each bit in the real memory is mapped to a word address in the alias memory.

In the Bit Band regions, each bit of real RAM or Peripheral register is mapped to a word address in the bit band alias region (Fig. 3.18), and by writing 1's and 0's to the alias word address, we can set and clear the real memory bit location. Similarly, if we write a word to the real memory location, we can read the bit band alias address to check the current state of a bit in that word. To use bit banding, you simply need to calculate the word address in the bit band region that maps to the bit location in the real memory that you want to modify. Then create a pointer to the word address in the bit band region. Once this is done, you can control the real bit memory location by reading and writing to the alias region via the pointer. The calculation for the word address in the bit band alias region is as follows:

Bit Band Word Address = bit band alias base + (byte offset from peripheral base address x 32) + (bit number x 4)

So, for example, if we wanted to read and write to bit 8 of the GPIO B port register on a typical Cortex-M microcontroller, we can calculate the bit band alias address as follows:

GPIO B data register address = 0x40010C0C
Peripheral base address 0x400000000

Register byte offset from peripheral base address = 0x40010C0C − 0x40000000
= 0x00010C0C

Bit Band Word Address = 0x42000000 + (0x000010C0C * 0x20)) + (0x8 *0x4)
= 0x422181A0

We can define a pointer to this address

```
#define GPIO_PORTB_BIT8 = (*((volatile unsigned long *)0x422181A0))
```

Now by reading and writing to this word address, we can directly control the individual port bit.

```
GPIO_PORTB_BIT8 = 1    //set the port pin
```

This will compile to the following assembler instructions:

```
Opcode      Assembler
F04F0001    MOV   r0,#0x01
4927        LDRr  1,[pc,#156]   ;@0x080002A8
6008        STR   r0,[r1,#0x00]
```

This sequence uses one 32-bit instruction and two 16-bit instructions or a total of eight bytes. If we compare this to setting the port pin by using a logical OR to write directly to the port register.

```
GPIOB->ODR | = 0x00000100;  //LED on
```

We then get the following code sequence:

```
Opcode      Assembler
481         ELDR  r0,[pc,#120]   ;@0x080002AC
6800        LDR   r0,[r0,#0x00]
F4407080    ORR   r0,r0,#0x100
491         CLDR  r1,[pc,#112]   ;@0x080002AC
6008        STR   r0,[r1,#0x00]
```

This uses four 16-bit instructions and one 32-bit instruction or 12 bytes.

The use of bit banding gives us a win−win situation, smaller code size, and faster operation. So as a simple rule, if you are going to repetitively access a single bit location, you should use bit banding to generate the most efficient code.

You may find some compiler tools or Silicon Vendor software libraries that provide macro functions to support bit banding. You should generally avoid using such macros as they may not yield the most efficient code and may be tool chain dependent.

Exercise 3.2: Bit Banding

In this exercise, we will look at defining a bit band variable to toggle a port pin and compare its use to bitwise AND and OR instructions.

Open the Pack Installer.

Select the Boards tab and "The Designers Guide Tutorial Examples."

Select the example tab and Copy "Exercise 3.2: Bit Banding."

In this exercise, we want to toggle an individual port pin. We will use the PortB bit 8 pin as we have already done the calculation for the alias word address.

So now, in the "C" code, we can define a pointer to the bit band address

```
#define PortB_Bit8 (*((volatile unsigned long *)0x422181A0))
```

And in the application code, we can set and clear the port pin by writing to this pointer

```
PortB_Bit8 = 1;
PortB_Bit8 = 0;
```

Build the project and start the debugger.

Enable the timing analysis with the debug\execution profiling\Show Time menu (Fig. 3.19).

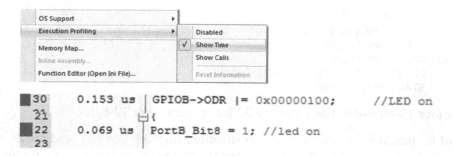

Figure 3.19
Enable the timing profile to show run time per line of code.

This opens an additional column in the debugger, which will display the execution time for each line. We can use this to compare the execution time for the bit band instruction to the AND and OR methods.

Open the disassemble window and examine the code generated for each method of setting the port pin.

The Bit Banding instructions are the best way to set and clear individual bits. You should use them in any part of your code that repetitively manipulates a bit.

Dedicated Bit Manipulation Instructions

In addition to the bit band support, the THUMB-2 instruction set has some dedicated bit-orientated instructions (Table 3.4). Some of these instructions are not directly accessible from the "C" language and are supported by compiler "intrinsic" calls.

Table 3.4: In addition to bit banding the Cortex-M3 processor has some dedicated bit manipulation instructions

BFC	Bit field clear
BFI	Bit field insert
SBFX	Signed bit field extract
SXTB	Sign extend a byte
SXTH	Sign extend a half-word
UBFX	Unsigned bit field extract
UXTB	Zero extend a byte
UXTH	Zero extend a half-word

SysTick Timer

All of the Cortex-M processors contain a standard timer. This is called the SysTick timer and is a 24-bit countdown timer with auto-reload (Fig. 3.20). Once started, the SysTick timer will count down from its initial value. When it reaches zero, it will raise an interrupt, and a new count value will be loaded from the reload register. The main purpose of this timer is to generate a periodic interrupt for a real-time operating system (RTOS) or other event-driven software. If you are not running an RTOS you can also use it as a simple timer peripheral.

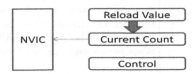

Figure 3.20
The SysTick timer is a 24-bit countdown timer with auto-reload. It is generally used to provide a periodic interrupt for an RTOS scheduler.

The default clock source for the SysTick timer is the Cortex-M CPU clock. It may be possible to switch to another clock source, but this will vary depending on the actual microcontroller you are using. While the SysTick timer is common to all the Cortex-M processors, its registers occupy the same memory locations within the Cortex-M3/M4 and M7. In the Cortex-M0 and M0 + , the SysTick registers are located in the SCB and have different symbolic names to avoid confusion. The SysTick timer interrupts line and all of the microcontroller peripheral lines are connected to the Nested Vector Interrupt Controller or NVIC.

Nested Vector Interrupt Controller

Aside from the Cortex-M CPU, the next major unit within the Cortex-M processor is the NVIC. The NVIC is the same to use between all Cortex-M processors. Once you have set up an interrupt on a Cortex-M3, the process is the same for Cortex-M0 through to Cortex-M7. The NVIC is designed for fast and efficient interrupt handling and is configured through a set of memory-mapped registers (Table 3.5). On a Cortex-M3 you will reach the first line of "C" code in your interrupt routine after 12 cycles for zero wait state memory system. This interrupt latency is fully deterministic, so from any point in the background (noninterrupt) code, you will enter the interrupt with the same latency. As we have seen, multicycle instructions can be halted with no overhead and then resumed once the interrupt has finished. On the Cortex-M3/M4 and Cortex-M7, the NVIC supports up to 240 interrupt sources. While the Cortex-M0 is limited to 32, the NVIC supports up to 256 interrupt priority levels on Cortex-M3, M4, and M7 and four priority levels on Cortex-M0.

Table 3.5: The nested vector interrupt controller consists of seven register groups that allow you to enable, set priority levels and monitor the user interrupt peripheral channels

Register	Maximum Size in Words[a]	Description
Set Enable	8	Provides an interrupt enable bit for each interrupt source
Clear Enable	8	Provides an interrupt disable bit for each interrupt source
Set Pending	8	Provides a set pending bit for each interrupt source
Clear Pending	8	Provides a clear pending bit for each interrupt source
Active	8	Provides an interrupt active bit for each interrupt source
Priority	60	Provides an 8-bit priority field for each interrupt source
Software trigger	1	Write the interrupt channel number to generate a software interrupt

[a]The actual number of words used will depend on the number of Interrupt channels implemented by the microcontroller manufacturer.

Operating Modes

While the Cortex-M CPU is executing background (noninterrupt code), the CPU is in an operating mode called Thread mode. When an interrupt is raised, the NVIC will cause the processor to jump to the appropriate Interrupt Service Routine. When this happens, the CPU changes to a new operating mode called Handler mode. In simple applications without an RTOS, you can use the default configuration of the Cortex-M processor out of reset, and there is no major functional difference in these operating modes. You are able to access all of the microcontroller's resources from either Thread or Handler mode. The Cortex-M processors can be configured with a more complex operating model that introduces operating differences between Thread and Handler mode that limit access to some processor registers and instructions. For now, we will look at basic interrupt handling and then introduce the more advanced configuration options in Chapter 5, Advanced Architecture Features.

Interrupt Handling—Entry

When a microcontroller peripheral raises an interrupt line, the NVIC will cause two things to happen in parallel. First, the exception vector is fetched over the I-CODE bus. This is the address of the entry point into the Interrupt Service Routine. This address is pushed into R15, the Program Counter forcing the CPU to jump to the start of the interrupt routine. In parallel, the CPU will automatically PUSH key registers onto the stack and POP them back at the end of the ISR. This stack frame consists of the following registers, xPSR, PC, LR, R12, R3, R2, R1, R0 (Fig. 3.21). This stack frame preserves the state of the processor and provides R0—R3 for use by the Interrupt Service Routine. If the Interrupt Service Routine needs to use more CPU registers, it must PUSH them onto the stack and POP them on exit.

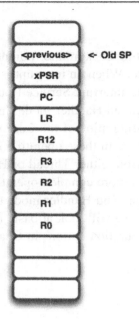

Figure 3.21
When an interrupt or exception occurs the CPU will automatically push a stack frame. This consists of the xPSR, PC, LR, R12, and registers R0–R3. At the end of the interrupt or exception the stack frame is automatically unstacked.

The interrupt entry process takes 12 cycles on the Cortex-M3/M4/M7 and 16 cycles on the Cortex-M0. All of these actions are handled by microcode in the CPU. Any additional entry code will be provided by the compiler. In practice, you simply write the service routine as a "C" function.

The exception vectors are stored in an Interrupt Vector Table. The Interrupt Vector Table is located at the start of the address space 0x00000000. The first four bytes are used to hold the initial value of the MSP. The starting value of the stack pointer is calculated by the compiler and linker. The resulting value will be placed at the start of the image when the program is built. After reset this value will be automatically loaded into R13 to initialize the MSP. The Interrupt Vector Table then has address locations every four bytes growing upward through the address space. The first four bytes of the vector table hold the address of the reset handler. This is followed by 10 processor exception addresses and then addresses for each of the microcontroller Interrupt Service Routines. The vector table for each microcontroller comes predefined as part of the startup code, as we will see in the next chapter. A label for each interrupt service routine is stored at each interrupt vector location. To create a matching interrupt service routine, you simply need to declare a void "C" function using the same name as the interrupt vector label. A sample vector table is shown below:

```
        AREA RESET, DATA, READONLY
        EXPORT__Vectors
__Vectors DCD__initial_sp    ; Top of Stack
        DCD Reset_Handler        ; Reset Handler
      ; External Interrupts
        DCD WWDG_IRQHandler      ; Window Watchdog
        DCD PVD_IRQHandler       ; PVD through EXTI Line detect
        DCD TAMPER_IRQHandler    ; Tamper
        DCD RTC_IRQHandler       ; Real time clock
        DCD FLASH_IRQHandler     ; Flash
        DCD RCC_IRQHandler       ; RCC
```

So to create the "C" routine to handle an interrupt from the real-time clock, we create a "C" function named as follows.

```
void RTC_IRQHandler(void) {
    ....
}
```

When the project is built the linker will resolve the address of the "C" routine and locate it in the vector table in place of the label. If you are not using this particular interrupt in your project the label still has to be declared to prevent an error during the linking process. Following the interrupt vector table, a second table declares all of the interrupt service routine addresses. These are declared as WEAK labels. This means that the declaration can be overwritten if the label is declared elsewhere in the project. In this case, they act as a "backstop" to prevent any linker errors if the interrupt routine is not formally declared in the project source code.

```
EXPORT RTC_IRQHandler    [WEAK]
B.
ENDP
```

Interrupt Handling—Exit

Once the Interrupt Service Routine has finished its task, it will force a return from the interrupt to the background code. However, the THUMB-2 instruction set does not have a Return or Return From Interrupt instruction. The Interrupt Service Routine will use the same return method as a noninterrupt routine, namely, a branch on the R14, the link register. During normal operation, the link register will contain the valid return address. However, when we entered the interrupt, the current contents of R14 were pushed onto the stack and in its place the CPU entered a special value (Table 3.6) into the R14, link register, this value is not a valid return address as it is out of the executable region defined in the memory map template. When the CPU detects this value, instead of doing a normal branch, it is forced to restore the stack frame and resume normal processing.

Table 3.6: At the start of an exception or interrupt R14 (link register) is pushed onto the stack

Interrupt Return Value	Meaning
0xFFFFFFF9	Return to Thread mode and use the Main Stack Pointer
0xFFFFFFFD	Return to Thread mode and use the Process Stack Pointer
0xFFFFFFF1	Return to Handler mode

The CPU then places a control word in R14. At the end of the interrupt the code will branch on R14. The Control word is not a valid return address and will cause the CPU to retrieve a stack frame and return to the correct operating mode.

Exiting Interrupt Routines Important!

The interrupt lines that connect the user peripheral interrupt sources to the NVIC interrupt channels can be level-sensitive or edge sensitive. In most microcontrollers, the default is level sensitive. Once an interrupt has been raised, it will be asserted on the NVIC until it is cleared. This means that if you exit an interrupt service routine with the interrupt still asserted on the NVIC, a new interrupt will be raised. You must clear the interrupt status flags in the user peripheral before exiting the Interrupt Service Routine to cancel the interrupt. If the peripheral generates another interrupt while its interrupt line is asserted, a further interrupt will not be raised. So it is best to clear the interrupt status flags at the beginning of the interrupt routine, then any further interrupts from the peripheral will be served. To further complicate things, some peripherals will automatically clear some of their status flags. For example, an ADC conversion complete flag may be automatically cleared when the ADC data register is read, thus deasserting the interrupt. Keep this in mind when you are reading the Microcontroller user manual.

Exercise 3.3: SysTick Interrupt

This project demonstrates setting up a first interrupt using the SysTick timer.

Open the Pack Installer.

Select the Boards tab and "The Designers Guide Tutorial Examples."

Select the example tab and Copy "Exercise 3.3: SysTick Interrupt."

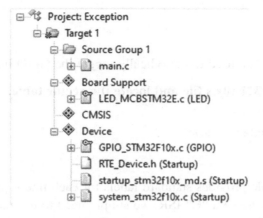

Figure 3.22
SysTick interrupt project layout.

This project (Fig. 3.22) consists of the minimum amount of code necessary to get the Cortex-M processor running and to generate a SysTick timer interrupt.

Open the main.c file

```
#include "stm32f10x.h"
#define SYSTICK_COUNT_ENABLE       1
#define SYSTICK_INTERRUPT_ENABLE   2
int main (void)
{
GPIOB->CRH = 0x33333333;
SysTick->VAL = 0x9000;
SysTick->LOAD = 0x9000;
SysTick->CTRL = SYSTICK_INTERRUPT_ENABLE | SYSTICK_COUNT_ENABLE;
while(1);
}
```

The main function configures a bank of port pins as outputs. Next, we load the SysTick timer, and SySTick reload register and then enable the timer and its interrupt line to the NVIC. Once this is done, the background code sits in a while(1) loop doing nothing.

When the timer counts down to zero, it will generate an interrupt which will run the SysTick ISR

```
void SysTick_Handler (void)
{
static unsigned char count = 0,ledZero = 0x0F;
   if(count + + >0x60)
   {
      ledZero = ledZero ^ 0xFF;
      LED_SetOut(ledZero);
```

```
        count = 0;
    }
}
```

The interrupt routine is then used to periodically toggle the GPIO lines.

Open the Device::STM32F10x.s file and locate the vector table.

```
SysTick_Handler PROC
    EXPORT SysTick_Handler    [WEAK]
    B.
    ENDP
```

The Default Handler table provides standard labels for each interrupt source created as "weak" declarations. To create a "C" ISR, we simply need to use the label as the name for a void function. The "C" function will then override the assembled stub and be called when the interrupt is raised.

Build the project and start the debugger.

Without running the code, open the Register window and examine the state of the registers.

In particular, note the value of the stack pointer (R13), the link register (R14), and the PSR. We can also see the operating mode of the CPU (Fig. 3.23).

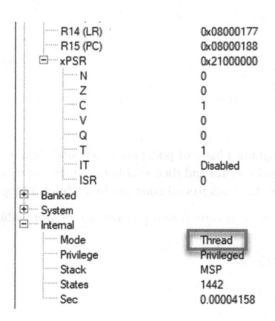

Figure 3.23
CPU register values at the start of the code.

Set a breakpoint in the interrupt routine and start the code running (Fig. 3.24).

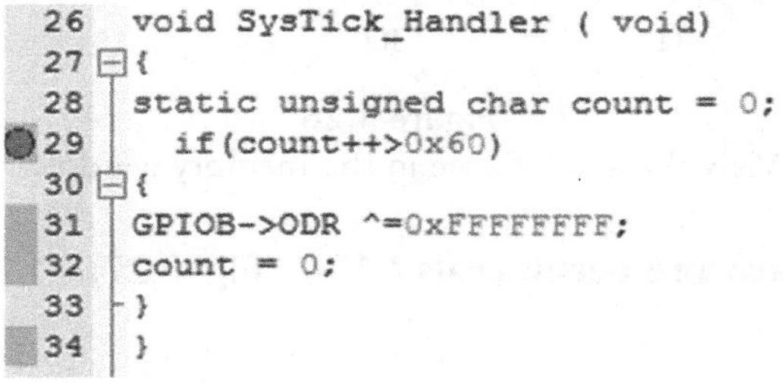

```
26  void SysTick_Handler ( void)
27 ⊟{
28   static unsigned char count = 0;
●29     if(count++>0x60)
30 ⊟{
31   GPIOB->ODR ^=0xFFFFFFFF;
32   count = 0;
33 ├}
34   }
```

Figure 3.24
A breakpoint at the start of the SysTick Handler function.

When the code hits the breakpoint again, examine the register window (Fig. 3.25).

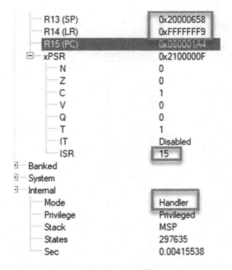

Figure 3.25
The Link register (R14) now holds a return code that forces the CPU to return from interrupt at the end of the Interrupt service Function.

Now, R14 has the interrupt return code in place of a normal return address, and the stack pointer has been decremented by 32 words.

Open a memory window at the current main stack pointer address and decode the stack frame (Fig. 3.26).

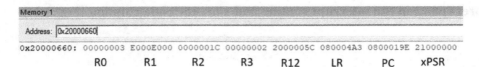

Figure 3.26
View the stack frame in the memory window.

Now open the peripherals\core peripherals\NVIC (Fig. 3.27).

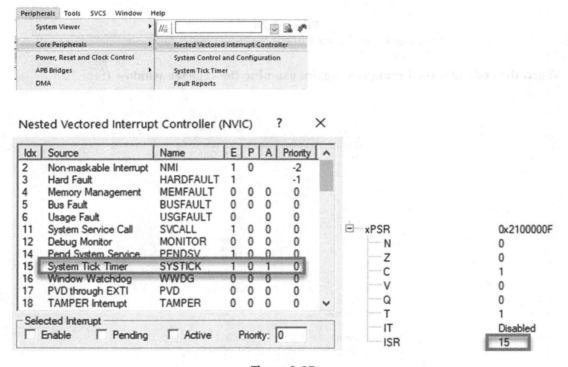

Figure 3.27
The NVIC peripheral window and the xPSR register view both show the SysTick timer as the active interrupt.

The NVIC peripheral window shows the state of each interrupt channel. Channel 15 is the SysTick timer. Its state is given by the Enabled (E), Active (A), and Pending (P) Columns. The idx column indicates the NVIC channel number and ties up with the ISR channel number in the PSR.

Now set a breakpoint on the closing brace of the interrupt function and run the code (Fig. 3.28).

```
     34   void SysTick_Handler ( void)
     35 ⊟ {
     36 |  static uint8_t count = 0;
     37 |  static uint32_t ledState = 0;
     38 |
     39 |    if(count++>0x60)
     40 ⊟  {
     41 |      ledState = ledState ^ LED_PATTERN;
     42 |      GPIOB->ODR = ledState;
     43 |      count = 0;
     44 |    }
     45 | }
```

Figure 3.28
Breakpoints on the ISR entry and exit points.

Now open the disassembly window and view the return instruction (Fig. 3.29).

```
   . . .
0x080001D4 7001      STRB      r1,[r0,#0x00]
   45: }
0x080001D6 4770      BX        lr
```

Figure 3.29
The return from an interrupt is a normal branch instruction.

The return instruction is a branch instruction, the same as if you were returning from a subroutine. However, the value in the Link register R14 will force the CPU to unstack and return from the interrupt. Single step this instruction (F11) and observe the return to the background code and the stacked values return to the CPU registers.

Cortex-M Processor Exceptions

In addition to the peripheral interrupt lines, the Cortex-M Processor has some internal exceptions. These occupy the first 15 locations of the vector table (Fig. 3.30).

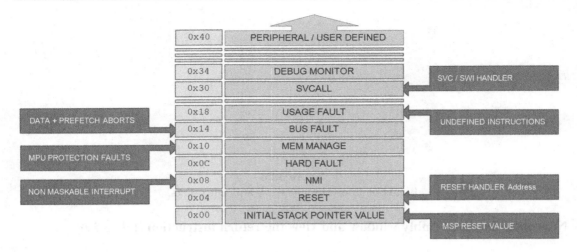

Figure 3.30
The first four bytes of memory hold the initial stack value. The vector table starts from 0x00000004. The first 10 vectors are for the Cortex processor while the remainder are for user peripherals.

The first location in the Vector Table is the Reset Handler. When the Cortex-M processor is reset, the address stored here will be loaded into the Cortex-M Program Counter, forcing a jump to the start of your application code. The next location in the Vector Table is for a nonmaskable interrupt. How this is implemented will depend on the specific microcontroller you are using. It may, for example, be connected to an external pin on the microcontroller or to a peripheral such as a watchdog within the microcontroller. The next four exceptions are for handling faults that may occur during the execution of the application code. All of these exceptions are present on the Cortex-M3/M4 and Cortex-M7. The type of faults that can be detected by the processor are Usage Fault, Bus Fault, Memory Manager Fault, and Hard Fault. The Cortex-M0 only implements the Hard Fault handler.

Usage Fault

A usage fault occurs when the application code has incorrectly used the Cortex-M processor. The typical cause is when the processor has been given an invalid Op code to execute. Most Arm compilers can generate code for a range of Arm processor cores. So, it is possible to incorrectly configure the compiler and to produce code that will not run on a Cortex-M processor. Other causes of a usage fault are shown in Table 3.7.

Table 3.7: Possible causes of the usage fault exception

Undefined Instruction
Invalid interrupt return address
Unaligned memory access using load and store multiple instructions
Divide by zero*
Unaligned memory access*

*This feature must be enabled in the system control block "Configurable Fault Usage" register.

Bus Fault

A bus fault is raised when an error is detected on an internal microcontroller bus (more about the bus matrix in Chapter 5: Advanced Architecture Features). The most common reason for a bus fault is that you are trying to access memory that does not exist. The potential reasons for this fault are shown in Table 3.8.

Table 3.8: Possible causes of the bus fault exception

Invalid memory region
Wrong size of transfer ie byte write to a word only peripheral register
Wrong processor privilege level (We will look at privilege levels in Chapter 5: Advanced Architecture Features)

Memory Manager Fault

The Memory Protection Unit (MPU) is an optional Cortex-M processor peripheral that can be added when the microcontroller is designed. It is available on all variants except the Cortex-M0. The Memory Protection Unit is used to control access to different regions of the Cortex-M address space depending on the operating mode of the processor. We will look at the MPU hardware in more detail in Chapter 5, Advanced Architecture Features, and its use within safety systems in Chapter 11, RTOS Techniques. For now, the Memory Protection Unit can raise an exception for the cases shown in Table 3.9.

Table 3.9: Possible causes of the memory manager fault exception

Accessing an MPU region with the wrong privilege level
Writing to a read-only region
Accessing a memory location outside of the defined MPU regions
Program execution from a memory region that is defined as nonexecutable

Hard Fault

A Hard fault can be raised in two ways: first, if a bus error occurs when the vector table is being read; second, the Hard Fault exception is also reached through fault escalation. This means that if the Usage, Memory Manager, or Bus fault exceptions are active but their dedicated Interrupt Service Routines are not enabled, or if the exception service does not have a sufficient priority level, then the fault will escalate to a Hard fault.

Enabling Fault Exceptions

The Hard Fault handler is always enabled after reset and can only be disabled by setting the CPU FAULTMASK register. The other fault exceptions must be enabled in the SCB, "System Handler Control and State" register (SCB->SHCSR). The SCB->SHCSR register also contains Pend and Active bits for each Fault Exception.

We will look at the Fault Exceptions and tracking faults in Chapter 5, Advanced Architecture Features.

Priority and Preemption

The NVIC contains a group of priority registers with an 8-bit field for each interrupt source. In its default configuration, the top 7 bits of the priority register allow you to define the preemption level (Fig. 3.31). The lower the preemption level, the more important the interrupt. So, if an interrupt is being served and a second interrupt is raised with a lower preemption level, then the state of the current interrupt will be saved, and the processor will serve the new interrupt. When it is finished, the processor will resume serving the original interrupt provided a higher priority interrupt is not pending.

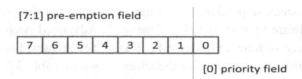

Figure 3.31
Each peripheral priority register consists of a configurable preemption field and a subpriority field.

The least significant bit is the subpriority bit. If two interrupts are raised with the same preemption level, the interrupt with the lowest subpriority level will be served first. This means we have 128 preemption levels, each with two subpriority levels.

Active priority field

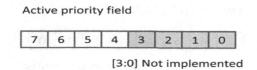

[3:0] Not implemented

Figure 3.32
Each priority register is 8 bits wide. However, the silicon manufacturer may not implement all of the priority bits. The implemented bits always extend from the MSB toward the LSB.

When the microcontroller is designed, the manufacturer can define the number of active bits in each of the priority registers (Fig. 3.32). For Cortex-M3/M4 and Cortex-M7, this can be a minimum of three and up to a maximum of eight. For Cortex-M0, M0 + , and M1, it is always two bits. Reducing the number of active priority bits reduces the NVIC gate count and hence its power consumption. If the manufacturer does not implement the full eight bits of the priority register, the least significant bits will be disabled, this makes it safer to port code between microcontrollers with different numbers of active priority bits. You will need to check the manufacturer's datasheet to see how many bits of the priority register are active.

Groups and Subgroup

After a processor reset, the first seven bits of the priority registers define the preemption level, and the least significant bit defines the subpriority level. This split between preemption group and priority subgroup can be modified by writing to the NVIC "Priority Group" Field in the "Application Interrupt and Reset Control" register. This register allows us to change the size of the preemption group field and priority subgroup as shown in Table 3.10. On reset, this register defaults to Priority Group Zero.

Table 3.10: Priority group and subgroup values

Priority Group	Preempt Group Bits	Subpriority Group Bits
0	7−1	0
1	7−2	1−0
2	7−3	2−0
3	7−4	3−0
4	7−5	4−0
5	7−6	5−0
6	7	6−0
7	None	7−0

So for example, if our microcontroller has four active priority bits, we could select Priority group 5, which would give us four levels of preemption, each with four levels of subpriority (Fig. 3.33).

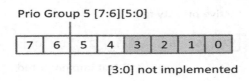

Figure 3.33
A priority register with four active bits and Prio group five. This yields four preempt levels and four priority levels.

Processor Exceptions

The highest preemption level for a user exception is zero. However, some of the Cortex-M processor exceptions have negative priority levels that allow them to always preempt a user peripheral interrupt. The processor exceptions and their default priority settings are shown in Table 3.11.

Table 3.11: Processor exceptions and user interrupt table

Group	Exception	Name	Priority	Description
Startup & Fault	1	Reset	-3	Reset Vector
	2	NMI	-2	Nonmaskable Interrupt
	3	Hard Fault	-1	Default Fault if no other fault handler enabled
	4	Mem Manager Fault	User Defined	MPU exception, illegal memory access
	5	Bus Fault	User Defined	AHB/APB bus error
	6	Usage Fault	User Defined	Instruction or program error
System	11	SV Call	User Defined	System SerVice call
	12	Debug	User Defined	Coresight debug exception
	14	Pend SV	User Defined	Pendable Service Request (see Chapter 10: Using a Real-Time Operating System)
	15	SysTick	User Defined	System Tick Timer exception
peripherals	16	Interrupt 0	User Defined	NVIC interrupt channel 0
	......	......	......	......
	255	Interrupt 255	User Defined	NVIC interrupt channel 255

Run Time Priority Control

There are three CPU registers that may be used to dynamically disable interrupt sources within the NVIC: PRIMASK, FAULTMASK, and BASEPRI registers, as shown in Table 3.12.

Table 3.12: The CPU PRIMASK, FAULTMASK, and BASEPRI registers are used to dynamically disable interrupts and exceptions

CPU Mask Register	Description
PRIMASK	Disables all exceptions except Hard Fault and NMI
FAULTMASK	Disables all exceptions except NMI
BASEPRI	Disables all exceptions at the selected Preemption level and lower Preemption level

These registers are not memory mapped. They are CPU registers and may only be accessed with the MRS and MSR instructions. When programming in "C" they may be accessed by dedicated compiler intrinsic instructions. We will look at accessing these instructions more closely in Chapter 4, Common Microcontroller Software Interface Standard.

Exception Model

When the NVIC serves a single interrupt, there is a delay of 12 cycles until we reach the interrupt service routine (Fig. 3.34) and a further 10 cycles at the end of the interrupt service routine until the Cortex-M processor resumes execution of the background code. This gives us fast deterministic handling of interrupts in a small microcontroller system with a limited number of active interrupt sources.

Data bus | Push the stack frame

Instruction bus | Fetch ISR address from vector table

12 Cycles

Figure 3.34
When an exception is raised a stack frame is pushed in parallel with the ISR address being fetched from the vector table. On Cortex-M3 and M4 this is always 12 cycles. On Cortex-M0 it takes 16 cycles. M0 + takes 15 cycles.

In more complex systems, there may be many active interrupt sources all demanding to be served as efficiently as possible. The NVIC has been designed with a number of optimizations to ensure fast interrupt handling in such a heavily loaded system. All of the interrupt handling optimizations described below are an integral part of the NVIC and do not require any configuration by the application code.

NVIC Tail Chaining

In a very interrupt-driven design, we can often find that while the CPU is serving a high-priority interrupt, a lower priority interrupt is also pending. In the earlier Arm CPUs and many other processors, it was necessary to return from the interrupt by POPing the CPU context from the stack back into the CPU registers and then performing a fresh stack PUSH before running the pending interrupt service routine. This is quite wasteful in terms of CPU cycles as it performs two redundant stack operations.

When this situation occurs on a Cortex-M processor, the NVIC uses a technique called tail chaining (Fig. 3.35) to eliminate the unnecessary stack operations. When the Cortex-M processor reaches the end of the active interrupt service routine, and there is a pending interrupt, then the NVIC simply forces the processor to vector to the pending interrupt service routine. This takes a fixed six cycles to fetch the start address of the pending interrupt routine, and then execution of the next interrupt service routine can begin. Any further pending interrupts are dealt with in the same way. When there are no further interrupts pending, the stack frame will be POPed back to the processor registers, and the CPU will resume execution of the background code. As you can see from the above diagram, tail chaining can significantly improve the latency between interrupt routines.

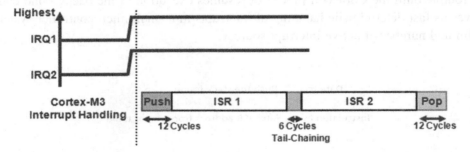

Figure 3.35

If an interrupt ISR is running and a lower priority interrupt is raised it will automatically "tail chained" to run six cycles after the initial interrupt has terminated.

NVIC Late Arriving

Another situation that can occur is a "late-arriving" high-priority interrupt. In this situation, a low-priority interrupt is raised, followed almost immediately by a high-priority interrupt. Most microcontrollers will handle this by preempting the initial interrupt. This is undesirable because it will cause two stack frames to be pushed and delay the high-priority interrupt.

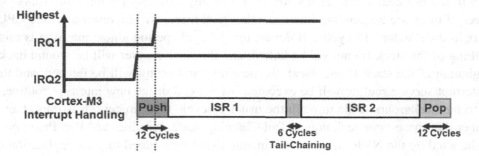

Figure 3.36
If the Cortex-M processor is entering and ISR and a higher priority interrupt is raised the NVIC will automatically switch to serve the high-priority interrupt. This will only happen if the initial interrupt is in its first 12 cycles.

If this situation occurs on a Cortex-M Processor (Fig. 3.36) and the high-priority interrupt arrives within the initial 12-cycle microcode PUSH of the low-priority stack frame, then the NVIC will switch to serving the high-priority interrupt, and the low-priority interrupt will be tail chained to execute once the high-priority interrupt is finished. For the "late-arriving" switch to happen, the high-priority interrupt must occur in the initial 12-cycle period of the low-priority interrupt. If it occurs any later than this, then it will preempt the low-priority interrupt, which requires the normal stack PUSH and POP.

NVIC POP Preemption

The final optimization technique used by the NVIC is called POP preemption (Fig. 3.37). This is kind of a reversal of the late-arriving technique discussed above.

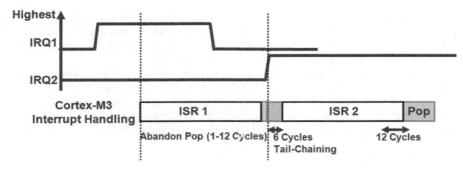

Figure 3.37
If an interrupt is raised while is in its exiting 12 cycles the processor will "rewind" the stack and serve the new interrupt with a minimum 6-cycle delay.

When a typical microcontroller reaches the end of an interrupt service routine, it always has to restore the stack frame regardless of any pending interrupts. As we have seen above, the NVIC will use tail chaining to efficiently deal with any currently pending interrupts. However, if there are no pending interrupts the stack frame will be restored to the CPU registers in the standard 12 cycles. If during this 12-cycle period a new interrupt is raised, the POPing of the stack frame will be halted, and the stack pointer will be wound back to the beginning of the stack frame. Next, the new interrupt vector will be fetched, and the new interrupt service routine will be executed. At the end of the new interrupt routine, we return to the background code through the usual 10-cycle POP process. It is important to remember that these three techniques, Tail Chaining, Late Arriving, and Pop Preemption, are all handled by the NVIC without any instructions being added to your application code.

Exercise 3.4: Working with Multiple Interrupts

This exercise extends our original SysTick exception exercise to enable an ADC and hardware timer interrupts. We can use these three interrupts to examine the behavior of the NVIC when it has multiple interrupt sources.

Open the Pack Installer.

Select the Boards tab and "The Designers Guide Tutorial Examples."

Select the example tab and Copy "Exercise 3.4: Multiple Interrupts."

Open main.c and locate the main() function.

```
uint32_t BACKGROUND = 0;
uint32_t ADC = 0;
uint32_t SYSTICK = 0;
uint32_t TIMER = 0;

int main (void)
{
GPIOB->CRH = 0x33333333;                // Configure the Port B LED pins
init_ADC();                             //initialize the ADC and enable its EoC interrupt
TIM1_Init ();                           //initialize a hardware timer and enable a
                                        compare interrupt
SysTick_Config(SystemCoreClock / 100);  // Configure the Systick and enable interrupt
NVIC_EnableIRQ(ADC1_2_IRQn);            // Enable the ADC Interrupt
NVIC_SetPriorityGrouping (5);           // Set the priority grouping to 5.3
NVIC_SetPriority (SysTick_IRQn,4);      // Set the Systick and ADC priority
NVIC_SetPriority (ADC1_2_IRQn,4);       // Set the ADC and Timer Priorities to the same as
                                        the SysTick

NVIC_SetPriority (TIM1_UP_IRQn,4);
```

```
//NVIC_SetPriority (ADC1_2_IRQn,3);      //uncomment this line to raise the priority of the
                                         ADC
//NVIC_SetPriority (SysTick_IRQn,3);     //uncomment this line to raise the priority of the
                                         SysRick
while(1)
{
    BACKGROUND = 1;                           // Set the background and sleep toggle bits
}
}
```

We initialize the SysTick timer with a standard helper function. In addition the ADC peripheral is also configured. In this example we are using some additional standard helper functions to setup the NVIC registers. We will have a look at these functions in more detail in the next chapter.

We have also added four variables BACKGROUND, TIMER, ADC, and SYSTICK. These will be set to logic one when the matching region of code is executing and zero at other times. This allows us to track execution of each region of code using the debugger system analyzer.

```
void ADC_IRQHandler (void){
int i;
BACKGROUND              = 0;
SYSTICK                 = 0;
for (i = 0;i<0x1000;i + +){
   ADC = 1;
}
ADC1->SR & = ~(1 << 1);    /* clear EOC interrupt    */
ADC = 0;
}
```

The ADC interrupt handler sets the execution region variables and then sits in a delay loop. Before exiting, it also writes to the ADC status register to clear the end of conversion flag. This deasserts the ADC interrupt request to the NVIC.

The SysTick and timer interrupt contain similar code that sets their variables to a logic one during execution of the ISR and then clears them on exit.

Build the project and start the simulator.

Add each of the execution variables to the system analyzer and start the code running (Fig. 3.38).

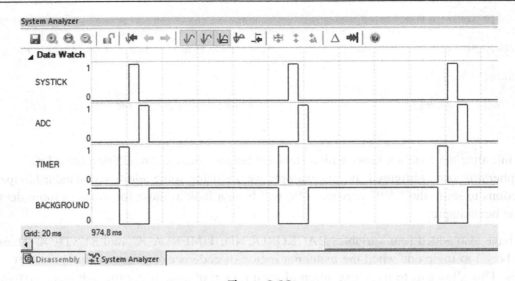

Figure 3.38
The SysTick interrupt is executed (logic high) then the ADC interrupt is tail chained and will run when the SysTick ends.

The SysTick interrupt is raised, which starts the ADC conversion. The ADC finishes conversion and raises its interrupt before the SysTick interrupt completes so it enters a Pending state. When the SysTick interrupt completes, the ADC interrupt is tail chained and begins execution without returning to the background code.

Exit the debugger and comment out the line of code that clears the ADC end of conversion flag.

```
//ADC1->SR &= ~(1 << 1);
```

Build the code and restart the debugger and observe the execution of the interrupts in the system analyzer window (Fig. 3.39).

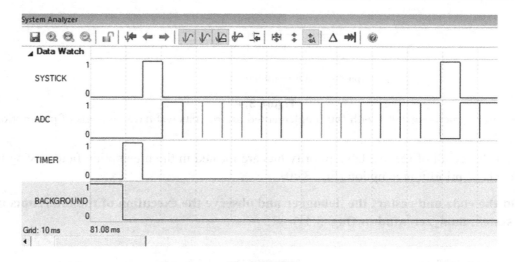

Figure 3.39
The ADC tail chains the SysTick and runs multiple times because the ADC status flag has not been cleared.

After the first ADC interrupt has been raised, the interrupt status flag has not been cleared, and the ADC interrupt line to the NVIC stays asserted. This causes continuous ADC interrupts to be raised by the NVIC, blocking the activity of the background code. The SysTick interrupt has the same priority as the ADC, so it will be tail Chained to run after the current ADC interrupt has finished. Neglecting to clear interrupt status flags is the most common mistake made when first starting to work with the Cortex-M processors.

Exit the debugger and uncomment the end of conversion code.

```
ADC1->SR &= ~(1 << 1);
```

Add the following lines to the background initializing code.

```
NVIC_SetPriority    (ADC1_2_IRQn,3);
```

This programs the user peripheral NVIC "Interrupt Priority" registers to set the ADC priority level and the "System Handler Priority" registers to set the SysTick priority level. These are both byte arrays that cover the 8-bit priority field for each exception source. However, on this microcontroller, the manufacturer has implemented four priority bits out of the possible eight. The priority bits are located in the upper nibble of each byte. On reset the PRIGROUP is set to zero which creates a 7-bit preemption field and 1-bit priority field.

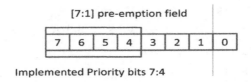

[7:1] pre-emption field

| 7 | 6 | 5 | 4 | 3 | 2 | 1 | 0 |

Implemented Priority bits 7:4

Figure 3.40

After reset a microcontroller with four implemented priority bits will have 16 levels of preemption.

On our device all of the available priority bits are located in the preemption field, giving us 16 levels of priority preemption (Fig. 3.40).

Build the code and restart the debugger and observe the execution of the interrupts in the system analyzer window (Fig. 3.41).

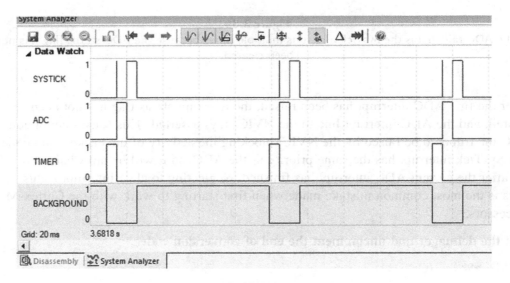

Figure 3.41

The ADC is at a higher priority than the SysTick so it preempts the SysTick interrupt.

The ADC now has the lowest preemption value and is, therefore, the most important interrupt, as soon as its interrupt is raised, it will preempt the SysTick interrupt. When it completes, the SysTick interrupt will resume and complete before returning to the background code.

Exit the debugger and uncomment the following lines in the background initialization code.

The AIRC register cannot be written too freely. It is protected by a key field which must be programmed with the value 0x5FA before a write is successful. Here we are writing directly to the register. The next chapter will introduce some helper functions which make this much easier.

```
temp = SCB->AIRCR;
temp & = ~(SCB_AIRCR_VECTKEY_Msk | SCB_AIRCR_PRIGROUP_Msk);
temp = (temp|((uint32_t)0x5FA << 16) | (0x05 << 8));
SCB->AIRCR = temp;
```

This programs the PRIGROUP field in the AIRC register to a value of five, which means a two-bit preemption field and a six-bit priority field. This maps onto the available four-bit priority field, giving four levels of preemption, each with four priority levels (Fig. 3.42).

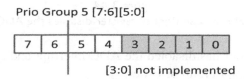

Figure 3.42
Program the PRIGROUP to adjust the preemption and priority bit size.

Build the code and restart the debugger and observe the execution of the interrupts in the system analyzer window (Fig. 3.43).

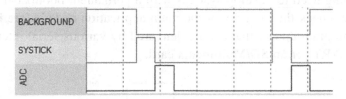

Figure 3.43
The SysTick and ADV have different priority levels but the same preempt level. Now the ADC cannot preempt the SysTick.

The ADC interrupt is no longer preempting the SysTick timer despite them having different values in their priority registers. This is because they now have different values in the priority field but the same preempt value.

Exit the debugger and change the interrupt priorities as shown below:

```
NVIC->IP[18] = (2<<6 | 2<<4);
SCB->SHP[11] = (1<<6 | 3<<4);
```

Set the base priority register to block The ADC preempt group.

```
__set_BASEPRI (2<<6);
```

Build the code and restart the debugger and observe the execution of the interrupts in the system analyzer window (Fig. 3.44).

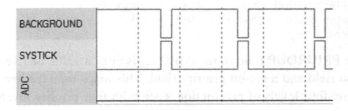

Figure 3.44
The setting of the Base Priority register disables the ADC interrupt.

Setting the BASEPRI register has disabled the ADC interrupt and any other interrupts that are on the same level Preempt group or lower.

Bootloader Support

The interrupt vector table is located at the start of memory when the Cortex-M processor is reset. However, it is possible to relocate the vector table to a different location in memory at runtime. As software embedded in small microcontrollers becomes more sophisticated, there is an increasing need to develop systems with a permanent bootloader program (Fig. 3.45) that can check the integrity of the main application code before it runs and then check for a program update that can be delivered by various serial interfaces (Ethernet, USB, UART) or an SD/Multimedia card.

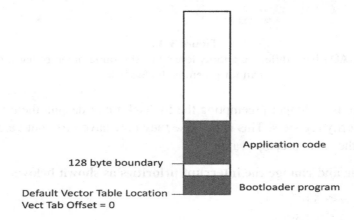

Figure 3.45
A bootloader program can be placed into the first sector in the Flash memory. It will check if there is an update to the application code before starting the main application program.

Once the bootloader has performed its checks, and if necessary, updated the application program, it will start the application code running by forcing the program counter to its reset handler. To operate correctly, the application code requires the hardware vector table to be mapped to the start address of the application code (Fig. 3.46).

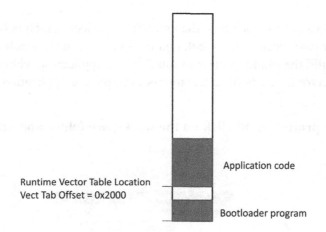

Figure 3.46
When the application code starts to run it must relocate the vector table to the start of the application code by programming the NVIC vector Table Offset register.

The Vector table can be relocated by writing to a register in the NVIC called the "Vector Table Offset" register. This register allows you to relocate the vector table to any 128-byte boundary in the Cortex-M processor memory map.

Exercise 3.5: Bootloader

This exercise demonstrates how a bootloader and application program can both be resident on the same Cortex-M microcontroller and how to load both programs into the debugger.

Open the Pack Installer.

Select the Boards tab and "The Designers Guide Tutorial Examples."

Select the example tab and Copy "Exercise 3.5: Bootloader."

This will open a multiproject workspace which is a more advanced feature of the μVision IDE that allows you to view two or more projects at the same time (Fig. 3.47).

Figure 3.47
The bootloader and Blinky project in a multiproject workspace.

The workspace consists of two projects, the bootloader project, which is built to run on the Cortex-M processor reset vector as normal, and the blinky project, which is our application. First, we need to build the blinky project to run from an application address which not in the same FLASH sector as the bootloader. In this example the application address is chosen to be 0x80002000.

Expand the blinky project, right click on the workspace folder and set it as the active project (Fig. 3.48).

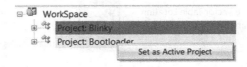

Figure 3.48
Select a project and right click to make it the active project.

Now click on the blinky project folder and open the options for target\target tab (Fig. 3.49).

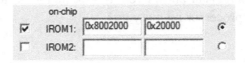

Figure 3.49
The blinky project must have its code offset to the 0x2000 start address.

The normal start address for this chip is 0x8000000 and we have increased this to 0x8002000 so the linker will now start the application image from this higher address.

Open the system_stm32F10x.c file and locate line 128.

```
#define VECT_TAB_OFFSET 0x02000
SCB->VTOR = FLASH_BASE | VECT_TAB_OFFSET;
```

This contains a #define for the Vector Table Offset register. Normally this is zero, but if we set this to 0x2000, the vector table will be remapped to match our application code when it starts running.

Some startup code may adjust the table offset automatically by using the following code:

```
extern void *__Vectors;
SCB->VTOR = (uint32_t) &__Vectors;
```

where the label Vectors is placed at the start of the interrupt vector table.

Build the blinky project.

Expand the bootloader project and set it as the active project (Fig. 3.50).

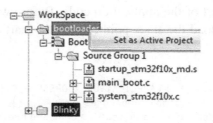

Figure 3.50
Select the bootloader as the active project.

Open main_boot.c.

The bootloader program demonstrates how to jump from one program to the start of another. We need to define the start address of our second program. This must be a multiple of 128 bytes (0x200). In addition, a void function pointer is also defined:

```
#define APPLICATION_ADDRESS 0x2000
typedef void (*pFunction)(void);
pFunction Jump_To_Application;
uint32_t JumpAddress;
```

When the bootloader code enters main(), it would perform any custom checks on the application code FLASH, such as a checksum or hash signature, and could also test other critical aspects of the hardware. The bootloader would then check to see if there is a new application ready to be programmed into the application area. This could be in response to a command from an upgrade utility via a serial interface, for example. If the application program checks fail or a new update is available, we would enter into the main bootloader code.

```
int main(void) {
uint32_t bootFlags;
    /* check the integrity of the application code */
    /* Check if an update is available */
    /* if either case is true set a bit in the bootflags register */
    bootFlags = 0;

    if (bootFlags! = 0) {

    //enter the Flash update code here
    }
```

If the application code and the hardware is ok, then the bootloader will hand over to the application code. The reset vector of the application code is now located at the application address + 4, and this can be loaded into the function pointer, which can then be executed, resulting in a jump to the start of the application code. Before we jump to the application code, it is also necessary to load the stack pointer with the start address expected by the application code

```
else {
    JumpAddress = *(__IO uint32_t*) (APPLICATION_ADDRESS + 4);
    Jump_To_Application = (pFunction) JumpAddress;
    // read the first four bytes of the application code and program this value into the stack
    pointer:
    //This sets the stack ready for the application code
    __set_MSP(*(__IO uint32_t*) APPLICATION_ADDRESS);
    Jump_To_Application();
}}
```

Build the project.

Open the bootloader options for target\debug tab and open the loadApp.ini file (Fig. 3.51).

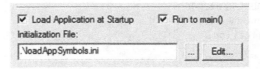

Figure 3.51
The debug script is used to load the application project symbols. This allows you to debug the bootloader and application code simultaneously.

```
Load "\\Blinky\\Blinky.AXF" incremental
```

This script file can be used with the simulator or the hardware debugger. It is used to load the blinky application code as well as the bootloader code. This allows us to debug seamlessly between the two separate programs.

Start the debugger.

Single step the code through the bootloader checking that the correct stack pointer address is loaded into the MSP and that the blinky reset handler address is loaded into the function pointer. You can see these values in the debugger by using the memory, register, and watch windows (Fig. 3.52).

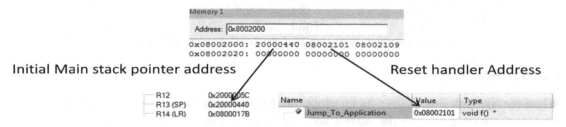

Initial Main stack pointer address   Reset handler Address

Figure 3.52
The first eight bytes of the application image hold the initial stack pointer address and the reset handler address.

Open The blinky.c file in the blinky project and set a breakpoint on main() (Fig. 3.53).

```
35 int main (void) {
36    uint32_t ad_avg = 0;
37    uint16_t ad_val = 0, ad_val_ = 0xFFFF;
```

Figure 3.53
Set a breakpoint on main() in the blinky project.

Run the code.

Now the Cortex-m processor has left the bootloader program and entered the blinky application program. The startup system code has programmed the Vector Table Offset register, now the hardware vector table matches the blinky software.

Open the peripherals\core peripherals\nested vector interrupt table (Fig. 3.54).

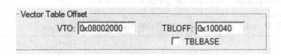

Figure 3.54
By programming the VTO the hardware interrupt table has been moved to match the software vector table.

Now the vector table is at 0x8002000 to match the blinky code.

Open the IRQ.C file and set a breakpoint on the SysTick interrupt handler (Fig. 3.55).

```
 27  void SysTick_Handler (void) {
 28      static unsigned long ticks = 0;
 29      static unsigned long timetick;
 30      static unsigned int  leds = 0x01;
```

Figure 3.55
Set a breakpoint at the start of the SysTick interrupt handler.

Run the code.

When the SysTick timer raises its interrupt, the handler address will be fetched from the blinky vector table and not the default address at the start of the memory, then the correct SysTick handler routine will be executed. Now the blinky program is running happily at its offset address. In a real-world project the application code may download a new candidate image to a staging area within the FLASH memory. Once the candidate has been successfully stored, we need to force a reset so that the bootloader runs and updates the candidate image into the active image slot. It is possible to perform a software reset by writing to the "System Control Block Application Interrupt and Reset Control" register. While you can write directly to the AIRC register there is also a dedicated function

```
NVIC_SystemReset();
```

That we will meet in the next chapter.

Power Management

While the Cortex-M0 and M0 + are specifically designed for low-power operation, the Cortex-M3 and M4 still have remarkably low-power consumption. While the actual power consumption will depend on the manufacturing process used by the silicon vendor, the figures in Table 3.13 give an indication of the expected power consumption for each processor.

Table 3.13: Power consumption figures by processor variant in 90 nm LP (low-power) process

Processor	Dynamic Power Consumption (μW/MHz)	Details
Cortex-M0 +	11	Excludes debug units
Cortex-M0	16	Excludes debug units
Cortex-M3	32	Excludes MPU and debug units
Cortex-M4	33	Excludes FPU, MPU, and debug units

The Cortex-M processors are capable of entering low-power modes called SLEEP and DEEPSLEEP. When the processor is placed in SLEEP mode, the main CPU clock signal is stopped, which halts the Cortex-M processor. The rest of the microcontroller clocks and peripherals will still be running and can be used to wake up the CPU by raising an interrupt.

The DEEPSLEEP mode is an extension of the SLEEP mode. When DEEPSLEEP is entered, the CPU will enter its SLEEP mode but will also assert a signal to the surrounding microcontroller hardware. This signal is typically connected to the power management unit (PMU) within the microcontroller. This allows the silicon vendor to implement a range of low-power modes that can be triggered when the CPU enters DEEPSLEEP.

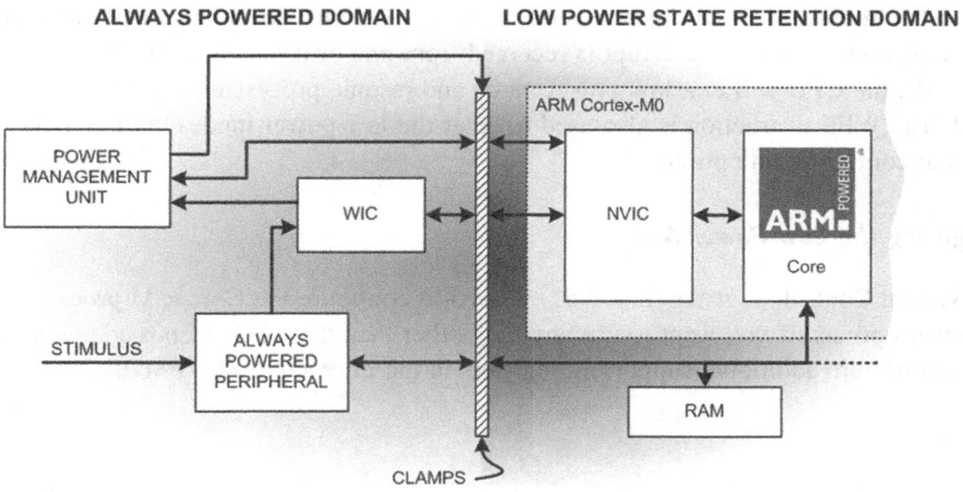

Figure 3.56
The wake-up controller is a small area of gates which do not require a clock source. The WIC can be located on a different power domain to the Cortex-M processor. This allows all the processor clocks to be halted. The range of available power modes is defined by the silicon manufacturer.

When a Cortex-M processor has entered a low-power sleep mode, it can be woken up by a microcontroller peripheral raising an interrupt to the NVIC. However, the NVIC needs a clock to operate, so if all the clocks are stopped, we need another hardware unit to tell the PMU to restore the clocks before the NVIC can respond. The Cortex-M processors can be fitted with an optional unit called the Wake-up Interrupt Controller (WIC) (Fig. 3.56). The WIC handles interrupt detection when the CPU is stopped and allows the Cortex-M processor to fully enter low-power modes. The wake-up controller can also be placed on a different power domain to the main Cortex-M processor. This allows the microcontroller manufacturers to design a device that can have low-power modes where most of the chip is switched off while keeping key peripherals alive to wake up the processor.

Entering Low-Power Modes

The THUMB-2 instruction set contains two dedicated instructions, shown in Table 3.14, that will place the Cortex-M processor into SLEEP or DEEPSLEEP mode.

Table 3.14: Low-power entry instructions

Instruction	Description	CMSIS-Core Intrinsic
WFI	Wait for Interrupt	__WFI()
WFE	Wait for Event	__WFE

As its name implies, the WFI instruction will place the Cortex-M processor in the selected low-power mode. When an interrupt is received from one of the microcontroller peripherals, the CPU will exit low-power mode and resume processing the interrupt as normal. The WFE instruction is also used to enter the low-power modes but has some additional configuration options.

Configuring the Low-Power Modes

The "System Control" register (Fig. 3.57) is used to configure the Cortex-M processor low-power options. If you want to use anything other than the CPU sleep mode, you will need to configure additional supporting registers in the microcontroller PMU.

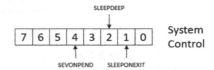

Figure 3.57
The system control register contains the Cortex-M processor low-power configuration bits.

The WFE instruction places the Cortex-M processor into its sleeping mode. The processor may be woken by an interrupt in the same manner as the WFI instruction (Fig. 3.58). However, the WFE instruction has an internal event latch. If the event latch is set to one, the processor will clear the latch but not enter low-power mode. If the latch is zero, it will enter low-power mode. On a typical microcontroller, the events are the peripheral interrupt signals. So, if there are pending interrupts, this latching will prevent the processor from sleeping. The SEVONPEND bit is used to change the behavior of the WFE instruction.

If this bit is set, the peripheral interrupt lines can be used to wake the processor even if the interrupt is disabled in the NVIC. This allows you to place the processor into its sleep mode. Then when a peripheral interrupt occurs, the processor will wake and resume execution of the instruction following the WFE instruction rather than jumping to an interrupt routine.

The Sleep on Exit bit is used in conjunction with the WFI instruction. Setting the Sleep on Exit bit will force the microcontroller to enter its sleep mode when it reaches the end of an interrupt. This allows you to design a system that wakes up in response to an interrupt, runs the required code, and will then automatically return to sleep. In such a system no stack management is required (except in the case of preempted interrupts) during the interrupt entry/exit sequence and no background code will be executed.

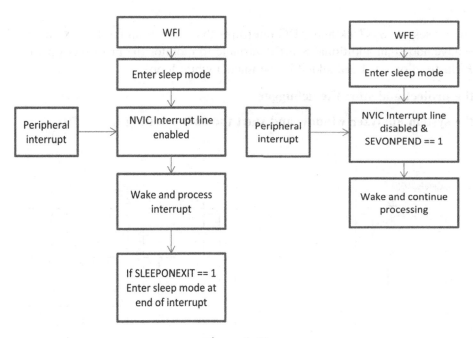

Figure 3.58
The two low-power entry instructions place the Cortex processor into its low-power mode. Both modes use a peripheral interrupt to wake the processor but their wake-up behavior is different.

An interrupt disabled in the NVIC cannot be used to exit a sleep mode entered by the WFI instruction. However, the WFE instruction will respond to activity on any interrupt line even if it is disabled or temporarily disabled by the processor mask registers (BASEPRI, PRIMASK, and FAULT MASK).

Exercise 3.6: Low-Power Modes

In this exercise, we will use the exception project to experiment with the Cortex-M low power modes.

In the pack installer, select "Exercise 3.6: Low-power modes" and press the copy button.

```
while(1)
{
   SLEEP = 1;
   BACKGROUND = 0;
   __wfe();
   BACKGROUND = 1;
   SLEEP = 0;
}
```

This project uses the SysTick and ADC interrupts that we saw in the last example. This time we have added an additional SLEEP variable to monitor the processor operating state. The WFI instruction has been added in the main() while loop.

Build the project and start the debugger.

Open the system analyzer window and start the code running (Fig. 3.59).

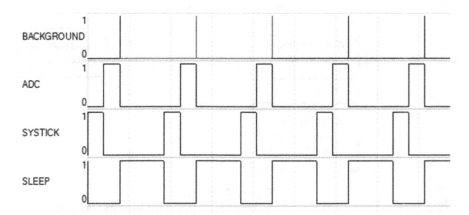

Figure 3.59
In our simple project the background code only executes the__wfi() instruction forcing the CPU to enter sleep state.

Here, we can see that the background code executes the__WFI() instruction and then goes to sleep until an interrupt is raised. When the interrupts have been completed, we return to the background code, which will immediately place the processor in its low-power mode.

Exit the debugger and change the code in main() to match the lines below.

```
SCB->SCR = 0x2;
SLEEP = 1;
BACKGROUND = 0;
__wfi();
while(1)
```

The first line of code sets the SLEEPONEXIT flag, which forces the processor into a low-power mode when it completes an interrupt.

Build the code and restart the debugger and observe the execution of the interrupts in the system analyzer window (Fig. 3.60).

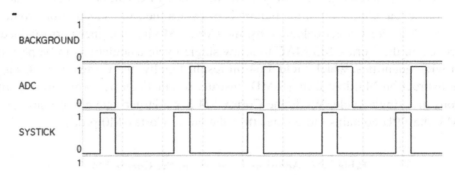

Figure 3.60
Once sleep on exit has been enabled, we no longer execute code in the background (noninterrupt code).

Here, we can see that the interrupts are running, but after the initializing code has run, the background loop never executes, so the background and sleep variables are never updated. In the debugger, you will also be able to see from the coverage monitor that the main() while loop is never executed. This feature allows the processor to wake up, run some critical code and then sleep with minimum overhead.

Moving From the Cortex-M3

In this chapter, we have concentrated on learning the Cortex-M3 processor. Now that you are familiar with how the Cortex-M3 works, we can examine the differences between the M3 and the other Cortex-M variants. As we will see, these are mainly architectural differences, and if you can use the Cortex-M3, you can easily move up to a Cortex-M4 or down to a Cortex-M0(+)-based microcontroller. Increasingly silicon manufacturers make a microcontroller family where variants have the same package pinout and peripherals but can be selected with either a Cortex-M0 or Cortex-M3 processor, allowing you to seamlessly switch devices trading off performance versus cost.

Cortex-M4

The Cortex-M4 is most easily described as a Cortex-M3 with additional FPU and Digital Signal Processing (DSP) instructions as shown in Table 3.15. We will look at these features in Chapter 9, Practical DSP for Cortex-M4 and Cortex-M7, but here we will take a tour of the main differences between the Cortex-M3 and M4. The Cortex-M4 offers the same processing power of 1.25 DMIPS/MHz as the Cortex-M3 but has a much greater maths capability. This is delivered in three ways. The hardware FPU can perform calculations in as little as one cycle compared to the hundreds of cycles the same calculation would take on the Cortex-M3. For integer calculations, the Cortex-M4 has a higher performance MAC that improves on the Cortex-M3 MAC to allow single-cycle calculations to be performed on 32-bit wide quantities, which yield a 64-bit result. Finally, the Cortex-M4 adds a group of Single Instruction Multiple Data (SIMD) instructions that can perform multiple integer calculations in a single cycle. While the Cortex-M4 has a larger gate count than the Cortex-M3, the FPU contains more gates than the entire Cortex-M0 processor.

Table 3.15: Additional features in the Cortex-M4

Feature	Comments
Floating-point unit	See Chapter 9, Practical DSP for Cortex-M Microcontrollers
DSP SIMD instructions	See Chapter 9, Practical DSP for Cortex-M Microcontrollers
GE field in xPSR	Additional condition code flag in PSR for "Greater than or equal"
Extended integer MAC unit	The Cortex-M4 extends the integer MAC to support single-cycle execution of 32-bit multiplies which yield a 64-bit result

Cortex-M0

The Cortex-M0 is a reduced version of the Cortex-M3; it is intended for low-cost and low-power microcontrollers. However, the Cortex-M0 has a benchmark processing power of 0.84 DMIPS/MHz which actually makes it a very capable processor for a small microcontroller. While the Cortex-M0 is capable of running at high clock frequencies, it is often designed into low-cost devices with simple memory systems. A typical device based on a Cortex-M0 will have a core clock frequency of 50–60 MHz. While it essentially has the same programmers model as a Cortex-M3 there are some limitations shown in Table 3.16.

Table 3.16: Features not included in the Cortex-M0

Feature	Comments
3 Stage pipeline	Same pipeline stages as the Cortex-M3 but no speculative branch target fetch
Von Newman bus interface	Instruction and data use the same bus port, which can have a slower performance compared to the Harvard architecture of the M3/M4
No conditional IF THEN blocks	Conditional branches are always used, which cause a pipeline flush
No Saturated maths instructions	
SYSTICK timer is optional	However, so far, every Cortex-M0 microcontroller has it fitted
No Memory protection unit	The MPU is covered in Chapter 5, Advanced Architecture Features
32 NVIC channels	A limited number of interrupt channels compared to the M3/M4; however, in practice, this is not a real limitation the M0 is intended for small devices with a limited number of peripherals
Four programmable priority levels	Priority level registers only implement 2 bits (4 levels), and there is no priority group setting
Hard Fault exception only	No Usage fault, Memory management fault, or Bus fault exception vectors
16-cycle interrupt latency	Same deterministic interrupt handling as the M3 but with a four cycle overhead
No priority group	Limited to a fixed 4 levels of preemption
No BASEPRI register	
No Faultmask register	
Reduced debug features	The M0 has fewer debug features compared to the M3/M4 see Chapter 8, Debugging With CoreSight, for more details
No exclusive access instructions	The exclusive access instructions are covered in Chapter 5, Advanced Architecture Features
No reverse bit order or count leading zero instructions	These instructions are often used to implement DSP algorithms
Reduced number of registers in the system control block	See below
Vector table cannot be relocated	The NVIC does not include the Vector table offset register
All code executes at privileged level	The Cortex-M0 does not support the unprivileged operating mode This is important for safety and security applications

The SCB contains a reduced number of features compared to the Cortex-M3/M4 as shown in Table 3.17. Also, the SysTick timer registers have been moved from the NVIC to the SCB.

Table 3.17: Registers in the Cortex-M0 system control block

Register	Size in Words	Description
SysTick Control and Status	1	Enables the timer and its interrupt
SysTick Reload	1	Holds the 24-bit reload value
SysTick Current Value	1	Holds the current 24-bit timer value
SysTick Calibration	1	Allows trimming the input clock frequency
CPU ID	1	Hardwired ID and revision numbers from Arm and the Silicon Manufacturer
Interrupt Control and State	1	Provides Pend bits for the SysTick and NMI interruts and extended interrupt Pending\active information
Application Interrupt and Reset Control	1	Contains the same fields as the Cortex-M3 minus the PRIGROUP field
Configuration and Control	1	
System Handler Priority	2	These registers hold the 8-bit priority fields for the configurable Processor exceptions

Cortex-M0 +

The Cortex-M0 + is an enhanced version of the Cortex-M0. As such, it boasts lower power consumption figures combined with greater processing power. The Cortex-M0 + also brings the Memory Protection Unit and real-time debug capability to very low-end devices. The Cortex-M0 + also introduces a fast IO port which speeds up access to peripheral registers, typically GPIO ports, to allow fast switching of port pins. As we will see in Chapter 8, Debugging With CoreSight, the debug system is fitted with a new trace unit called the micro trace buffer. This allows you to capture a history of executed code with a low-cost development tool. The key features of the Cortex-M0 + are shown in Table 3.18.

Table 3.18: Cortex-M0 + features

Feature	Comments
Code compatible with the Cortex-M0	The Cortex-M0 + is code compatible with the Cortex-M0 and provides higher performance with lower power consumption
Two-stage pipeline	This reduces the number of Flash accesses and hence power consumption
I/O port	The I/O port provides single cycle access to GPIO and peripheral registers
Vector table can be relocated	This supports more sophisticated software designs. Typically a Bootloader and separate application
Supports 16-bit flash memory accesses	This allows devices featuring the Cortex-M0 + to use low-cost memory
Code can execute at privileged and unprivileged levels	The Cortex-M0 + has the same operating modes as the Cortex-M3/M4
Memory protection unit	The Cortex-M0 + has a similar MPU to the Cortex-M3/M4
Micro trace buffer	This is a "snapshot" trace unit that can be accessed by low-cost debug units

Conclusion

In this chapter, we have covered the key features of the Cortex-M processor family. To develop successfully with a Cortex-M-based device, you will need to be completely familiar with all the topics covered in this chapter and Chapter 2, Developing Software for the Cortex-M Family. Now that we have a basic understanding of the Cortex-M processor, we will look at the more advanced processor features plus a range of software development methods and techniques.

Common Microcontroller Software Interface Standard

Introduction

The widespread adoption of the Cortex-M processor into general-purpose microcontrollers has led to two rising trends within the electronics industry. First of all, the same processor is available from a wide range of vendors, each with their own family of microcontrollers. In most cases, each vendor creates a family of microcontrollers that span a range of requirements for embedded systems developers. This proliferation of devices means that as a developer, you can select a suitable microcontroller from several thousand devices while still using the same tools and skills regardless of the Silicon Vendor. This explosive growth in Cortex-M-based microcontrollers has made the Cortex-M processor the de facto industry standard for 32-bit microcontrollers, and there are currently no real challengers.

The flip side of the coin is differentiation. It would be possible for a microcontroller vendor to design their own proprietary 32-bit processor. However, this is expensive to do and also requires an ecosystem of affordable tools and software to enable mass adoption. It is more cost-effective to license the Cortex-M processor from Arm and then use their own expertise to create a microcontroller with innovative peripherals. There are now more than 17 (up from 10 when I first wrote this book 3 years ago) Silicon Vendors shipping Cortex-M-based microcontrollers. While the Cortex-M processor is the same in all devices, each manufacturer of the final silicon seeks to offer a unique set of user peripherals (Fig. 4.1) for a given range of applications. This can be a microcontroller designed for low-power applications, motor control, communications, or graphics. This way, a silicon vendor can offer a microcontroller with a state-of-the-art processor which has a wide ecosystem of development tools and software while at the same time using their skill and knowledge to develop a microcontroller featuring an innovative set of peripherals.

The Designer's Guide to the Cortex-M Processor Family.
DOI: https://doi.org/10.1016/B978-0-323-85494-8.00005-X

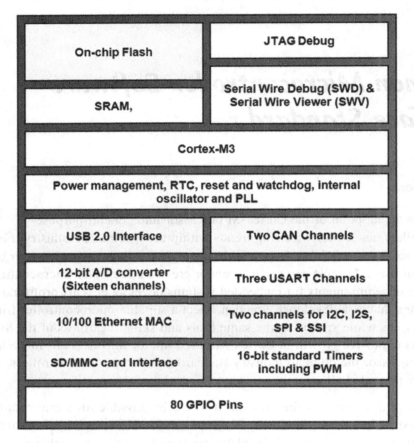

Figure 4.1
Cortex-based microcontrollers can have a number of complex peripherals on a single chip. To make these work you will need to use some form of third-party code. CMSIS is intended to allow stacks from different sources to integrate together easily.

These twin factors have led to a vast "cloud" of standard microcontrollers with increasingly complex peripherals. As well as typical microcontroller peripherals such as USART, I2C, ADC, and DAC, a modern high-end microcontroller could well have a Host/Device USB controller, Ethernet MAC, SDIO controller, LCD interface. The software to drive any of these peripherals is effectively a project in itself, so gone are the days of a developer using an 8/16-bit microcontroller and writing all of the application code from the reset vector. To release any kind of sophisticated product, it is almost certain that you will be using some form of third-party code in order to meet project deadlines. The third-party code may take the form of example code, an open-source or commercial stack, or a library provided by the silicon vendor. Both of these trends have created a need to make "C" level code more portable between different development tools and different microcontrollers. There is also a need to be able to easily integrate code taken from a variety of sources into a single project.

Figure 4.2
CMSIS-compliant software development tools and middleware stacks are allowed to carry the CMSIS logo.

In order to address these issues, a consortium of silicon vendors and tools vendors has developed a set of standards called CMSIS (seeMsys, Fig. 4.2). This originally stood for "Cortex Microcontroller Software Interface Standard" but more recently has been rebranded as "Common Microcontroller Software Interface Standard."

CMSIS Specifications

The main aim of CMSIS is to improve software portability and reusability across different microcontrollers and toolchains. This allows software from different sources to integrate seamlessly together. Once learnt, CMSIS helps to speed up software development through the use of standardized software functions.

At this point, it is worth being clear about exactly what CMSIS is. CMSIS consists of 10 interlocking specifications that support code development across all Cortex-M-based microcontrollers. The 10 specifications are as follows: CMSIS-Core, CMSIS-RTOS, CMSIS-DSP, CMSIS-NN, CMSIS-Driver, CMSIS-Zone, CMSIS-Pack, CMSIS-SVD, CMSIS-DAP, and CMSIS-Build (Fig. 4.3). As we will see, some of these standards will directly affect how you write "C" code, while others work more "under the hood" to help align the development ecosystem to prevent work from being endlessly duplicated.

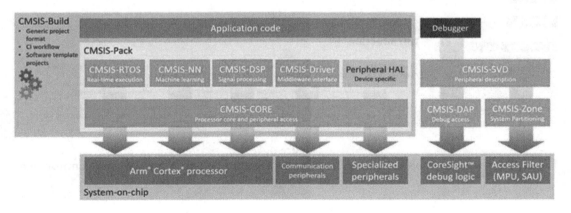

Figure 4.3
CMSIS consists of a several separate specifications (CORE, DSP, RTOS, SVD, DRIVER, DAP, and PACK) which make source code more portable between tools and devices.

It is also worth being clear about what CMSIS is not. CMSIS is not a complex abstraction layer that forces you to use a complex and bulky library. Rather the CMSIS-Core specification takes a very small amount of resources about 1k of code and just 4 bytes of RAM. It is used to standardize the way you access the Cortex-M processor and microcontroller registers. Furthermore, CMSIS does not really affect the way you develop code or force you to adopt a particular methodology. It simply provides a framework that helps you to develop your project, integrate third-party code, and reuse code on future projects. Once you are familiar with the key specifications, we will look at a software architecture that uses CMSIS to boost productivity and code reuse. Each of the CMSIS specifications is not that complicated and can be learnt easily through the course of this book.

The full documentation for each of the CMSIS specifications can be downloaded from the URL http://www.keil.com/cmsis. Each of the CMSIS specifications is integrated into the MDK-Arm toolchain and the CMSIS documentation is available by opening the Run-Time Environment and clicking on the CMSIS link in the description column (Fig. 4.4).

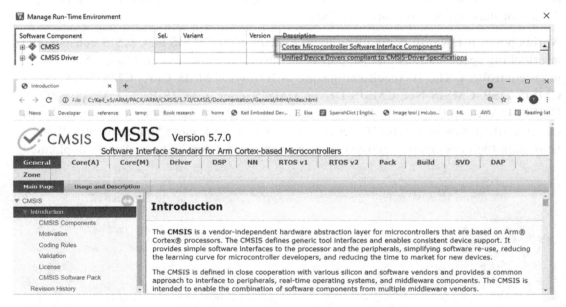

Figure 4.4
The CMSIS documentation is accessed through the Description link in the Run Time Environment Manager.

CMSIS-Core

The Core specification provides a minimal set of functions and macros to access the key Cortex-M processor registers. The Core specification also defines a function to configure the microcontroller oscillators and clock tree in the startup code, so the device is ready for use when you reach main(). The Core specification also standardizes the naming convention for the device peripheral registers. The CMSIS-Core specification includes support for the Instrumentation Trace (ITM) during debug sessions.

CMSIS-RTOS

The CMSIS-RTOS2 specification provides a standard API for a Real-Time Operating System RTOS. This is, in effect, a set of wrapper functions that translate the CMSIS-RTOS2 API to the API of the specific RTOS that you are using. We will look at the use of an RTOS in general and the CMSIS-RTOS2API in Chapter 10, Using a Real Time Operating System. The Keil RTX RTOS was the first RTOS to support the CMSIS-RTOS2 API, and it has been released as an open-source reference implementation. RTX can be compiled with the Keil/Arm, GCC, and IAR compilers. It is licensed with a three-clause BSD license that allows its unrestricted use in commercial and noncommercial applications.

CMSIS-DSP

As we have seen in Chapter 3, Cortex-M Architecture, the Cortex-M4 is a "Digital Signal Controller" with a number of enhancements to support DSP algorithms. Developing a real-time DSP system is best described as a "nontrivial pass time" and can be quite daunting for all but the simplest systems. To help mere mortals include DSP algorithms in Cortex-M4/M7 and Cortex-M3 projects, CMSIS includes a DSP library that provides over 60 of the most commonly used DSP mathematical functions. These functions are optimized to run on the Cortex-M4 and Cortex-M7 but can also be compiled to run on the Cortex-M3. We will have a look at using this library in Chapter 9, Practical DSP for Cortex-M Microcontrollers.

CMSIS-Driver

The CMSIS-Driver specification defines a standard API for a range of peripherals common to most microcontroller families. This includes peripherals such as USART, SPI, and I2C as well as more complex peripherals such as Ethernet MAC and USB. The CMSIS-Drivers are intended to provide a standard target for middleware libraries. For example, this would allow a third-party developer to create a USB library that uses the CMSIS USB driver. Such a library could then be deployed on any device that has a CMSIS USB driver. This greatly speeds up support for new devices and allows library developers to concentrate on adding features to their products rather than continually having to spend time developing support for new devices. It is important to note here that the development of CMSIS-Drivers is really the responsibility of the Silicon Vendor. They should provide a set of drivers when the microcontroller is released.

CMSIS-SVD and DAP

A key problem for software toolchain vendors is to provide debug support for new devices as soon as they are released. One of the main areas that must be customized in the debugger is the "Peripheral View" windows that show the developer the current state of the microcontroller peripherals. With the growth in both the number of Cortex-M vendors and the rising number and complexity of on-chip peripherals, it is becoming all but impossible for any given tools vendor to maintain support for all available microcontrollers. To overcome this hurdle, the CMSIS-SVD specification defines a "System Viewer Description" (SVD) file. This file is provided and maintained by the Silicon Vendor and contains a complete description of the microcontroller peripheral registers in an XML format. This file is then imported by the development tool, which uses it to automatically construct the peripheral debug windows for the microcontroller. This approach allows full debugger support to be available as soon as new microcontrollers are released. Like the CMSIS drivers, the development of the SVD files is the responsibility of the Silicon Vendor.

The CMSIS-DAP specification defines the interface protocol for a hardware debug adapter that sits between the host PC and the Debug Access Port DAP of the microcontroller (Fig. 4.5). This allows any software debugger that supports CMSIS-DAP to connect to any hardware debug adapter that also supports the CMSIS-DAP protocol. There are an increasing number of very low-cost evaluation boards that contain an integrated debugger that connects to a PC using USB. In many cases, this hardware debugger supports the CMSIS-DAP protocol so that it can be connected to any compliant toolchain.

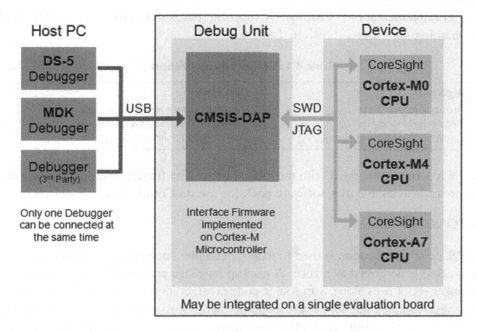

Figure 4.5
The CMSIS-DAP debug adapter provides a standard serial protocol that supports multiple IDEs and debuggers.

CMSIS-Pack

The CMSIS-Core and Driver specifications can be used to develop reusable software components that are portable across device families. The CMSIS-Pack specification defines a method of bundling all of the component elements (software files, examples, help files, templates) into a software Pack. Such a Pack can be downloaded and installed into your toolchain. The Pack also contains information on the software component dependencies, that is, the other files that need to be present when the component is used. This allows you to quickly integrate software from many sources to build a platform on which to develop your application code. As the complexity of Cortex-M microcontrollers is ever-increasing, this is a very important technology to increase a developer's productivity and code reliability.

CMSIS-NN

CMSIS Neural Net is a collection of building block functions that support the efficient design of Neural Net algorithms on Cortex-M microcontrollers. They aim to minimize the overall memory footprint whilst providing around a five-times performance increase over other typical implementations. CMSIS-NN supports (but is not limited to) the design of the Machine learning algorithms shown in Table 4.1.

Table 4.1: CMSIS NN algorithms

Algorithm	Description
Artificial Neural Net	Used for pattern recognition
Recurrent Neural Net	Pattern recognition in a time series
Convolutional Neural Net	DSP + ANN for image recognition

CMSIS-NN has been integrated with the Tensor Flow Lite framework, and the Tensor Flow Lite interpreter is available as a software pack for use on Cortex-M microcontrollers.

CMSIS-ZONE

The CMSIS-Zone specification is used to manage complex memory maps where each execution region is described as a zone. A markup language and an external utility are used to generate linker script files along with source code configuration files. CMSIS-Zone can be used to manage complex memory maps found in multiprocessor projects, defining execution regions for safety-critical software and also partitioning memory for security applications.

CMSIS-Build

CMSIS-Build defines a generic project file format that allows projects to be shared between different IDEs and build systems. The CMSIS-Build specification also defines workflows for continuous integration servers based on software components supplied in the CMSIS-Pack format. CMSIS-Build also includes the concept of software layers that can be retargeted to different hardware platforms.

Overview of CMSIS-Core

In the remainder of this chapter, we will look at the CMSIS-Core specification and then cover the remaining CMSIS specifications throughout the rest of this book. The CMSIS-Core specification provides a standard set of low-level functions, macros, and peripheral register definitions that allows your application code to easily access the Cortex-M processor and microcontroller peripheral registers. This framework needs to be added to your code at the start of a project. This is actually very easy to do as the CMSIS-Core functions are very much part of the compiler toolchain.

Coding Rules

While CMSIS is important for providing a standardized software interface for all Cortex-M microcontrollers, it is also interesting for embedded developers because it is based on a consistent set of "C" coding rules called MISRA-C. When applied, these coding rules generate clear, unambiguous "C" code, this approach is worth studying as it embodies many of the best practices that should be adopted when writing the "C" source code for your own application software.

MISRA-C

The MISRA-C coding standard is maintained and published by MIRA. MIRA stands for "Motor Industry Research Agency." MIRA is located near Rugby in England and is responsible for many of the industry standards used by the UK motor industry. In 1998 its software division released the first version of its coding rules formally called "MISRA guidelines for the use of C in-vehicle electronics" (Fig. 4.6).

Figure 4.6
The CMSIS source code has been developed using MISRA-C as a coding standard.

The original MISRA-C specification contained 127 rules which attempted to prevent common coding mistakes and resolve gray areas of the ANSI C specification when applied to embedded systems. Although initially intended for the automotive industry, MISRA-C has found acceptance in the wider embedded systems community. In 2004 a revised edition of MISRA-C was released with the title "MISRA-C Guidelines for the use of C in critical systems." This change in the title reflects the growing adoption of MISRA-C as a coding standard for general embedded systems. There have been two further updates to the MISRA-C standard in 2008 and 2012 which have expanded the number of rules to 143 with 20 directives. One of the other key attractions of MISRA-C is that it was written by engineers and not computer scientists. This has resulted in a clear, compact, and easy-to-understand set of rules. Each rule is clearly explained with examples of good coding practice. This means that the entire coding standard is contained in a slim volume that can easily be read in an evening. A typical example of a MISRA-C rule is shown below:

Rule 13.6 (required) Numeric variables being used within a for loop for iteration counting shall not be modified in the body of the loop

Loop counters shall not be modified in the body of the loop. However, other loop control variables representing logical values may be modified in the loop. For example, a flag to indicate that something has been completed, which is then tested in the for statement.

```
Flag = 1;
For ((I = 0;(i<5) && (flag= =1);i+ +)
{
/*.....*/
Flag = 0;/* Compliant — allows early termination of the loop */
i = i + 3;          /*Not Compliant — altering the loop counter */
}
```

Where possible, the MISRA-C rules have been designed so that they can be statically checked either manually or by a dedicated tool. The MISRA-C standard is not an open standard and is published in paper and electronic form on the MIRA website. Full details of how to obtain the MISRA Standard are available in Appendix A.

In addition to the MISRA-C guidelines, CMSIS enforces some additional coding rules. To prevent any ambiguity in the compiler implementation of standard "C" types CMSIS uses the data types defined in the ANSI C header file stdint.h as shown in Table 4.2.

Table 4.2: CMSIS variable types

Standard ANSI C Type	Misra C Type
signed char	int8_t
Signed short	int16_t
Signed int	int32_t
Signed __int64	int64_t
unsigned char	unit8_t
unsigned short	uint16_t
unsigned int	uint32_t
unsigned __int64	uint64_t

The typedefs ensure that the expected data size is mapped to the correct C language data type for a given compiler. Using typedefs like this is good practice as it avoids any ambiguity about the underlying variable size, which may vary between compilers, particularly if you are migrating code between different processor architectures and compiler tools.

CMSIS also specifies IO type qualifiers for accessing peripheral variables, as shown in Table 4.3. These are typedefs that make clear the type of access each peripheral register allows.

Table 4.3: CMSIS IO qualifiers

Misra-C IO Qualifier	ANSI C Type	Description
#define __I	volatile const	Read Only
#define __O	volatile	Write Only
#define __IO	volatile	Read and Write

While this does not provide any extra functionality for your code, it provides a common mechanism that can be used by static checking tools to ensure that the correct access is made to each peripheral register.

Much of the CMSIS documentation is autogenerated using a tool called Doxygen. This is a free download released under a GPL license. While Doxygen cannot actually write the documentation for you, it does do much of the dull, boring stuff for you (leaving you to do the exciting documentation work). Doxygen works by analyzing your source code and extracting declarations and specific source code comments to build up a comprehensive "object dictionary" for your project. The default output format for Doxygen is a browsable HTML, but this can be converted to other forms if desired.

The CMSIS source code comments contain specific tags prefixed by the @ symbol for example @brief. These tags are used by Doxygen to annotate descriptions of the CMSIS functions.

```
/**
* @brief Enable Interrupt in NVIC Interrupt Controller
* @param IRQn interrupt number that specifies the interrupt
* @return none.
* Enable the specified interrupt in the NVIC Interrupt Controller.
* Other settings of the interrupt such as priority are not affected.
*/
```

When the Doxygen tool is run, it analyses your source code and generates a report containing a dictionary of your functions and variables based on the comments and source code declarations.

CMSIS-Core Structure

The CMSIS-Core functions can be included in your project through the addition of three files (Fig. 4.7). These include the default startup code with the CMSIS standard vector table. The second file is the system_ < device>.c file that contains the necessary code to initialize the microcontroller system peripherals. Finally, the <device>.h header file, which imports the CMSIS header files that contain the CMSIS-Core functions and macros. Generally, these files will be part of the initial project configuration when you create a new project.

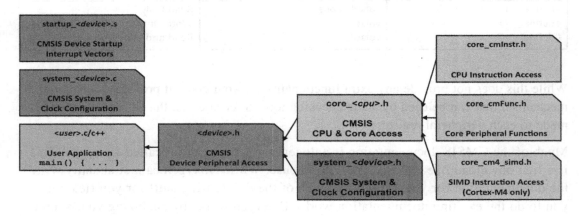

Figure 4.7

The CMSIS-Core standard consists of the device startup, system C code, and a device header. The device header defines the device peripheral registers and pulls in the CMSIS header files. The CMSIS header files contain all of the CMSIS-Core functions.

Startup code

The startup code provides the reset vector, initial stack pointer value, and a symbol for each of the interrupt vectors.

```
__Vectors  DCD __initial_sp      ;Top of Stack
           DCD  Reset_Handler     ;Reset Handler
           DCD  NMI_Handler       ;NMI Handler
           DCD  HardFault_Handler ;Hard Fault Handler
           DCD  MemManage_Handler ;MPU Fault Handler
```

When the processor starts, it will initialize the main stack pointer by loading the value stored in the first four bytes of the vector table. Then it will jump to the reset handler;

```
Reset_Handler PROC
  EXPORT Reset_Handler    [WEAK]
  IMPORT __main
  IMPORT SystemInit
      LDR  R0, =SystemInit
      BLX  R0
      LDR  R0, =__main
      BX   R0
      ENDP
```

System Code

The reset handler calls the SystemInit() function which is located in the CMSIS system_ < device>.c file. This code is delivered by the silicon manufacturer and it provides all the necessary code to configure the microcontroller after it leaves the reset vector. Typically, this includes setting up the internal phase-locked loops, configuring the microcontroller clock tree and internal bus structure, and enabling the external bus if required. The configuration of the initializing functions is controlled by a set of #defines located at the start of the module. This allows you to customize the basic configuration of the microcontroller system peripherals. Since the SystemInit() function is run when the microcontroller leaves reset the microcontroller system peripherals and the Cortex-M processor will be in a fully configured state when the program reaches main(). In the past, this system initializing code was something you would have had to write or crib from example code. On a new microcontroller, this would have been a few days' work, so the SystemInit() function does save you a lot of time and effort. The SystemInit() function also sets the CMSIS global variable SystemCoreClock to the CPU frequency. This variable can then be used by the application code as a reference value when configuring the microcontroller peripherals. CMSIS-Core also defines an additional function to update the SystemCoreClock variable if the CPU clock frequency is changed on the fly. The function SystemCoreClockUpdate(); is a void function that must be called if the CPU clock frequency is changed. This function is tailored to each microcontroller and will evaluate the clock tree registers to calculate the new CPU operating frequency and change the SystemCoreClock variable accordingly.

Once the SystemInit() function has run and we reach the application code, we will need to access the CMSIS-Core functions. This framework is added to the application modules through the microcontroller-specific header file.

Device Header File

The header file first defines all of the microcontroller special function registers in a CMSIS standard format.

A typedef structure is defined for each group of special function registers on the supported microcontroller. In the code below, a general GPIO typedef is declared for the group of GPIO reregisters. This is a standard type def, but we are using the IO qualifiers to designate the type of access granted to a given register.

```
typedef struct
{
  __IO uint32_t MODER;     /*!< GPIO port mode register, Address offset: 0x00 */
  __IO uint32_t OTYPER;    /*!< GPIO port output type register, Address offset: 0x04 */
  __IO uint32_t OSPEEDR;   /*!< GPIO port output speed register, Address offset: 0x08 */
  __IO uint32_t PUPDR;     /*!< GPIO port pull-up/pull-down register, Address offset: 0x0C */
  __IO uint32_t IDR;       /*!< GPIO port input data register, Address offset: 0x10 */
  __IO uint32_t ODR;       /*!< GPIO port output data register, Address offset: 0x14 */
  __IO uint16_t BSRRL;     /*!< GPIO port bit set/reset low register, Address offset: 0x18 */
  __IO uint16_t BSRRH;     /*!< GPIO port bit set/reset high register, Address offset: 0x1A */
  __IO uint32_t LCKR;      /*!< GPIO port configuration lock register, Address offset: 0x1C */
  __IO uint32_t AFR[2];    /*!< GPIO alternate function registers, Address offset: 0x24-0x28 */
} GPIO_TypeDef;
```

Next #defines are used to layout the microcontroller memory map. First, the base address of the peripheral special function registers is declared and then offset addresses to each of the peripheral busses, and finally an offset to the base address of each GPIO port.

```
#define PERIPH_BASE              ((uint32_t)0x40000000)
#define APB1PERIPH_BASE    PERIPH_BASE

#define GPIOA_BASE    (AHB1PERIPH_BASE + 0x0000)
#define GPIOB_BASE    (AHB1PERIPH_BASE + 0x0400)
#define GPIOC_BASE    (AHB1PERIPH_BASE + 0x0800)
#define GPIOD_BASE    (AHB1PERIPH_BASE + 0x0C00)
```

The register symbols for each GPIO port can then be declared.

```
#define GPIOA    ((GPIO_TypeDef*) GPIOA_BASE)
#define GPIOB    ((GPIO_TypeDef*) GPIOB_BASE)
#define GPIOC    ((GPIO_TypeDef*) GPIOC_BASE)
#define GPIOD    ((GPIO_TypeDef*) GPIOD_BASE)
```

In the application code we can program the peripheral special function registers by accessing the structure elements.

```
void LED_Init (void) {
  RCC->AHB1ENR |= ((1UL << 3) );           /* Enable GPIOD clock */
  GPIOD->MODER   &= ~((3UL << 2*12) |
                  (3UL << 2*13) |
                  (3UL << 2*14) |
                  (3UL << 2*15) );         /* PD.12...15 is output */
  GPIOD->MODER |= ((1UL << 2*12) |
                  (1UL << 2*13)|
                  (1UL << 2*14)|
                  (1UL << 2*15));
```

The microcontroller <device>.h include file provides similar definitions for all of the on-chip peripheral special function registers. These definitions are created and maintained by the silicon manufacturer, and as they do not use any non-ANSI keywords in the include file may be used with any "C" compiler. This means that any peripheral driver code written to the CMSIS specification is fully portable between CMSIS-compliant tools.

The microcontroller include file also provides definitions of the interrupt channel number for each peripheral interrupt source.

```
WWDG_IRQn           = 0, /*!< Window WatchDog Interrupt*/
PVD_IRQn            = 1, /*!< PVD through EXTI Line detection Interrupt*/
TAMP_STAMP_IRQn     = 2, /*!< Tamper and TimeStamp interrupts through the EXTI line*/
RTC_WKUP_IRQn       = 3, /*!< RTC Wakeup interrupt through the EXTI line*/
FLASH_IRQn          = 4, /*!< FLASH global Interrupt*/
RCC_IRQn            = 5, /*!< RCC global Interrupt*/
EXTI0_IRQn          = 6, /*!< EXTI Line0 Interrupt*/
EXTI1_IRQn          = 7, /*!< EXTI Line1 Interrupt*/
EXTI2_IRQn          = 8, /*!< EXTI Line2 Interrupt*/
EXTI3_IRQn          = 9, /*!< EXTI Line3 Interrupt*/
EXTI4_IRQn          = 10, /*!< EXTI Line4 Interrupt*/
```

In addition to the register and interrupt definitions the Silicon Vendor may also provide a library of peripheral driver functions. Again as this code is written to the CMSIS standard, it will compile with any suitable development tool. Often, these libraries are very useful for getting a project working quickly and minimize the amount of time you have to spend writing low-level code. However, they are often very general libraries that do not yield the most optimized code. So if you need to get the maximum performance or minimal code size, you will need to rewrite the driver functions to suit your specific application. The microcontroller include file also imports up to five further include files. These are "stdint.h" a "CMSIS-Core" file for the Cortex-M processor you are using. A header file "system_<device>.h" is also included to give access to the functions in the system file. The CMSIS instruction intrinsic and helper functions are contained in two further files, "core_cminstr.h" and "core_cmfunc.h." If you are using the Cortex-M4 or Cortex-M7, an additional file "core_CM4_simd.h" is added to provide support for the Cortex-M4 SIMD instructions. As discussed earlier, the "stdint.h" file provides the MISRA-C types which are used in the CMSIS definitions and should be used through your application code.

CMSIS-Core Header files

Within the CMSIS-Core specification, there are a small number of defines that are set up for a given microcontroller (Table 4.4). These can be found in the <device>.h processor include file.

Table 4.4: CMSIS configuration values

CMSIS Define	Description
__CMx_REV	Core revision number
__NVIC_PRIO_BITS	Number of priority bits implemented in the NVIC priority registers
__MPU_PRESENT	Defines if an MPU is present (see Chapter 5: Advanced Architecture Features)
__FPU_PRESENT	Defines if an FPU is present (see Chapter 5: Advanced Architecture Features)
__Vendor_SysTickConfig	Defines if there is a vendor-specific SysTick Configuration

The processor include file also imports the CMSIS header files, which contain the CMSIS-Core helper functions. The helper functions are split into the groups shown in Table 4.5.

Table 4.5: CMSIS function groups

CMSIS-Core Function Groups
NVIC access functions
SysTick configuration
CPU register Access
CPU instruction intrinsics
Cortex-M4 SIMD intrinsics
ITM debug functions
FPU Functions
Level 1 Cache Functions (Cortex-M7 only)
Cortex-M Vector Extensions
Trust Zone for Armv8-M
MPU Functions for Armv6/v7-M
MPU Functions for Armv6/v7-M
PMU Functions for Armv8.1-M

The NVIC group provides all the functions necessary to configure the Cortex-M interrupts and exceptions. A similar function is provided to configure the SysTick timer and interrupt. The CPU register group allows you to easily read and write to the CPU registers using the

MRS and MSR instructions. Any instructions that are not reachable by the "C" language are provided with dedicated intrinsic functions and are contained in the CPU instructions group. An extended set of intrinsics are also provided for the Cortex-M4 and Cortex-M7 to access the SIMD instructions. Finally, some standard functions are provided to access the debug ITM.

Interrupts and Exceptions

Management of the NVIC registers may be done by the functions provided in the interrupt and exception group (Table 4.6). These functions allow you to set up an NVIC interrupt channel and manage its priority as well as interrogate the NVIC registers during run time.

Table 4.6: CMSIS Interrupt and exception group

CMSIS Function	Description
NVIC_SetPriorityGrouping	Set the priority grouping
NVIC_GetPriorityGrouping	Read the priority grouping
NVIC_EnableIRQ	Enable a peripheral interrupt channel
NVIC_DisableIRQ	Disable a peripheral interrupt channel
NVIC_GetPendingIRQ	Read the pending status of an interrupt channel
NVIC_SetPendingIRQ	Set the pending status of an interrupt channel
NVIC_ClearPendingIRQ	Clear the pending status of of an interrupt channel
NVIC_GetActive	Get the active status of an interrupt channel
NVIC_SetPriority	Set the active status of an interrupt channel
NVIC_GetPriority	Get the priority of an interrupt channel
NVIC_EncodePriority	Encodes the priority value in terms of priority Group and subgroup
NVIC_DecodePriority	Decodes the priority value in terms of priority Group and subgroup
NVIC_SystemReset	Forces a system reset
NVIC_ClearTargetState	Clears the Interrupt target field in the nonsecure NVIC (Armv8-m secure state only)
NVIC_SetTargetState	Sets the Interrupt target field in the nonsecure NVIC (Armv8-m secure state only)
NVIC_GetVector	Read the address of an interrupt service routine
NVIC_GetEnabledIRQ	Returns the current interrupt enable status of a specified Interrupt channel
NVIV_GetPendingIRQ	Returns the current pending status of a specified Interrupt channel

A configuration function is also provided for the SysTick timer (Table 4.7).

Table 4.7: CMSIS systick function

CMSIS Function	Description
SysTick_Config	Configures the timer and enables the interrupt

So, for example, to configure an external interrupt line, we first need to find the name for the external interrupt vector used in the startup code vector table.

```
DCD    FLASH_IRQHandler              ; FLASH
DCD    RCC_IRQHandler                ; RCC
DCD    EXTI0_IRQHandler              ; EXTI Line0
DCD    EXTI1_IRQHandler              ; EXTI Line1
DCD    EXTI2_IRQHandler              ; EXTI Line2
DCD    EXTI3_IRQHandler              ; EXTI Line3
DCD    EXTI4_IRQHandler              ; EXTI Line4
DCD    DMA1_Stream0_IRQHandler       ; DMA1 Stream 0
```

So, for external interrupt line 0, we simply need to create a void function duplicating the name used in the vector table:

```
void EXTI0_IRQHandler (void);
```

This now becomes our interrupt service routine. In addition, we must configure the microcontroller peripheral and NVIC to enable the interrupt channel. In the case of the external interrupt line, the following code will setup Port A pin 0 to generate an interrupt to the NVIC on a falling edge.

```
AFIO->EXTICR[0]    & = ~AFIO_EXTICR1_EXTI0;        /* clear used pin      */
AFIO->EXTICR[0]    | = AFIO_EXTICR1_EXTI0_PA;      /* set PA.0 to use     */
EXTI->IMR          | = EXTI_IMR_MR0;               /* unmask interrupt    */
EXTI->EMR          & = ~EXTI_EMR_MR0;              /* no event            */
EXTI->RTSR         & = ~EXTI_RTSR_TR0;             /* no rising edge trigger */
EXTI->FTSR         | = EXTI_FTSR_TR0;              /* set falling edge trigger */
```

Next, we can use the CMSIS functions to enable the interrupt channel.

```
NVIC_EnableIRQ(EXTI0_IRQn);
```

Here, we are using the defined enumerated type for the interrupt channel number. This is declared in the microcontroller header file <device>.h. Once you get a bit familiar with the CMSIS-Core functions, it becomes easy to intuitively work out the name rather than having to look it up or look up the NVIC channel number.

We can also add a second interrupt source by using the SysTick configuration function which is the only function in the SysTick group.

```
uint32_t SysTick_Config(uint32_t ticks)
```

This function configures the countdown value of the SysTick timer and enables its interrupt, so an exception will be raised when its count reaches zero. The SysTick timer input frequency is usually derived from the CPU clock frequency. This allows us to easily setup the SysTick timer to generate a desired periodic interrupt using the

SystemCoreClock frequency. So a one millisecond interrupt can be generated as follows:

```
SysTick_Config(SystemCoreClock/1000);
```

Again, we can look up the exception handler from the vector table.

```
DCD  0;                 Reserved
DCD  PendSV_Handler;    PendSV Handler
DCD  SysTick_Handler;   SysTick Handler
```

and create a matching 'C' function;

```
void SysTick_Handler (void);
```

Now that we have two interrupt sources, we can use other CMSIS interrupt and exception functions to manage the priority levels. The number of priority levels will depend on how many priority bits have been implemented by the silicon manufacturer. For all of the Cortex-M processors, we can use a simple "flat" priority scheme where zero is the highest priority. The priority level is set by

```
NVIC_SetPriority(IRQn_Type IRQn,uint32_t priority);
```

The set priority function is a bit more intelligent than a simple macro. It uses the IRQn NVIC channel number to differentiate between user peripherals and the Cortex-M processor exceptions. This allows it to program either the system handler priority registers in the system control block or the Interrupt priority registers in the NVIC itself. The NVIC_SetPriority() function also uses NVIC_PRIO_BITS definition to shift the priority value into the active priority bits which have been implemented by the Silicon Vendor.

```
_STATIC_INLINE void NVIC_SetPriority(IRQn_Type IRQn, uint32_t priority)
{
 if(IRQn < 0){
 SCB->SHP[((uint32_t)(IRQn) & 0xF)-4] = ((priority << (8 - __NVIC_PRIO_BITS)) & 0xff);
 } /* set Priority for Cortex-M System Interrupts */
else {
 NVIC->IP[(uint32_t)(IRQn)] = ((priority << (8 - __NVIC_PRIO_BITS)) & 0xff); }   /* set
 Priority for device specific Interrupts */
}
```

For Cortex-M3/M4 and Cortex-M7, we have the option to set priority Groups and subgroups as discussed in Chapter 3, Cortex-M Architecture. Depending on the number of priority bits defined by the manufacturer, we can configure priority groups and subgroups.

```
NVIC_SetPriorityGrouping();
```

To set the NVIC priority grouping, you must write to the "Application Interrupt and Reset Control" Register. As discussed in Chapter 3, Cortex-M Architecture, this register is protected by its VECTKEY field. In order to update this register, you must write "0x5FA" to the VECTKEY field. The SetPriorityGrouping() function provides all the necessary code to do this.

```
__STATIC_INLINE void NVIC_SetPriorityGrouping(uint32_t PriorityGroup)
{
  uint32_t reg_value;
  uint32_t PriorityGroupTmp = (PriorityGroup & (uint32_t)0x07);       /* only values 0...7
                                                                         are used         */
  reg_value = SCB->AIRCR;                            /* read old register configuration */
  reg_value &= ~(SCB_AIRCR_VECTKEY_Msk | SCB_AIRCR_PRIGROUP_Msk); /* clear bits to change  */
  reg_value = (reg_value| ((uint32_t)0x5FA << SCB_AIRCR_VECTKEY_Pos)| /* Insert write
  key and priorty group */
                  (PriorityGroupTmp << 8));
  SCB->AIRCR = reg_value;
}
```

The "Interrupt and Exception" group also provides a system reset function that will generate a hard reset of the whole microcontroller.

```
NVIC_SystemReset(void);
```

This function writes to bit two of the "Application Interrupt Reset Control" Register. This strobes a logic line out of the Cortex-M Core to the microcontroller reset circuitry, which resets the microcontroller peripherals and the Cortex-M processor. However, you should be a little careful here as the implementation of this feature is down to the microcontroller manufacturer and may not be fully implemented. So if you are going to use this feature, you need to test it first to ensure that the microcontroller peripherals do in fact go back to their reset configuration. Bit zero of the same register will do a warm reset of the Cortex-M processor. That is, force a reset of the Cortex-M processor but leave the microcontroller registers configured.

Exercise 4.1: CMSIS and User Code Comparison

In this exercise, we will revisit the multiple interrupts example and examine a rewrite of the code using the CMSIS-Core functions.

Open the Pack Installer.

Select the Boards:Designers Guide Tutorial.

Select the example tab and copy "Ex 4.1 CMSIS Multiple Interrupt."

Open main.c in both projects and compare the initializing code.

The SysTick timer and ADC interrupts can be initialized with the following CMSIS functions.

```
SysTick_Config(SystemCoreClock / 100);
NVIC_EnableIRQ                    (ADC1_2_IRQn);
NVIC_SetPriorityGrouping          (5);
NVIC_SetPriority                  (SysTick_IRQn,4);
NVIC_SetPriority                  (ADC1_2_IRQn,4);
```

We can compare this to the equivalent non CMSIS code. . ..

```
SysTick->VAL = 0x9000;                          //Start value for the sys Tick counter
SysTick->LOAD = 0x9000;                         //Reload value
SysTick->CTRL = SYSTICK_INTERRUPT_ENABLE
          |SYSTICK_COUNT_ENABLE;                //Start and enable interrupt
NVIC->ISER[0] = ( 1UL << 18);                   /* enable ADC Interrupt        */
NVIC->IP[18] = (2<<6 | 2<<4);
SCB->SHP[11] = (1<<6 | 3<<4);
Temp = SCB->AIRCR;
    Temp &= ~(SCB_AIRCR_VECTKEY_Msk | SCB_AIRCR_PRIGROUP_Msk);
    Temp = (Temp|((uint32_t)0x5FA << SCB_AIRCR_VECTKEY_Pos)|(5 << 8));
    SCB->AIRCR = Temp;
```

Although both blocks of code achieve the same thing, the CMSIS version is much faster to write, more readable, and far less prone to coding mistakes.

Build both projects and compare the size of the code produced.

The CMSIS functions introduce a small overhead, but this is an acceptable trade-off against ease of use, portability, and maintainability.

CMSIS-Core Register Access

The next group of CMSIS functions gives you direct access to the processor CPU registers (Table 4.8).

Table 4.8: CMSIS CPU register functions

Core Function	Description
__get_Control	Read the CPU CONTROL register
__set_Control	Write to the CONTROL register
__get_IPSR	Read the IPSR register
__get_APSR	Read the APSR register
__get_xPSR	Read the xPSR register
__get_PSP	Read the Process stack pointer
__set_PSP	Write to the process stack pointer
__get_PSPLIM	Read the value of the Process stack limit register(Armv8-M only)
__set_PSPLIM	Write a value to the Process stack limit register (Armv8-M only)
__get_MSP	Read the main stack pointer
__set_MSP	Write to the main stack pointer

(Continued)

Table 4.8: (Continued)

Core Function	Description
__get_MSPLIM	Read the value of the Main stack limit register(Armv8-M only)
__set_MSPLIM	Write a value to the Main stack limit register (Armv8-M only)
__get_PRIMASK	Read the PRIMASK
__set_PRIMASK	Write to the PRIMASK
__get_BASEPRI	Read the BASEPRI register
__set_BASEPRI	Write to the BASEPRI register
__set_BASEPRI_MAX	Writes a value to the BASEPRI mask register
__get_FAULTMASK	Read the FAULTMASK
__set_FAULTMASK	Write to the FAULTMASK
__get_FPSCR	Read the FPSCR
__set_FPSCR	Write to the FPSCR
__enable_irq	Enable interrupts and configurable fault exceptions
__disable_irq	Disable interrupts and configurable fault exceptions
__enable_fault_irq	Enables interrupts and all fault handlers
__disable_fault_irq	Disables interrupts and all fault handlers

These functions provide you with the ability to globally control the NVIC interrupts and set the configuration of the Cortex-M processor into its more advanced operating mode. First, we can globally enable and disable the microcontroller interrupts with the following functions.

```
__set_PRIMASK(void);
__set_FAULTMASK (void);
__enable-IRQ
__enable_Fault-irq
__set_BASEPRI()
```

While all of these functions are enabling and disabling interrupt sources, they all have slightly different effects. The __set_PRIMASK() function and the enable_IRQ/Disable_IRQ functions have the same effect in that they set and clear the PRIMASK bit, which enables and disables all interrupt sources except the Hard Fault Handler and the Non-Maskable interrupt. The __set_FAULTMASK() function can be used to disable all interrupts except the Non-Maskable Interrupt. We will see later how this can be useful when we want to bypass the Memory protection unit. Finally, the __set_BASEPRI() function sets the minimum active priority level for user peripheral interrupts. When the Base priority register is set to a nonzero level any interrupt at the same priority level or lower will be disabled.

These functions allow you to read the program status register and its aliases. You can also access the control register to enable the advanced operating modes of the Cortex-M processor as well as explicitly setting the stack pointer values. A dedicated function is also provided to access the Floating-point status and control register if you are using the Cortex-M4 or Cortex-M7. We will have a closer look at the more advanced operating modes of the Cortex-M processor in Chapter 5, Advanced Architecture Features.

CMSIS-Core CPU Intrinsic Instructions

The CMSIS-Core header also provides two groups of standardized intrinsic functions (Table 4.9). The first group is common to all Cortex-M processors, and the second provides standard intrinsic for the Cortex-M4 SIMD instructions.

Table 4.9: CMSIS instruction intrinsics

CMSIS Function	Description	More Information
__NOP	No Operation	See Chapter 3, Cortex-M
__WFI	Wait for interrupt	Architecture
__WFE	Wait for event	
__SEV	Send Event	
__ISB	Instruction synchronization barrier	
__DSB	Data synchronization barrier	
__DMD	Data Memory synchronization barrier	
__REV	Reverse byte order (32 bit)	See this Chapter for rotation
__REV16	Reverse byte order (16 bit)	instructions
__REVSH	Reverse byte order, signed short	
__RBIT	Reverse bit order (not for Cortex-M0)	
__ROR	Rotate right by n bits	
__LDREXB	Load exclusive (8 bits)	See Chapter 5, Advanced
__LDREXH	Load exclusive (16 bits)	Architecture Features, for
__LDREXW	Load exclusive (32 bits)	exclusive access instructions
__STREXB	Store exclusive (8 bits)	
__STREXH	Store exclusive (16 bits)	
__STREXW	Store exclusive (32 bits)	
__CLREX	Remove exclusive lock	
__SSAT	Signed saturate	See Chapter 3, Cortex-M
__USAT	Unsigned saturate	Architecture
__CLZ	Count leading zeros	See Chapter 3, Cortex-M Architecture

The CPU intrinsics provide direct access to Cortex-M processor instructions that are not directly reachable from the "C" language. Using an intrinsic will allow a dedicated single cycle instruction to replace multiple instructions generated by standard "C" code.

With the CPU intrinsics, we can enter the low-power modes using the __WFI() and _WFE() instructions. The CPU intrinsics also provide access to the saturated math's instructions that we met in Chapter 3, Cortex-M Architecture. The intrinsic functions also give access to the execution barrier instructions that ensure completion of a data write or instruction execution before continuing with the next instruction. The next group of instruction intrinsics is used to guarantee exclusive access to a memory region by one region of code. We will have a look at these in Chapter 5, Advanced Architecture Features. The remainder of the CPU intrinsics support single cycle data manipulation functions such as the rotate and reverse bit order instructions.

Exercise 4.2: Intrinsic Bit Manipulation

In this exercise, we will look at the data manipulation intrinsic supported in CMSIS.

Open the Pack Installer.

Select the Boards:Designers Guide Tutorial.

Select the example tab and Copy "Ex 4.2: CMSIS-Core Intrinsic."

The exercise declares an input variable and a group of output variables and then uses each of the intrinsic data manipulation functions.

```
outputREV          = __REV(input);
outputREV16        = __REV16(input);
outputREVSH        = __REVSH(input);
outputRBIT         = __RBIT(input);
outputROR          = __ROR(input,8);
outputCLZ          = __CLZ(input);
```

Build the project and start the debugger.

Add the input and each of the output variables to the watch window.

Step through the code and count the cycles taken for each function.

While each intrinsic instruction takes a single cycle, some surrounding instructions are required so the intrinsic functions take between 9 and 18 cycles.

Examine the values in the output variables to familiarize yourself with the action of each intrinsic.

Consider how you would code each intrinsic using standard "C" instructions.

CMSIS SIMD Intrinsics

The next group of CMSIS intrinsics provides direct access to the Cortex-M4 and Cortex-M7 SIMD instructions.

The SIMD instructions provide simultaneous calculations for two sixteen-bit operations or four eight-bit operations. This greatly enhances any form of repetitive calculation over a data set, as in a digital filter. We will take a close look at these instructions in Chapter 9, *Practical DSP for Cortex-M Microcontrollers.*

CMSIS-Core Debug Functions

The CMSIS-Core functions also provide enhanced debug support through the CoreSight ITM. The CMSIS standard has two dedicated debug specifications CMSIS-SVD and CMSIS-DAP which we will look at in Chapter 8, Debugging with CoreSight. However, the CMSIS-Core specification contains some helpful debug support.

Hardware Breakpoint

First of all, there is a dedicated intrinsic to add a hardware breakpoint to your code.

```
__BKPT(uint8_t value)
```

Using this intrinsic will place a hardware breakpoint instruction at this location in your code. When this point is reached, execution will be halted, and the "value" will be passed to the debugger. During development, the __BKPT() intrinsic can be used to trap error conditions and halt the debugger.

Instrumentation Trace

As part of its hardware debug system, the Cortex-M3, Cortex-M4, and Cortex-M7 provide an "ITM" unit. This can be thought of as a debug UART which is connected to a console window in the debugger. By adding debug hooks (Instrumenting) into your code it is possible to read and write data to and from the debugger while the code is running. We will look at using the ITM for additional debug and software testing in Chapter 8, Debugging with CoreSight. For now, there are a number of CMSIS functions that standardize communication with the ITM (Table 4.10).

Table 4.10: CMSIS debug functions

CMSIS Debug Function	Description
volatile int ITM_RxBuffer = ITM_RXBUFFER_EMPTY;	Declare one word of storage for receive flag
ITM_SendChar(c);	Send one character to the ITM
ITM_CheckChar()	Check if any data has been received
ITM_ReceiveChar()	Read one character from the ITM

CMSIS Core Functions for Corex-M7

With the release of the Cortex-M7 processor at the end of 2014, the CMSIS-Core specification was extended to provide some additional functions to support the new features introduced by the Corex-M7 (Table 4.11).

Table 4.11: Cortex-M7 CMSIS core support

CMSIS Cortex-M7 function	Description
Cache Functions	Eleven functions to support the Instruction and Data Caches
FPU Function	One function to support the FPU

As we will see in Chapter 6, Cortex-M7 Processor, the Cortex-M7 introduces Data and Instruction caches to the Cortex-M processor family. The CMSIS cache functions allow you to enable and disable the caches and manage them as your code executes. We will look at these functions in Chapter 6, Cortex-M7 Processor.

MPU Support

All of the Cortex-M processors may be fitted with a Memory protection Unit when the microcontroller is designed. The CMSIS core specification defined a set of support functions that are used to configure and manage the MPU. However, there are two versions of the MPU an original version used by the Armv7-M-based devices (M0 + /M3/M4 and M7) and a later version used by Armv8.x-M-based devices (M23/M33/M55/M85). A set of functions are provided for each MPU version. We will look at the CMSIS support for Armv7-M devices in Chapter 5, Advanced Architecture Features, and support for Armv8.x-M devices in Chapter 7, Armv8-M Architecture.

Armv8-M Support

The CMSIS core specification provides additional support for new Armv8.x-M features (Table 4.12). We will cover these features in Chapter 7, Armv8-M Architecture.

Table 4.12: Armv8-M CMSIS core support

Feature	Description
MVE	Cortex-M Vector Extensions for Armv8.1
PMU	Performance Monitoring Unit for Armv8.1
TrustZone	Support for the Armv8.x security extension

Conclusion

A good understanding of each CMSIS specification is key to effectively developing applications for any Cortex-M-based microcontroller. In this chapter we have introduced each CMSIS specification and taken a detailed look at the CMSIS-Core specification. We will look at the remaining CMSIS specifications throughout the rest of this book.

CHAPTER 5

Advanced Architecture Features

Introduction

In the last few chapters, we have covered most of what you need to know to develop with a Cortex-M-based microcontroller. In this chapter, we will look at some of the more advanced features of the Cortex-M processor. All of the features discussed in this chapter are included in the Cortex-M0 + /M3/M4 and M7. In this chapter, we will look at the different operating modes built into each of the Cortex-M processors and some additional instructions that are designed to support the use of a Real-Time Operating System. We will also have a look at the optional Memory Protection Unit (MPU), which can be fitted to the Cortex-M0 + , M3, M4, and M7, and how this can partition the memory map into regions with different execution privileges. To round the chapter off, we will have a look at the bus interface between the Cortex-M processor and the microcontroller system.

Cortex Processor Operating Modes

When the Cortex-M processor comes out of reset, it is running in a simple "flat" mode where all of the application code has access to the entire processor address space and unrestricted access to the CPU and NVIC registers. While this is ok for many applications, the Cortex-M processor has a number of features that let you place the processor into a more advanced operating model that is suitable for high-integrity software and also supports the use of a Real-Time Operating System.

As a first step to understanding the more advanced operating modes of the Cortex-M processor, we need to review its basic operation. The CPU can be running in two different modes, Thread mode and Handler mode (Fig. 5.1). When the processor is executing background code (i.e., noninterrupt code), it is running in Thread mode. When the processor is executing interrupt code, it is running in Handler mode.

The Designer's Guide to the Cortex-M Processor Family.
DOI: https://doi.org/10.1016/B978-0-323-85494-8.00004-8

		Operations (privilege out of reset)	Stacks (Main out of reset)
Modes (Thread out of reset)	Handler - Processing of exceptions	**Privileged execution Full control**	**Main Stack Used by OS and Exceptions**
	Thread - No exception is being processed - Normal code execution	**Privileged or Unprivileged**	**Main or Process**

Figure 5.1

Each Cortex-M Processor has two execution modes, (interrupt) Handler and (background) Thread. It is possible to configure these modes to have privileged and unprivileged access to memory regions. It is also possible to configure a two-stack operating mode.

The operating state of the Cortex-M CPU is managed through the CONTROL register. Since this register is a CPU register rather than a memory-mapped address and can only be accessed by the MRS and MSR instructions. The CMSIS core specification provides dedicated functions to read and write to the CONTROL register.

```
void __set_CONTROL(uint32_t value);
uint32_t __get_CONTROL(void);
```

When the processor starts to run out of reset there is no operating difference between Thread and Handler mode. Both modes have full access to all features of the CPU, this is known as privileged execution. By programming the Cortex-M processor CONTROL register (Fig. 5.2), it is possible to place the Thread mode into unprivileged execution by setting the Thread Privilege Level (TPL) bit.

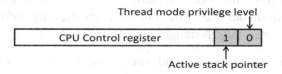

Figure 5.2

The Control register is a CPU register which can only be accessed by the MRS and MSR instructions. It contains two bits that configure the thread mode privilege level and activation of the process stack.

In unprivileged mode, the MRS, MSR, and Change Processor State (CPS) instructions are disabled for all CPU registers except the register file (R0−R15) and the Application Program Status Register (APSR) alias of the Program Status Register (PSR). This prevents the Cortex-M processor from accessing the CONTROL, FAULTMASK, PRIMASK registers, and the PROGRAM STATUS register (except the APSR). In unprivileged mode, it is also not possible to access the SysTick timer registers, NVIC, or the System Control Block. These limitations attempt to prevent unprivileged code from accidentally disturbing the operation of the Cortex-M processor. If Thread Mode has been limited to unprivileged access, it is only possible to clear the TPL bit from Handler Mode. Once the TLP bit has been set, the application code running in Thread Mode can no longer influence the operation of the Cortex-M processor. When the processor responds to an exception or interrupt, it moves into Handler mode, which always executes code in Privileged Mode with full access to all the processor resources.

The CONTROL register also contains an additional bit, the Active Stack Pointer Selection. When set this bit enables an additional stack pointer called the Process Stack Pointer (PSP).

The Process Stack is a banked R13 stack pointer which is used by code running in Thread mode. When the Cortex-M processor responds to an exception it enters Handler mode. This causes the CPU to switch stack pointers. In this configuration, we now have two distinct operating modes each with their own stack. The Thread mode that can be used for application code and the Handler mode which is used for interrupts and RTOS code.

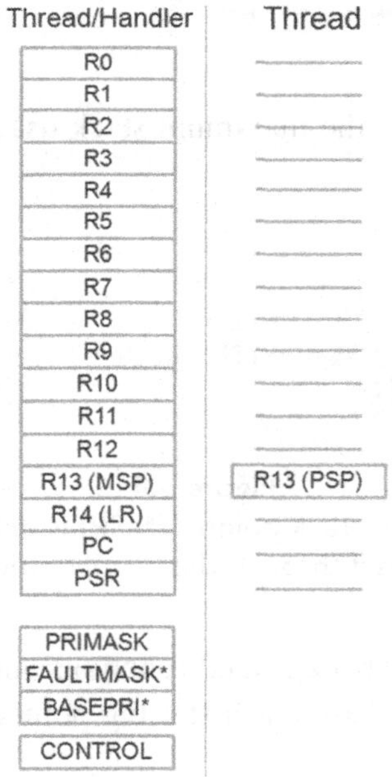

Thread/Handler	Thread
R0	—
R1	—
R2	—
R3	—
R4	—
R5	—
R6	—
R7	—
R8	—
R9	—
R10	—
R11	—
R12	—
R13 (MSP)	R13 (PSP)
R14 (LR)	—
PC	—
PSR	—

PRIMASK
FAULTMASK*
BASEPRI*

CONTROL

Figure 5.3
At reset R13 is the main stack pointer and is automatically loaded with the initial stack value. The CPU CONTROL register can be used to enable a second banked R13 register. This is the process stack which is used in thread mode. The application code must load an initial stack value into this register.

As we have seen in Chapter 3, Cortex-M Architecture, at reset the Main Stack Pointer (MSP) will be loaded with the value stored in the first four bytes of memory (Fig. 5.3) and the total stack space is defined in the startup code (Fig. 5.4). However, the Process Stack is not automatically initialized and must be set up by the application code before it is enabled. Fortunately, the CMSIS-Core specification contains functions to configure the Process Stack.

```
void __set_PSP(uint32_t TopOfProcStack);
uint32_t __get_PSP(void);
```

So, if you have to manually set the initial value of the Process Stack, what should it be? There is not an easy way to answer this, but the compiler produces a report file that details the static calling tree for the project. This file is created each time the project is built and is called < project name >. htm.

Static Call Graph for image .\OperatingModes.axf

#<CALLGRAPH># ARM Linker, 5060960: Last Updated: Tue Aug 10 10:18:04 2021

Maximum Stack Usage = 144 bytes + Unknown(Functions without stacksize, Cycles, Untraceable Function Pointers)

Call chain for Maximum Stack Depth:

__rt_entry_main ⇒ main ⇒ __2printf ⇒ _printf_char_file ⇒ _printf_char_common ⇒ __printf

The report file includes a value for the maximum stack usage and a calling tree for the longest call chain.

```
Stack_Size     EQU     0x00000100

         AREA     STACK, NOINIT, READWRITE, ALIGN=3
Stack_Mem     SPACE     Stack_Size
__initial_sp
```

Option	Value
⊟ Stack Configuration	
Stack Size (in Bytes)	0x0000 0100
⊟ Heap Configuration	
Heap Size (in Bytes)	0x0000 0200

Figure 5.4
The stack size allocated to the main stack pointer (MSP) is defined in the startup code and can be configured through the configuration wizard.

This calling tree is likely to be for background functions and will be the maximum value for the PSP. This value can also be used as a starting point for the MSP when the PSP is not being used.

Exercise 5.1: Stack Configuration

In this exercise, we will have a look at configuring the operating mode of the Cortex-M processor so the thread mode is running with unprivileged access and uses the PSP.

Open the Pack Installer.

Select the Boards::Designers Guide Tutorial.

Select the example tab and copy "Ex 5.1 Process Stack Configuration."

Build the code and start the debugger and run to main().

This is a version of the blinky project we used earlier, with some code added to configure the processor operating mode. The new code includes a set of #defines.

```
#define USE_PSP_IN_THREAD_MODE  (1<<1)
#define THREAD_MODE_IS_UNPRIVILIGED 1
#define PSP_STACK_SIZE    0x200
```

The first two declarations define the location of the bits which need to be set in the CONTROL register to enable the process stack and switch the thread mode into unprivileged access. Then, we define the size of the process stack space. At the start of main(), we can use the CMSIS functions to configure and enable the process stack. Finally, there is an instruction barrier to ensure that all the functions have been completed before the code continues. We can also examine the operating modes of the processor in the register window (Fig. 5.5).

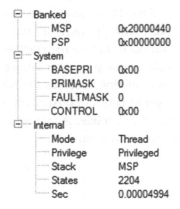

Figure 5.5
The Cortex-M processor is in Thread/privileged mode using the main stack. The PSP is not initialized.

```
_initalPSPValue = __get_MSP() + PSP_STACK_SIZE;
_set_PSP(initalPSPValue);
__set_CONTROL(USE_PSP_IN_THREAD_MODE);
__ISB();
```

Figure 5.6
Now the processor is in Thread/privileged mode but is using the process stack which has been initialized with a stack space of 200 h bytes.

When you reach the main() function, the processor is in thread mode with full privileged access to all features of the microcontroller. Also, only the MSP is being used. If you step through the three configuration lines, the code first reads the contents of the MSP. This will be at the top of the main stack space. To get the start address for the PSP, we simply add the desired stack size in bytes (Fig. 5.6). This value is written by the process stack before enabling it in the CONTROL register. Always configure the stack before enabling it in case there is an active interrupt that could occur before the stack is ready. Next, we need to execute any code that needs to configure the processor before switching the thread mode to unprivileged access. The ADC_Init() function accesses the NVIC to configure an interrupt and the SysTick timer is also configured (Fig. 5.7). Accessing these registers will be prohibited when we switch to unprivileged mode. Again an instruction barrier is used to ensure the code completes before execution continues.

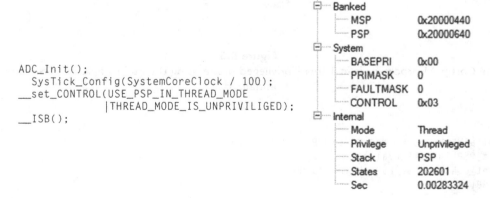

```
ADC_Init();
  SysTick_Config(SystemCoreClock / 100);
__set_CONTROL(USE_PSP_IN_THREAD_MODE
            |THREAD_MODE_IS_UNPRIVILIGED);
__ISB();
```

Figure 5.7
Now the processor has been set into Thread/unprivileged mode.

Set a breakpoint in the IRQ.c module line 32 (Fig. 5.8).

This is in the SysTick interrupt handler routine. Now run the code and it will hit the breakpoint when the SysTick handler interrupt is raised.

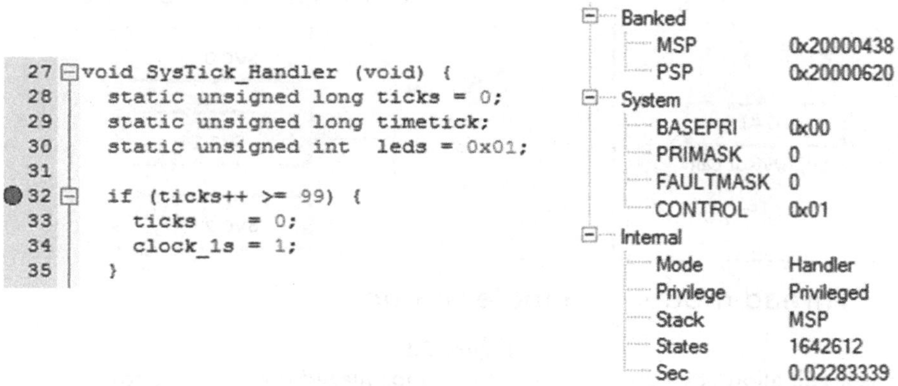

```
27 ⊟void SysTick_Handler (void) {
28      static unsigned long ticks = 0;
29      static unsigned long timetick;
30      static unsigned int  leds = 0x01;
31
32 ⊟    if (ticks++ >= 99) {
33        ticks    = 0;
34        clock_1s = 1;
35      }
```

⊟ Banked		
── MSP	0x20000438	
── PSP	0x20000620	
⊟ System		
── BASEPRI	0x00	
── PRIMASK	0	
── FAULTMASK	0	
── CONTROL	0x01	
⊟ Internal		
── Mode	Handler	
── Privilege	Privileged	
── Stack	MSP	
── States	1642612	
── Sec	0.02283339	

Figure 5.8
During the exception the processor enters Handler/privileged mode.

Now that the processor is serving an interrupt, it has moved into interrupt handler mode with privileged access to the Cortex-M processor and is using the main stack.

Supervisor Call

Once configured, this more advanced operating mode provides a partition between the exception/interrupt code running in Handler mode and the background application code running in Thread mode. Each operating mode can have its own code region, ram region, and stack. This allows the interrupt handler code full access to the chip without the risk that its operation may be corrupted by the application code. However, at some point, the application code will need to access features of the Cortex-M processor that are only available in the Handler mode with its full privileged access. To allow this to happen, the Thumb-2 instruction set has an instruction called Supervisor Call (SVC). When this instruction is executed, it raises a supervisor exception which moves the processor from executing the application code in Thread/Unprivileged mode to an exception routine in Handler/Privileged mode (Fig. 5.9). The SVC has its own location within the vector table and behaves like any other exception.

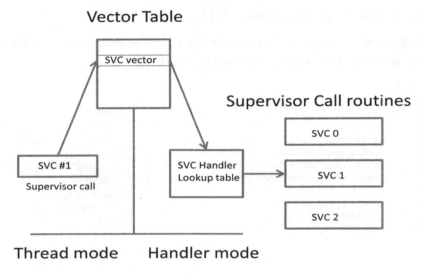

Figure 5.9
The supervisor call allows execution to move from unprivileged thread mode to privileged handler mode and gain full unrestricted access to the cortex processor. The SVC instruction is used by RTOS API calls.

The SVC instruction may also be encoded with an 8-bit value called an ordinal. When the SVC call is executed, this ordinal value can be read and used as an index to call one of 256 different supervisor functions (Fig. 5.10).

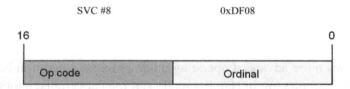

Figure 5.10
The unused portion of the SVC instruction can be encoded with an ordinal number. On entry to the SVC handler this number can be read to determine which SVC functions to execute.

The compiler toolchain provides a dedicated SVC support function that is used to extract the ordinal value and then call the appropriate function. First, the SVC support function reads the link register to determine the operating mode. It then reads the value of the saved PC from the appropriate stack. We can then read the memory location holding the SVC instruction and extract the ordinal value. This number is then used as an index into a lookup table to load the address of the function which is being called. The function is then called and will be executed in privileged mode before we return back to the application code running in unprivileged Thread mode. This mechanism may seem an overly complicated way of calling a function, but it provides the basis of a supervisor/user split where an operating system is running in privileged mode and acts as a scheduler to application threads running in unprivileged Thread mode. This way the individual threads do not have access to critical processor features except by making API calls to the RTOS.

Exercise 5.2: SVC

In this exercise, we will look at calling some functions with the SVC instruction rather than branching to the routine as in a standard function call.

Open the Pack Installer.

Select the Boards::Designers Guide Tutorial.

Select the example tab and Copy "Ex 5.2: Supervisor Call."

First, let's have a look at the project structure.

Figure 5.11
SVC instructions are supported by adding the additional svc_user.c module. This module provides code to "decode" the SVC instruction + ordinal.

The project consists of the standard project startup file and the initializing system file. The application source code is in the file test.c. There is an additional source file svc_user.c (Fig. 5.11) that provides support for handling SVC exceptions. The svc_user.c file contains the SVC exception handler. This is a standard support file that is provided with the Arm compiler and is included as a template when you add a new file to the project. We will have a closer look at its operation later. The application code in test.c is calling two simple functions that in turn call routines to perform basic arithmetic operations.

```
int main (void) {
    test_a();
    test_t();
while(1);
}
void test_a (void) {
    res = add (74, 27);
    res + = mul4(res);
```

```
}
void test_t (void) {
   res = div (res, 10);
   res = mod (res, 3);
}
```

Each of the arithmetic functions is designed to be called with an SVC instruction so that all of these functions run in Handler mode rather than Thread mode. In order to convert the arithmetic functions from standard functions to software interrupt functions, we need to change the way the function prototype is declared. The way this is done will vary between compilers, but in the Arm compiler, there is a function qualifier__svc. This is used as shown below to convert the function to be an SVC and allows you to pass up to four parameters and get a return value. So the add() function is declared as follows:

```
int__svc(0) add (int i1, int i2);
int__SVC_0 (int i1, int i2) {
   return (i1 + i2);
}
```

The__svc qualifier defines this function as an SVC and defines the ordinal number of the function. The ordinals used must start from zero and grow upwards contiguously to a maximum of 256. To enable each SVC, it is necessary to build a lookup table in the SVC.c file.

```
#define USER_SVC_COUNT 4 // Number of user SVC functions
extern void__SVC_0 (void);
extern void__SVC_1 (void);
extern void__SVC_2 (void);
extern void__SVC_3(void);
extern void * const osRtxUserSVC[1 + USER_SVC_COUNT];
         void * const osRtxUserSVC[1 + USER_SVC_COUNT] = {
         (void *)USER_SVC_COUNT,
         (void *)__SVC_0,
         (void *)__SVC_1,
         (void *)__SVC_2,
         (void *)__SVC_3,
};
```

When the code is compiled, the labels will be replaced by the entry address of each function.

In the project build the code and start the simulator. Step the code until you reach line 61 the call to the add function.

The following code is displayed in the disassembly window and in the register window we can see that the processor is running in thread mode (Fig. 5.12):

```
        61:  res = add (74, 27);
        0x0800038A 211B    MOVS    r1,#0x1B
        0x0800038C 204A    MOVS    r0,#0x4A
        0x0800038E DF03    SVC     0x03
```

Internal	
Mode	Thread
Privilege	Privileged
Stack	MSP
States	2120
Sec	0.00004878

Figure 5.12

Prior to the SVC instruction the processor is running in Thread mode.

The function parameters are loaded into the parameter passing registers R0 and R1, and the normal branch instruction is replaced by an SVC instruction. The SVC instruction is encoded with an ordinal value of 3. If you make the disassembly window, the active window and step through these instructions the SVC exception will be raised and you will enter the SVC Handler in SVC.c In the registers window, you can also see that the processor is now running in Handler mode (Fig. 5.13).

```
__asm void SVC_Handler (void) {

    TST    LR,#4            ; Called from Handler Mode?
    MRSNE  R12,PSP          ; Yes, use PSP
    MOVEQ  R12,SP           ; No, use MSP
    LDR    R12,[R12,#24]    ; Read Saved PC from Stack
    LDRH   R12,[R12,#-2]    ; Load Halfword
    BICS   R12,R12,#0xFF00  ; Extract SVC Number
```

Internal	
Mode	Handler
Privilege	Privileged
Stack	MSP
States	2150
Sec	0.00004919

Figure 5.13

Once the SVC instruction has been executed, the processor will be running in Handler mode.

The first section of the SVC_Handler code works out which stack is in use and then reads the value of the program counter saved on the stack. The program counter value is the return address, so the load instruction deducts two to get the address of the SVC instruction. This will be the address of the SVC instruction that raised the exception. The SVC instruction is then loaded into R12 and the ordinal value is extracted. The code is using R12 because the Arm binary interface standard defines R12 as the "Intra Procedure Call Scratch Register"; this means it will not contain any program data and is free for use.

```
PUSH    {R4,LR}                           ; Save Registers
        LDR    LR, = SVC_Count
        LDR    LR,[LR]
        CMP    R12,LR
        BHS    SVC_Dead                   ; Overflow
        LDR    LR, = SVC_Table
        LDR    R12,[LR,R12,LSL #2]        ; Load SVC Function Address
        BLX    R12                        ; Call SVC Function
```

The next section of the SVC exception handler prepares to jump to the add() function. First, the link register and R4 are pushed onto the stack. The size of the SVC table is loaded into the link register. The SVC ordinal is compared to the table size to check it is less than the SVC table size and hence a valid number. If it is valid, the function address is loaded into R12 from the SVC Table and the function is called. If the ordinal number has not been added to the table the code will jump to a trap called SVC_DEAD. Although R4 is not used in this example it is preserved on the stack as it is possible for the called function to use it.

```
POP     {R4,LR}
TST     LR,#4
MRSNE   R12,PSP
MOVEQ   R12,SP
STM     R12,{R0-R3}     ; Function return values
BX      LR              ; RETI
```

Once the SVC function has been executed, it will return back to the SVC exception handler to clean up before returning to the background code in handler mode.

PEND_SV Exception

The Cortex-M0 + /M3, M4, and M7 have an additional processor exception called the PENDSV exception. The PENDSV exception has been added to the Cortex-M processor primarily to support a Real-Time Operating System. We will take a closer look at how an RTOS uses the PENDSV exception in Chapter 10, Using a Real-Time Operating System, but for now, we will look at how it works. The PEND exception can be thought of as an NVIC interrupt channel that is connected to a processor register rather than a microcontroller peripheral. A PENDSV exception can be raised by the application software writing to the PEND register, the PEND exception is then handled in the same way as any other exception.

Exercise 5.3: Pend_SV

In this example, we will examine how to use the Pend_SV system service call interrupt.

Open the Pack Installer.

Select the Boards::Designers Guide Tutorial.

Select the example tab and Copy "Ex 5.3: Pend Exception."

Build the project and start the debugger.

The code initializes the ADC and enables its "End of Conversion Interrupt." It also changes the PENDSV and ADC interrupt priority from their default options. The system code routine uses an SVC instruction to raise an exception and move to Handler mode.

```
NVIC_SetPriority(PendSV_IRQn,2);    //set interrupt priorities
NVIC_SetPriority(ADC1_2_IRQn,1);
NVIC_EnableIRQ(ADC1_2_IRQn);        //enable the ADC interrupt
ADC1->CR1   |=(1UL≪5);              //switch on the ADC and start a conversion
ADC1->CR2   |=(1UL≪0);
ADC1->CR2   |=(1UL≪22);
systemCode();                       //call some system code with an SVC interrupt
```

The SVC has priority zero (highest) while the ADC has priority 1 and the PENDV interrupt has priority 2 (lowest).

Set a breakpoint on the systemCode() function and run the code (Fig. 5.14).

```
58 | ADC1->CR2    |=  (1UL << 22);
●59 | systemCode();
60 |
```

Figure 5.14
Execute the code up to the systemCode() function.

Open the peripherals/core peripherals/Nested vector interrupt controller window and check the priority levels of the SVC, PENDSV, and ADC interrupts (Fig. 5.15).

Idx	Source	Name	E	P	A	Priority
11	System Service Call	SVCALL	1	0	0	0
14	Pend System Service	PENDSV	1	0	0	2
34	ADC Global Interrupt	ADC	1	0	0	1

Figure 5.15
The Peripherals/NVIC window shows that the interrupts are enabled, none are active or pending.
We can also see their priority levels.

Now step into the systemCode routine (F11) until you reach the C function.

```
void__svc(0) systemCode (void);
void__SVC_0    (void){
unsigned int i, pending;

pending=NVIC_GetPendingIRQ(ADC1_2_IRQn);
if(pending==1){
```

```
      SCB->ICSR |=1<<28;        //set the pend pend
  }else{
      Do_System_Code();
  }
  }
```

Inside the systemCode() routine, there is a short loop that represents the critical section of code that must be run. While this loop is running the ADC will finish conversion and as it has a lower priority than the SCV interrupt, it will enter a pending state. When we exit the loop, we test the state of any critical interrupts by reading their pending bits. If a critical interrupt is pending then the remainder of the system code routine can be delayed. To do this we set the PENDSVSET bit in the Interrupt Control and State register and quit the SVC handler.

Set a breakpoint on the exit brace (}) of the systemCode() routine and run the code (Fig. 5.16).

```
 38  |  Do_System_Code();
 39  | }
 40  | }
```

Figure 5.16
Run the Do_System_Code() routine.

Now use the NVIC debug window to examine the state of the interrupts (Fig. 5.17).

Idx	Source	Name	E	P	A	Priority
11	System Service Call	SVCALL	1	0	1	0
14	Pend System Service	PENDSV	1	1	0	2
34	ADC Global Interrupt	ADC	1	1	0	1

Figure 5.17
Now the SVC exception is active with the ADC and PENDSV exceptions pending.

Now the SVC call is active with the ADC and PendSV system service call in a pending state.

Single step out of the System Service Call until you enter the next interrupt.

Both of the pending interrupts will be tail chained on to the end of the system service call. The ADC has the highest priority so it will be served next (Fig. 5.18).

Idx	Source	Name	E	P	A	Priority
11	System Service Call	SVCALL	1	0	0	0
14	Pend System Service	PENDSV	1	1	0	2
34	ADC Global Interrupt	ADC	1	0	1	1

Figure 5.18
Now the ADC interrupt is active; when this ends, the PENDSV routine will be served and the system code routine will resume.

Step out of the ADC handler and you will immediately enter the PendSV System service interrupt which allows you to resume execution of the system code that was requested to be executed in the System Service Call interrupt.

Interprocessor Events

The Cortex-M processors are designed so that it is possible to build multiprocessor devices. An example would be to have a Cortex-M4 and a Cortex-M0 + within the same microcontroller. The Cortex-M0 will typically manage the user peripherals, while the Cortex-M4 runs the intensive portions of the application code. Alternatively, there are devices that have a Cortex-A processor which can run Linux and manage a complex user interface. On the same chip, there may be one or more Cortex-M processors which manage the real-time code. These more complex "system-on-chip" designs require methods of signaling activity between the different processors. The Cortex-M processors can be chained together by an external event signal. The event signal is set by using a set event instruction. This instruction can be added to your C code using the __SEV() intrinsic provided by the CMSIS core specification. When a __SEV() instruction is issued if will wake up the target processor if it has entered a low power mode. If the target processor is running, the event latch will be set so that when the target processor executes the __WFE() instruction, it will reset the event latch and keep running without entering the low power mode.

Exclusive Access Instructions

One of the key features of an RTOS is multitasking support. As we will see in chapter 10 Using a Real-Time Operating System, this allows you to develop your code as independent threads that conceptually are running in parallel on the Cortex-M processor. As your code develops, the program threads will often need to access common resources, be it SRAM or peripherals. An RTOS provides mechanisms called semaphores and mutexes (Fig. 5.19) which are used to control access to peripherals and common memory objects.

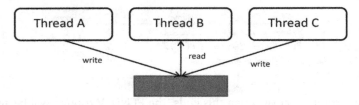

Figure 5.19
In a multiprocessor or multithread environment, it is necessary to control access to shared resources or errors such as read before write can occur.

While it is possible to design "memory lock" routines on any processor, the Cortex-M3, M4, and M7 provide a set of instructions that can be used to optimize exclusive access routines (Table 5.1).

Table 5.1: Exclusive access instructions

__LDREXB	Load exclusive (8 bits)
__LDREXH	Load exclusive (16 bits)
__LDREXW	Load exclusive (32 bits)
__STREXB	Store exclusive (8 bits)
__STREXH	Store exclusive (16 bits)
__STREXW	Store exclusive (32 bits)
__CLREX	Remove exclusive lock

In earlier Arm processors like the ARM7 and ARM9 the problem of exclusive access was answered by a swap instruction that could be used to exchange the contents of two registers. This instruction took four cycles, but it was an atomic instruction meaning that once started it could not be interrupted and was guaranteed exclusive access to the CPU to carry out its operation. As Cortex-M processors have multiple busses, it is possible for read and write accesses to be carried out on different busses and even by different bus masters which may themselves be additional Cortex-M processors. On the Cortex-M processor the new technique of exclusive access instructions has been introduced to support multitasking and multiprocessor environments.

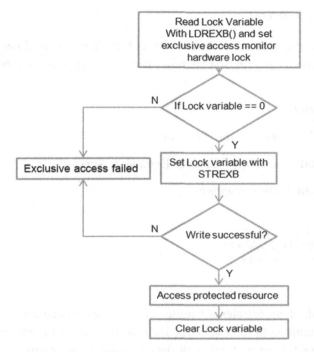

Figure 5.20
The Load and store exclusive instructions can be used to control access to a memory recourse.
They are designed to work with single and multiprocessor devices.

The exclusive access system works by defining a lock variable to protect the shared resource (Fig. 5.20). Before the shared resource can be accessed, the locked variable is checked using the exclusive read instruction; if it is zero, then the shared resource is not currently being accessed. Before we access the shared resource, the lock variable must be set using the exclusive store instruction. Once the lock variable has been set, we now have control of the shared resource and can write to it. If our process is preempted by an interrupt or another thread that also performs an exclusive access read, then a hardware lock in the exclusive access monitor is set, preventing the original exclusive store instruction from writing to the lock variable. This gives exclusive control to the preempting process.

When we are finished with the shared resource, the lock variable must be written to zero. This clears the variable and also removes the lock. If your code starts the exclusive access process but needs to abandon it, there is a clear exclusive (CLREX) instruction that can be used to remove the lock. The exclusive access instructions control access between different processes running on a single Cortex-M processor, but the same technique can be extended to a multiprocessor environment provided that the silicon designer includes the additional monitor hardware bus signals.

Exercise 5.4: Exclusive Access

In this exercise, we will create an exclusive access lock that is shared between a background thread process and an SVC handler routine to demonstrate the lock and unlock process.

Open the Pack Installer.

Select the Boards::Designers Guide Tutorial.

Select the example tab and Copy "Ex 5.4: Exclusive Access."

Build the code and start the debugger.

```
int main (void) {
  if(__LDREXB(&lock_bit) == 0){
    if (!STREXB(1,& lock_bit == 0)){
      semaphore+ + ; lock_bit = 0;
    }
  }
}
```

The first block of code demonstrates a simple use of the exclusive access instructions. We first test the lock variable with the exclusive load instruction. If the resource is not locked by another process, we set the lock bit with the exclusive store instruction. If this is successful, we can then access the shared memory resource called semaphore. Once this variable has been updated, the lock variable is written to zero, and the clear exclusive instruction releases the hardware lock.

Step through the code to observe its behavior.

```
if(__LDREXB(&lock_bit) == 0){
  thread_lock();
  if (!__STREXB(1,&lock_bit)){
    semaphore++;
  }
}
```

The second block of code does exactly the same thing except between the exclusive load and exclusive store, the function thread lock is called. This is an SVC routine that will enter handler mode and jump to the SVC0 routine.

```
void __svc(0) thread_lock (void);
void __SVC_0 (void) {
  __LDREXB(&lock_bit);
}
```

The SVC routine simply does another exclusive read of the lock variable that will set the hardware lock in the exclusive access monitor. When we return to the original routine and try to execute the exclusive store instruction, it will fail because any exception that happens between LDREX and STREX will cause the STREX to fail. The local exclusive access monitor is cleared automatically at exception entry/exit.

Step through the second block of code and observe the lock process.

Memory Protection Unit

The Cortex-M0 + , M3, M4, and M7 processors have an optional MPU which may be included in the processor core by the Silicon Manufacturer when the microcontroller is designed (Fig. 5.21). The MPU allows you to extend the privileged/unprivileged code model. By defining regions within the memory map of the Cortex-M processor. Each region can be configured to grant different access rights. If for example, the processor is running in unprivileged mode and tries to access an address within a privileged region, a memory protection exception will be raised and the processor will vector to the memory protection ISR. This allows you to detect and correct run time memory errors. The Cortex-M7 MPU has an additional role where it is used to configure the internal processor Instruction and Data Caches. In the next section, we will look at how to use the MPU with the Cortex-M0 + , M3, and M4. Then, Chapter 6, Cortex-M7 Processor, will look at the Cortex-M7 cache configuration features.

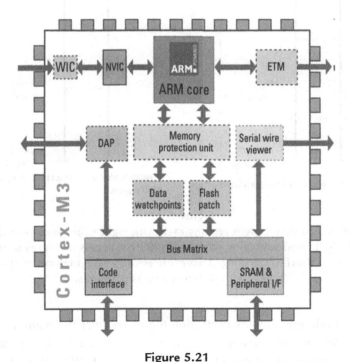

Figure 5.21
The Memory protection unit is available on the Cortex-M0 + , M3, M4, and M7. It allows you to place a protection template over the processor memory map.

In practice, the MPU allows you to define eight memory regions within the Cortex-M processor address space where each region has defined access rights (Fig. 5.22). These regions can then be further subdivided into eight equally sized subregions, which can, in turn, be granted their own access rights. There is also a default background region that covers the entire 4 Gbyte address space. When this region is enabled, it makes access to all memory locations privileged. If the background region is not enabled, access to memory outside the defined MPU regions is prohibited. To further complicate things memory regions may be overlapped with the highest numbered region taking precedent.

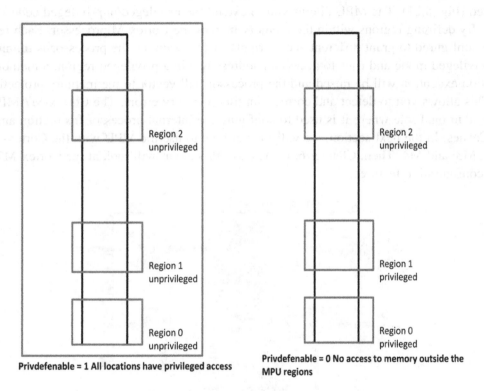

Region 2
unprivileged

Region 1
unprivileged

Region 0
unprivileged

Privdefenable = 1 All locations have privileged access

Region 2
unprivileged

Region 1
privileged

Region 0
privileged

Privdefenable = 0 No access to memory outside the MPU regions

Figure 5.22
The MPU allows you to define eight regions each with eight subregions over the processor memory map. Each region and subregion can grant different access privileges to its address range. It is also possible to set a default privileged access over the whole memory map and then create "holes" with different access privileges.

So, it is possible to build up complex protection templates over the memory address space. This allows you to design a protection regime that helps build a robust operating environment but also gives you enough rope to hang yourself. We will look at using the MPU in a safety system in Chapter 10, Using a Real-Time Operating System.

Configuring the MPU

The MPU is configured through a group of memory-mapped registers located in the Cortex-M processor system block (Fig. 5.23). These registers may only be accessed when the Cortex processor is operating in privileged mode.

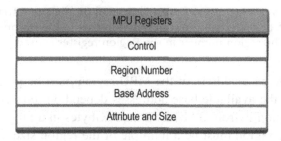

Figure 5.23
Each MPU region is configured through the region, base address, and Attribute registers. Once each region is configured the CONTROL register makes the MPU regions active.

The CONTROL register contains three active bits that affect the overall operation of the MPU (Fig. 5.24). These are the PRIVDEFENABLE bit that enables the background region and grants privileged access over the whole 4 Gbyte memory map. The next bit is the HFNMIENA bit. When set this bit enables the operation of the MPU during a hard fault, NMI, or FAULTMASK exception. The final bit is the MPU ENABLE bit. When set this enables the operation of the MPU. Typically, when configuring the MPU, the last operation performed is to set this bit. After reset, all of these bits are cleared to zero.

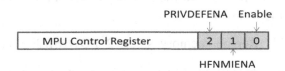

Figure 5.24
The CONTROL register allows you to enable the global privileged region (PRIVDEFENA). The enable bit is used to make the configured MPU regions active and HFMIENA is used to enable the regions when the Hard Fault, NMI, or Faultmask exceptions are active.

The remaining registers are used to configure the eight MPU regions. To configure a given region (0–7), first select the region by writing its region number into the region number register. Once a region has been selected, it can then be configured by the base address (Fig. 5.25) and the attribute and size register (Fig. 5.26). The base address register, contains a 27-bit address field along with a valid bit and also a repeat of the MPU region number.

Figure 5.25
The address region of the Base Address register allows you to set the start address of an MPU region. The address values that can be written will depend on the size setting in the Attribute and size register. If Valid is set to one, then the region number set in the region field is used otherwise the region number in the Region register is used.

As you might expect, the base address of the MPU region must be programmed into the address field. However, the available base addresses depend on the size defined for the region. The size of a region is from 32 bytes up to 4 Gbytes in a range of fixed sizes. The base address of an MPU region must be a multiple of the region size. Programming the address field sets the selected regions base address. You do not need to set the valid bit until you are ready to activate the region. Programming the attribute and size register finishes the configuration of an MPU region.

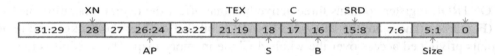

Figure 5.26
The Attribute and size register allows you to define the MPU region size from 32 bytes to 4G bytes. It also configures the memory attributes and access control options.

The size field defines the address size of the memory protection region in bytes. The region size is calculated using the formula:

$$\text{MPU Region memory size} = 2\ \text{POW(SIZE} + 1)$$

This gives us a minimum size starting at just 32 bytes. As noted above, the selected size also defines the range of possible base addresses. Next, it is possible to set the region attributes and access privileges. Like the Cortex-M processor, the MPU is designed to support multiprocessor systems. Consequently, it is possible to define regions as being shared between Cortex-M processors or as being exclusive to the given processor. It is also possible to define the cache policy for the area of memory covered by the MPU region. Currently, the vast majority of microcontrollers only have a single Cortex-M processor though asymmetrical multiprocessor devices have started to appear (typically, a Cortex-M4 and Cortex-M0). The Cortex-M7 introduces Instruction and Data Caches that are configured through the MPU. We will look at this in more detail in Chapter 6, Cortex-M7 Processor. This means that the MPU attributes can seem confusing when applied to a single-core microcontroller. The MPU attributes are defined by the TEX, C, B, and S bits, and suitable settings for most microcontrollers are shown in Table 5.2.

Table 5.2: Memory region attributes

Memory Region	TEX	C	B	S	Attributes
Flash	000	1	0	0	Normal memory, nonshareable, write through
Internal SRAM	000	1	0	1	Normal memory, shareable, write through
External SRAM	000	1	1	1	Normal memory, shareable, write back write allocate
Peripherals	000	0	1	1	Device memory, shareable

When working with the MPU, we are more interested in defining the access permissions. These are defined for each region in the AP field as shown in Table 5.3.

Table 5.3: Memory access rights

AP	Privileged	Unprivileged	Description
000	No Access	No Access	All accesses generate a permission fault
001	RW	No Access	Access form privileged code only
010	RW	RO	Unprivileged writes cause a permission fault
011	RW	RW	Full access
100	Unpredictable	Unpredictable	Reserved
101	RO	No Access	Reads by privileged code only
110	RO	RO	Read only for privileged and unprivileged code
111	RO	RO	

Once the size, attributes, and access permissions are defined, the enable bit can be set to make the region active. When each of the required regions has been defined, the MPU can be activated by setting the global enable bit in the CONTROL register. When the MPU is active and the application code makes an access that violates the permissions of a region, an MPU exception will be raised. Once you enter an MPU exception, there are a couple of registers that provide information to help diagnose the problem. The first byte of the Configurable Fault Status register located in the System Control Block is aliased as the Memory Manager Fault Status register (Fig. 5.27).

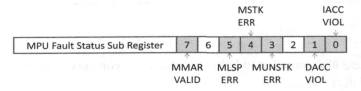

Figure 5.27

The MPU fault status register is a subsection of the configurable fault status register in the system control block. It contains error flags that are set when an MPU exception occurs. The purpose of each flag is shown below.

The meaning of each fault status flag is shown in Table 5.4.

Table 5.4: Fault status register flag descriptions

Flag	Description
IACCVIOL	Instruction Access Violation status flag
DACCVIOL	Data Access Violation status flag
MUNSTKERR	Memory Manager Fault on unstacking
MSTKERR	Memory Manager fault on stacking
MLSPERR	Memory Manager FPU Lazy Stacking error Cortex-M4 only (see Chapter 7: Armv8-M)
MMARVALID	Memory manager fault address valid

Depending on the status of the fault conditions, the address of the instruction that caused the memory fault may be written to a second register, the Memory Manager Fault Address Register. If this address is valid, it may be used to help diagnose the fault.

CMSIS Core MPU Support

The CMSIS Core specification provides a set of typdefs, macros, and functions to make configuration of the MPU a more manageable process. The CMSIS MPU functions provide a high-level interface to manage individual MPU regions, while a load function allows you to configure multiple regions by loading their configuration from a table stored in memory (Table 5.5).

Table 5.5: CMSIS MPU helper functions

Function	Description
ARM_MPU_Enable	Enable the memory protection unit
ARM_MPU_Disable	Disable the memory protection unit
ARM_MPU_ClrRegion	Clear and disable a given MPU region
ARM_MPU_SetRegion	Configure an MPU region (region number in RBAR value)
ARM_MPU_SetRegionEx	Configure an MPU region (explicit definition of region number)
ARM_MPU_OrderedMemcpy	Memcopy with strictly ordered memory access (used by ARM_MPU_Load)
ARM_MPU_Load	Load given MPU regions from a configuration table

A set of macros are also provided to help you define the contents of the region base address and attribute registers (Table 5.6).

Table 5.6: CMSIS MPU macro support

Macro	Description
ARM_MPU_RBAR	Used to construct an RBAR value
ARM_MPU_RASR	Used to construct a Region Attribute and Size Register Value
ARM_MPU_RASR_EX	As above but uses word wide "Access permissions" to define access attributes

To configure a region, we must use a structure provided by CMSIS core with the following typedef:

```
typedef struct {
  uint32_t RBAR;
  uint32_t RASR;
} ARM_MPU_Region_t;
```

We can declare a region structure.

```
ARM_MPU_Region_t threadRam;
```

And then use the ARM_MPU_RBAR macro to define a value for the base address register:

```
threadRam.RBAR = ARM_MPU_RBAR(1,0x10000000);
```

Here, we pass the region number (1) and the base address of the memory region (0x10000000).

Next, we can define the region attributes. In this case the region is SRAM.

```
threadRam.RASR = ARM_MPU_RASR(DISABLE_EXEC,
                 ARM_MPU_AP_FULL,          //full access for priv and un priv
                 0x0,                      //Type Extension
                 1,                        //SHAREABLE,
                 1,                        //CACHABLE,
                 0,                        //NON_BUFFERABLE,
                 0x00,                     //SUB_REGION_DISABLED,
                 ARM_MPU_REGION_SIZE_4KB);
```

Once the structure has been populated, we can program the selected region.

```
ARM_MPU_SetRegion (threadRam.RBAR, threadRam.RASR);
```

Once we have finished configuring the necessary regions, the MPU must be enabled.

```
ARM_MPU_ENABLE();
```

It is possible to configure a number of regions using the ARM_MPU_Load() function. First, we need to define a table that contains the configuration values for each MPU region that we need to use.

```
const ARM_MPU_Region_t mpuTable[3][4] = {
{
{.RBAR = ARM_MPU_RBAR(0UL, 0x08000000UL),
.RASR = ARM_MPU_RASR(0UL, ARM_MPU_AP_FULL, 0UL, 0UL, 1UL, 1UL, 0x00UL,
ARM_MPU_REGION_SIZE_1MB) },

{.RBAR = ARM_MPU_RBAR(1UL, 0x20000000UL),
.RASR = ARM_MPU_RASR(1UL, ARM_MPU_AP_FULL, 0UL, 0UL, 1UL, 1UL, 0x00UL,
ARM_MPU_REGION_SIZE_32KB) },

{.RBAR = ARM_MPU_RBAR(2UL, 0x40020000UL),
.RASR = ARM_MPU_RASR(1UL, ARM_MPU_AP_FULL, 2UL, 0UL, 0UL, 0UL, 0x00UL,
ARM_MPU_REGION_SIZE_8KB) },

{.RBAR = ARM_MPU_RBAR(3UL, 0x40022000UL),
.RASR = ARM_MPU_RASR(1UL, ARM_MPU_AP_FULL, 2UL, 0UL, 0UL, 0UL, 0xC0UL,
ARM_MPU_REGION_SIZE_4KB) }
},
```

Then, we can use the MPU_load() function to configure each of the MPU regions.

Exercise 5.5: MPU Configuration

In this exercise, we will configure the MPU to work with the Blinky project.

Open the Pack Installer.

Select the Boards::Designers Guide Tutorial.

Select the example tab and Copy "Ex 5.5: MPU Configuration."

This time the microcontroller used is an NXP LPC1768 which has a Cortex-M3 processor fitted with the MPU. First, we have to configure the project so that there are distinct regions of code and data that will be used by the processor in thread and handler modes. When doing this, it is useful to sketch out the memory map of the application code and then define a matching MPU template (Figs. 5.28 and 5.29).

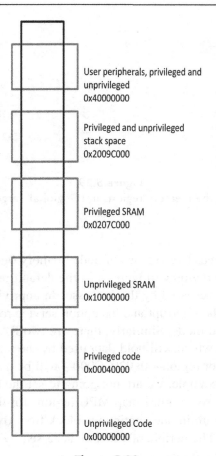

User peripherals, privileged and
unprivileged
0x40000000

Privileged and unprivileged
stack space
0x2009C000

Privileged SRAM
0x0207C000

Unprivileged SRAM
0x10000000

Privileged code
0x00040000

Unprivileged Code
0x00000000

Figure 5.28
The memory map of the blinky project can be split into six regions: Region 0: unprivileged
application code; Region 1: Privileged system code; Region 2: Unprivileged SRAM; Region 3:
Privileged SRAM; Region 4: Privileged and unprivileged stack space; Region 5: Privileged and
unprivileged user peripherals.

Open the options for the target/target menu (Fig. 5.30).

Here, we can set up the memory regions to match the proposed MPU protection template.

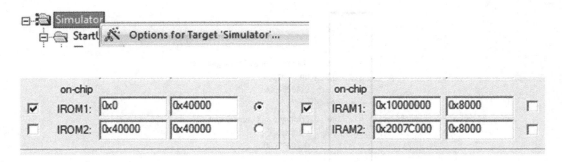

Figure 5.29
Configure the memory regions in the global Target dialog.

The target memory map defines two regions of code memory 0–0x3FFF and 0x4000–0x7FFF. The lower region will be used as the default region to hold the application code and will be accessed by the processor in unprivileged mode. The upper region will be used to hold the interrupt and exception service routines and will be accessed by the processor in privileged mode. Similarly, there are two RAM regions 0x10000000–0x100008000, which will hold data used by the unprivileged code and the system stacks, while the upper region—0x0207C000—will be used to hold the data used by the privileged code. In this example, we are not going to set the background privileged region (PRIVDEFENA = 0), so we must map MPU regions for the peripherals. All the peripherals except the GPIO are in one contiguous block from 0x40000000 while the GPIO registers sit at 0x000002C9. The peripherals will be accessed by the processor while it is in both privileged and unprivileged modes.

Select the IRQ.c file in the project window.

Open its local options and set the memory assignment as shown in Fig. 5.30.

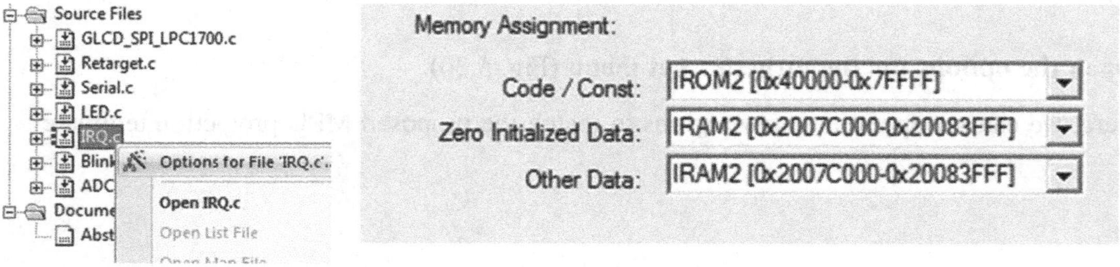

Figure 5.30
Use the Local memory assignment options to configure the memory regions used by the Interrupt code.

To prepare the code, we need to force the interrupt handler code into the regions which will be granted privileged access. In this example, all the code which will run in handler mode has been placed in one module. In the local options for this module, we can select the code and data regions that will be given privileged access rights by the MPU. All of the other code and data will be placed in the default memory regions, which will run in unprivileged mode. Once the project memory layout has been defined, we can add code to the project to set up the MPU protection template.

```
#define SIZE_FIELD                      1
#define ATTRIBUTE_FIELD                 16
#define ACCESS_FIELD                    24
#define ENABLE                          1
#define ATTRIBUTE_FLASH                 0x4
#define ATTRIBUTE_SRAM                  0x5
#define ATTRIBUTE_PERIPHERAL            0x3
#define PRIV_RW_UPRIV_RW                3
#define PRIV_RO_UPRIV_NONE              5
#define PRIV_RO_UPRIV_RO                6
#define PRIV_RW_UPRIV_RO                2
#define USE_PSP_IN_THREAD_MODE          2
#define THREAD_MODE_IS_UNPRIVILIGED     1
#define PSP_STACK_SIZE                  0x200
#define TOP_OF_THREAD_RAM               0x10007FF0
MPU->RNR    =    0x00000000;
MPU->RBAR   =    0x00000000;
MPU->RASR   =    (PRIV_RO_UPRIV_RO<<ACCESS_FIELD)
                 |(ATTRIBUTE_FLASH<<ATTRIBUTE_FIELD)
                 | (17<<SIZE_FIELD)|ENABLE;
```

The code shown above is used to set the MPU region for the unprivileged thread code at the start of memory. First, we need to set a region number followed by the base address of the region. Since this will be FLASH memory, we can use the standard attribute for this memory type. Next, we can define its access type. In this case, we can grant Read-Only access for both privileged and unprivileged modes. Next, we can set the size of the region, which is 256 K which must equal 2POW(SIZE + 1), which equates to 17. The enable bit is set to activate this region when the MPU is fully enabled. Each of the other regions is programmed in a similar fashion. Finally, the memory management exception and the MPU are enabled.

```
NVIC_EnableIRQ (MemoryManagement_IRQn);
MPU->CTRL = ENABLE;
```

Start the debugger, set a breakpoint on line 82, and run the code.

When the breakpoint is reached, we can view the MPU configuration via the peripherals/core peripherals/MPU (Fig. 5.31).

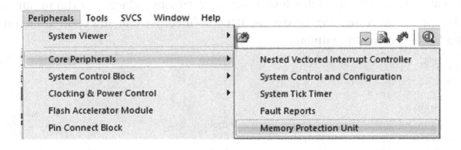

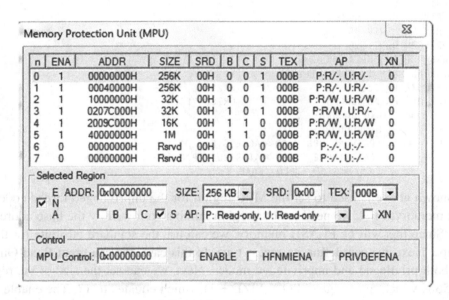

Figure 5.31
The debugger peripheral windows provide a detailed view of the configured MPU.

Here we can easily see the regions defined and the access rights that have been granted.

Now run the code for a few seconds and then halt the processor.

An MPU exception has been raised, and execution has jumped to the MemManager handler (Fig. 5.32).

```
147    MemManage_Handler\
148              PROC
149              EXPORT   MemManage_Handler        [WEAK]
150              B        .
151              ENDP
```

Figure 5.32
When an MPU exception occurs the code will vector to the default MemManager handler in the startup code.

The question is now, what caused the MPU exception? We can find this out by looking at the memory Manager Fault and Status register.

Open the Peripherals/Core Peripherals/Fault Reports window (Fig. 5.33).

Figure 5.33
The debugger also provides a condensed view of all the fault registers.

Here, we can see that the fault was caused by a data access violation to address 0x2007C008. If we now open the map file produced by the linker, we can search for this address and find what variable is placed at this location.

Highlight the project name in the project window and double click. This will open the map file. Now use the edit/find dialog to search the map file for the address 0x2007C008 (Fig. 5.34).

```
__initial_sp             0x10000220    Data       0  startup_lpc17xx.o(STACK)
clock_1s                 0x2007c000    Data       1  irq.o(.data)
```

Figure 5.34
You can view the linker MAP file by double clicking on the project root node in the project window. Then search through this file to find the symbol located at 0x2007C000.

This shows that the variable "clock_1s" is at 0x2007C000 and that it is declared in irq.c.

Clock_1s is a global variable that is also accessed from the main loop running in unprivileged mode. However, this variable is located in the privileged RAM region, so accessing it while the processor is running in unprivileged mode will cause an MPU fault.

Find the declaration of clock_1s in blinky.c and remove the extern keyword.

Now find the declaration for clock_1s in irq.c and add the keyword extern.

Build the code and view the updated map file (Fig. 5.35).

```
__stdin                           0x10000008   Data        4   retarget.o(.data)
clock_1s                          0x1000000c   Data        1   blinky.o(.data)
AD_done                           0x1000000e   Data        1   adc.o(.data)
```

Figure 5.35
Now the clock_1s variable is located in SRAM which can be accessed by both privileged and unprivileged code.

Now, clock_1s is declared in blinky.c and is located in the unprivileged RAM region so it can be accessed by both privileged and unprivileged code.

Restart the debugger, and the code will run without raising an MPU exception.

Memory Protection Unit Subregions

As we have seen, the MPU has a maximum of eight regions that may be individually configured with location size and access type. Any region which is configured with a size of 256 bytes or more will contain eight equally space subregions. When the region is configured, each of the subregions is enabled and has the default region attributes and access settings. It is possible to disable a subregion by setting a matching subregion bit in the SRD field of the MPU attribute and size register. When a subregion is disabled, a "hole" is created in the region. This "hole" inherits the attributes and access permission of any overlapped region. If there is no overlapped region, then the global privileged background region will be used. If the background region is not enabled, then no access rights will be granted, and an MPU exception will be raised if an access is made to an address in the subregion "hole" (Fig. 5.36).

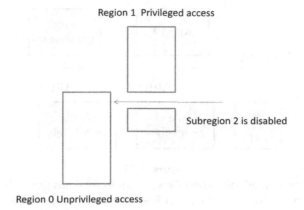

Region 1 Privileged access

Subregion 2 is disabled

Region 0 Unprivileged access

Figure 5.36
Each region has eight subregions. If a subregion is disabled, it inherits the access rights from an overlapped region or the global background region.

If we have two overlapped regions, the region with the highest region number will take precedence. In the case above, an unprivileged region is overlapped by a privileged region. The overlapped section will have privileged access. If a subregion is disabled in region 1, then the access rights in region 0 will be inherited and grant unprivileged access to the subregion range of addresses.

Memory Protection Unit Limitations

When designing your application software to use the MPU, it is necessary to realize that the MPU only monitors the activity of the Cortex-M processor. Many if not most Cortex-M-based microcontrollers have other peripherals, such as DMA units, which are capable of autonomously accessing memory and peripheral registers. These units are additional "Bus Masters," which arbitrate with the Cortex-M processor to gain access to the microcontroller resources. If such a unit makes an access to a prohibited region of memory, it will not trigger an MPU exception. This is important to remember as the Cortex-M processor has a bus structure that is designed to support multiple independent "Bus Master" devices.

AHB Lite Bus Interface

The Cortex-M processor family has a final important architectural improvement over the earlier generation of ARM7- and ARM9-based microcontrollers. In these first-generation Arm-based microcontrollers the CPU was interfaced to the microcontroller through two types of buss (Fig. 5.37). These were the Advanced High-Speed Bus (AHB) and the Advanced Peripheral Bus (APB).

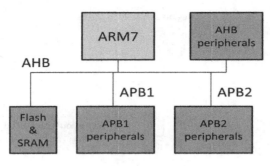

Figure 5.37
The first generation of Arm-based microcontrollers had an internal bus system based on the Advanced High Speed Bus and the Advanced Peripheral Bus. As multiple bus, masters (CPU, DMA) were introduced a bus arbitration phase had to be completed before a transfer could be made across the bus.

The AHB connects the CPU to the FLASH and SRAM memory, while the microcontroller peripherals were connected to one or more APB busses. The AHB bus also supported additional bus masters such as DMA units to sit alongside the ARM7 processor. While this system worked, the bus structure started to become a bottleneck, particularly as more complex peripherals such as Ethernet MAC and USB were added. These peripherals contained their own DMA units, which also needed to act as a bus master. This meant that there could be several devices (ARM7 CPU, General purpose DMA, and Ethernet MAC DMA) arbitrating for the AHB bus at any given point in time. As more and more complex peripherals are added, the overall throughput and deterministic performance became difficult to predict.

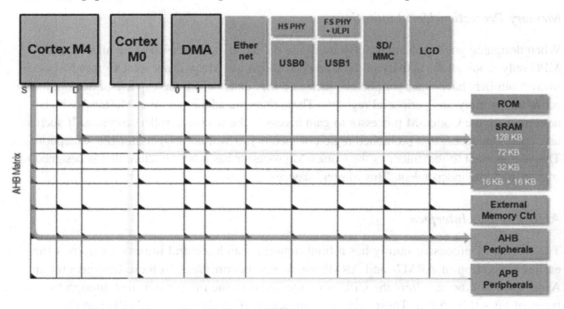

Figure 5.38
The Cortex-M processor family replaces the single AHB bus with a bus matrix that provides parallel paths for each bus master to each block of slave devices.

The Cortex-M processor family overcomes this problem by using an AHB bus matrix (Fig. 5.38). The AHB bus matrix consists of a number of parallel AHB busses that are connected to different device resources, such as a block of RAM or a group of peripherals on an APB bus. The mix of AHB busses and the layout of the device resources are assigned by the manufacturer when the chip is designed. Each region of device resources is a slave device. Each of these regions is then connected back to each of the bus masters through additional AHB busses to form the bus matrix. This allows manufacturers to design complex devices with multiple Cortex-M processors, DMA units, and advanced peripherals, each with parallel paths to the different device resources. The bus matrix is hardwired into the microcontroller and does not need any configuration by your application code. However, when you are designing the application code, you should pay attention to where different memory objects are located. For example, the memory used by the Ethernet controller should be placed in one block of SRAM, while the USB memory is located in a separate SRAM block. This allows the Ethernet and USB DMA unit to work in parallel, while the Cortex-M processor is accessing the FLASH and user peripherals. So by structuring the memory map of your application code, you can exploit this degree of parallelism and gain an extra boost in performance.

Conclusion

Many of the features in this chapter support more complex development frameworks such as an RTOS as we will see in Chapter 10, Using a Real-Time Operating System. This chapter also introduces the MPU which is essential for both Safety and Security applications.

CHAPTER 6

Cortex-M7 Processor

Introduction

In chapter three of this book, we have looked at the Cortex M0, M0 + , M3, and M4 processors which have a common programmer's model. In this chapter, we are going to examine the Cortex-M7, which is currently the highest performance Cortex-M processor available. It retains the same programmer's model as the other members of the family, so everything we have learned so far can be applied to the Cortex-M7. However, its internal architecture has some radical differences from the earlier Cortex-M processors that dramatically improve its performance (Fig. 6.1). The Cortex-M7 also has a more complex memory system which we need to understand and manage.

The Designer's Guide to the Cortex-M Processor Family.
DOI: https://doi.org/10.1016/B978-0-323-85494-8.00007-3

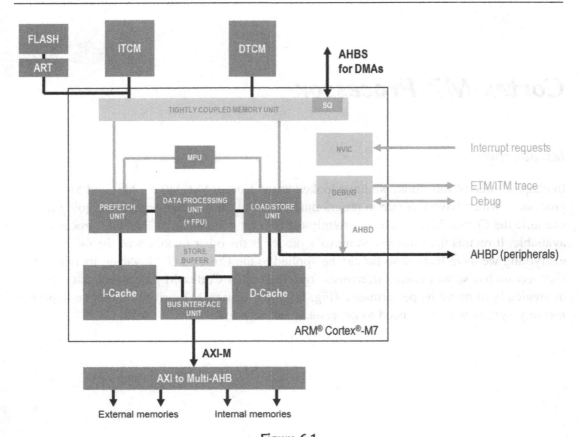

Figure 6.1
The Cortex-M7 has a more complex bus and memory structure but it retains the Cortex-M
programmer's model.

To a developer, the Cortex-M7 processor programmer's model looks the same as the other
Cortex-M processors. However, its performance is improved by a number of architectural
features that are very distinct from the rest of the family. These include a six-stage
dual-issue pipeline, an enhanced branch prediction unit, and a double-precision floating-
point math's unit. While these features boost the performance of the Cortex-M7, they are
largely transparent to a developer. For a developer, the Cortex-M7 really starts to differ
from the other Cortex-M processors in terms of its memory model. In order to achieve
sustained high-performance processing, the Cortex-M7 introduces a memory hierarchy that
consists of "Tightly Coupled Memories" (TCM) and a pair of on-chip caches in addition to
the main microcontroller FLASH and SRAM. The Cortex-M7 also introduces a new type of
bus interface, the AXI-M bus. The AXI-M bus allows 64-bit data transfers and also
supports multiple bus transactions.

Superscaler Architecture

The Cortex-M7 has a superscalar architecture and an extended pipeline. This means that it is able to "dual issue" instructions, and under certain conditions, the two instructions can be processed in parallel.

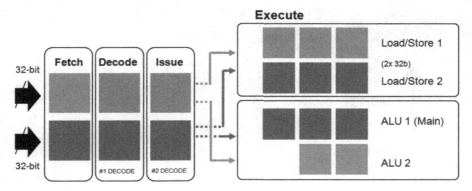

Figure 6.2
The cortex-M7 has a six-stage dual issue pipeline. The CPU has multiple processing "pipes" that allow different groups of instructions to be processed in parallel.

The Cortex-M7 pipeline has been increased to six stages (Fig. 6.2). The six-stage pipeline acts as two three-stage pipelines, each with a fetch decode and execute stage. The two pipelines are connected to the memory store via a 64-bit bus which allows two instructions to be fetched and injected into the pipelines in parallel. Both pipelines are, in turn, connected to several "pipes" within the processing unit. Each of the processor pipes is capable of executing a subset of instructions. This means that there are separate processing paths for Load and Store instructions, MAC instructions, ALU, and floating-point instructions. This makes it possible to parallel process instructions if we can issue the two instructions in the pipeline to different processing pipes. While this increases the complexity of the processor for a software developer, it all happens "under the hood" and is a challenge for the compiler to order the program instructions as interleaved memory accesses and data processing instructions.

Branch Prediction

The performance of the Cortex-M7 is further improved by adding a branch cache to the branch prediction unit. This is a dedicated cache unit with 64 entries, each 64 bits wide. During runtime execution, the Branch Target Address Cache (BTAC) will provide the correct branch address, so a branch instruction is always a single cycle (Fig. 6.3). If correctly predicted, the BTAC allows loop branches to be processed in a single cycle. The branch instructions can also be processed in parallel with another dual-issued instruction, further minimizing the branch overhead. This is a critical feature as it allows a general-purpose microcontroller to come close to matching the loop performance of a dedicated DSP device.

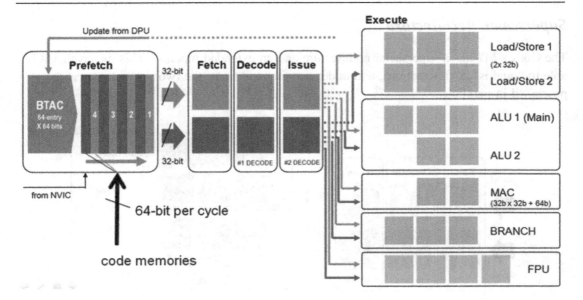

Figure 6.3
The Branch Target Address Cache ensures branch instructions are single cycle. This greatly reduces any overhead in program loops.

Exercise 6.1: Simple Loop

In this exercise, we are going to use some real hardware in the form of the STM32F7 Discovery board and the STM32F4 Discovery board in place of the software simulator. In the remainder of this chapter exercises, we will continue to use the STM32F7 Discovery board.

In this exercise, we will measure the execution time of a simple loop on both the Cortex-M4 and the Cortex-M7.

Open the Pack Installer.

Select the Boards::Designers Guide Tutorial.

Select the example tab and Copy "Ex 6.1 Cortex-M7 Simple Loop" (Fig. 6.4).

This is a multiproject workspace that contains projects for a Cortex-M4 project and a Cortex-M7. Each project contains the same code so that we can compare timings between processors.

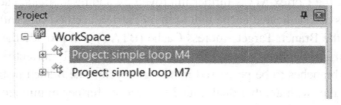

Figure 6.4
The two projects are opened in a multiproject workspace.

Batch build the project (Fig. 6.5).

Figure 6.5
Build both projects with the Batch Build command.

Select the M7 project (Fig. 6.6).

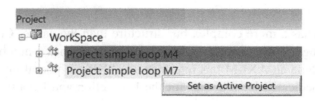

Figure 6.6
Select the M7 as the active project.

Start the debugger and set a breakpoint

Run the debugger until it hits the breakpoint instruction

Take the cycle count in the register window (Fig. 6.7).

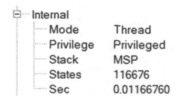

Figure 6.7
Read the cycle count for the M7.

Repeat for the M4 and compare (Fig. 6.8).

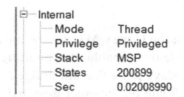

Figure 6.8
Read the cycle count for the M4.

With this simple example, we can see that code compiled for the Cortex-M7 runs much more efficiently than the Cortex-M4, even though this example does not use the caches and tightly coupled memory in the Cortex-M7. Combine these architectural improvements with a higher clock rate than a typical Cortex-M4 device. Then the Cortex-M7 offers a massive performance upgrade especially as real-world devices are now running in excess of 1 GHz. In the remainder of this chapter, we will look at optimizing the Cortex-M7 memory system to achieve even greater performance boosts.

Bus Structure

The Cortex-M7 also has a more complex bus structure than the earlier Cortex-M processors while still maintaining a linear memory map. The Cortex-M7 has a number of different bus interfaces which include an AXI-M bus interface, an AHB peripheral interface bus, an AHB slave bus, and also dedicated local busses for the Instruction and Data tightly coupled memories (Fig. 6.9).

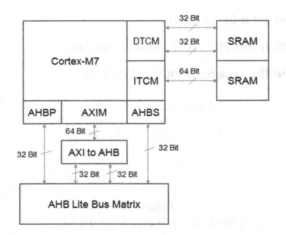

Figure 6.9
The Cortex-M7 introduces a number of new bus interfaces.

The main bus interface between the Cortex-M7 and the microcontroller is the Arm AXI-M bus. This is a 64-bit wide bus capable of supporting multiple transactions. The AXI bus provides a number of slave ports that are bridged onto an AHB lite bus. The AHB lite bus is the same as is found on the other members of the Cortex-M family. This helps silicon vendors reuse existing microcontroller designs with the Cortex-M7 processor. The Cortex-M7 also has an AHB peripheral bus (AHBP) port which is used to access the microcontroller peripherals via the AHB bus matrix. The AHBP bus is optimized for peripheral access, so peripheral register access will have a lower latency compared to other Cortex-M processors. The Cortex-M7 also has a second AHB port. This is the AHB slave (AHBS) port. This port allows other bus masters, typically DMA units, access to the instruction and data tightly coupled memories. A typical implementation of these busses within a microcontroller is shown in Fig. 6.10.

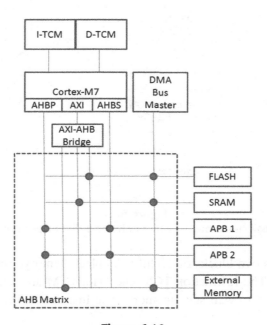

Figure 6.10
A Typical bus interface configuration for a Cortex-M7 microcontroller.

In this example, the AXI-M bus is connected to the AHB bus matrix via a bridge. However, the AHB bus-only connects the AXI-M bus ports to the microcontroller system memory. The AHB bus peripheral busses are connected back to the Cortex-M7 through the AHBP ports. Each of the DMA units can act as bus masters and have access to all of the system memory and peripherals. They can use the AHB Slave bus to access the Instruction and Data Tightly Coupled Memory. Although the AHBS bus is shown as being routed through the Cortex-M7, this bus will stay awake if the Cortex-M7 processor is placed into a low-power mode. This makes it possible for other bus masters to access the TCM memory even when the Cortex-M7 is fully asleep.

Memory Hierarchy

The more complex bus structure of the Cortex-M7 supports a memory map that contains a hierarchy of memory regions that need to be understood and controlled by a designer (Fig. 6.11).

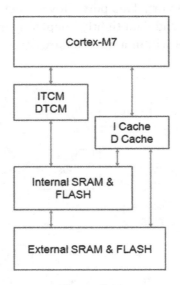

Figure 6.11
The Cortex-M7 has a memory hierarchy of different performance levels.

First, the Cortex-M7 has two blocks of Tightly Coupled Memory, one for data and one for instructions (Fig. 6.12). The Instruction TCM and Data TCM are connected to the processor via a 64-bit bus and are zero wait state memory. The Instruction TCM and Data TCM can be up to 16 MB.

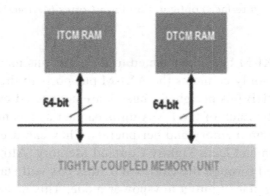

Figure 6.12
The tightly coupled memories are zero wait state memories interfaced directly to the CPU through dedicated 64-bit busses.

Code located in the Instruction TCM will run at zero wait states and will be highly deterministic. So this region of memory should be home to any critical routines and Interrupt Service Routines. The Data TCM is the fastest data RAM available to the processor and should be home to any frequently used data and is also a good location for stack and heap memory.

Exercise 6.2: Locating Code and Data into the TCM

This exercise demonstrates how to load code into the Instruction TCM and data into the Data TCM and compare its execution to code located in the system FLASH memory.

Open the Pack Installer.

Select the Boards::Designers Guide Tutorial.

Select the example tab and Copy "Ex 6.2−6.4 Cortex-M7 TCM and ICACHE."

The project contains benchmark code for the TCM and standard FLASH memory (Fig. 6.13).

Figure 6.13
The project has two loop modules which have their code located in the I-TCM and the AXI FLASH memory.

Open the options for target menu (Fig. 6.14).

In this project, we have two sets of loop functions that accumulate the values stored in a thousand element array. By default, the project will place the code in the flash memory which is located on the AXI-M bus at 0x08000000.

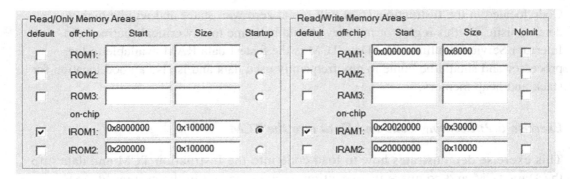

Figure 6.14
Memory layout with an additional region (RAM1) created to cover the ITCM.

However, we have also defined a RAM (RAM1) region at 0x00000000 for 32 K, which is the location of the Instruction TCM on this device. The default tick box is unchecked, so the linker will not place any objects in this region unless we specifically tell it to.

Open the "Options for file loop_TCM.c" (Fig. 6.15).

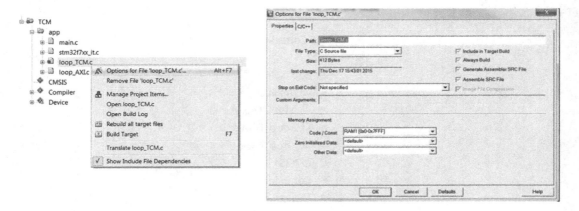

Figure 6.15
The module local options are used to map the code/const data into the RAM1 region. The Code located into the ITCM RAM region will be automatically booted out of the FLASH into the ITCM by the startup code.

In the memory assignment options, we can select the RAM1 region for the Code/Const segments in this file. This means that the code in the loop_TCM.c file will be stored in the FLASH memory but will be copied into the RAM1 region (The Instruction TCM SRAM) at startup. As a bonus, the debug symbols will also match the load address. This allows us to debug code loaded into the Instruction TCM without having to make a special build.

Open Main.c

The arrays which hold the data to be accumulated are defined at the start of main().

```
uint32_t array_DTCM[NUM_SAMPLES] __attribute__((at(0x20000000)));   //Locate in DTCM
                                                                     fastest data memory
uint32_t array_SDRAM[NUM_SAMPLES]__attribute__((at(0xC0000000)));   //Locate in SDRAM
                                                                     slowest data memory
```

The __attribute directive is used to force the linker to locate the array at an absolute address. In this case, we are forcing the first array into the D-TCM region and the second into the SRAM located on the AXI bus. In the main() application, we are configuring the SysTick timer to act as a stopwatch timer. Main() then calls identical loop functions which are executing from either the Instruction TCM or the AXI FLASH. The loop functions are also accessing an instance of the array that is located either in the Data TCM or the AXI-M SRAM.

Build the code and start the debugger.

Open the view/serial windows/debug printf window.

Run the code and observe the different run times displayed in the serial window.

Measured loop timing in AXI FLASH and ITCM using the DTCM for data.

AXI FLASH loop timing using DTCM RAM = 79156

ITCM loop timing using DTCM RAM = 24432

AXI FLASH loop timing using SDRAM = 89827

ITCM loop timing using SDRAM = 37306.

Cache Units

Outside of the tightly coupled memories, all program code and data will be held in the system memory. This will be the microcontroller internal FLASH and SRAM. The microcontroller may also have an external bus interface to access additional RAM and ROM memory. As we have seen with the Cortex-M3 and M4, it is necessary to provide some form of "Flash Memory Accelerator" to deliver instructions from the FLASH memory to the CPU in order to maintain processing at the full CPU clock speed. The Cortex-M7 is both a higher performance processor and is designed to run at higher clock frequencies, and this form of memory acceleration is no longer fully effective. In order to meet both its code and data throughput requirements, the Cortex-M7 can be implemented with both Data and Instruction Caches (Fig. 6.16).

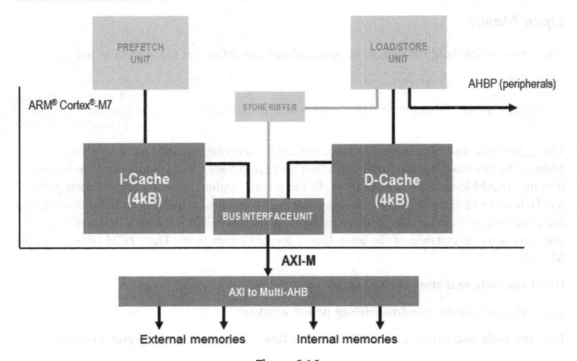

Figure 6.16
The I-Cache and D-Cache are used to cache memory located on the AXIM bus.

Both of the caches are used to hold copies of instructions and data fetched from memory accessed over the AXI-M bus. Once an instruction or data value has been loaded into the cache, it may be accessed far faster than the system FLASH and SRAM.

Cache Operation

Like the Tightly Coupled Memories, the Instruction and Data Caches are blocks of internal memory within the Cortex-M7 processor that are capable of being accessed with zero wait states. The Instruction Cache and the Data Cache may be up to 64 K bytes in size.

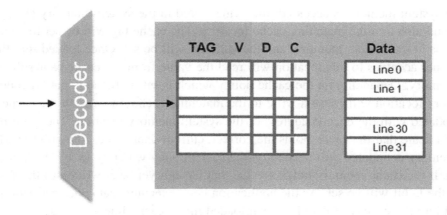

Figure 6.17
A Cache consists of a data line that is mapped to physical memory through a Tag. The C bits define the cache policy while the V and D bits are used to keep track of the data status.

The Instruction and Data Cache memory is arranged into a series of cache lines (Fig. 6.17) which are 32 bytes long. Each cache line has a cache tag. The cache tag contains the information which maps the cache line to the system memory. The tag also has a group of bits that detail the current status of the cache line. Both the instruction and data cache have a V bit which indicates that the cache line contains currently valid information. The data cache has the additional D and a pair of C bits. The Data cache D bit indicates that the data currently held in the cache is dirty. This means that the value held in the cache has been updated but the matching location in the system memory has not. The two C bits hold the cache attributes for this cache line, and we will discuss these in a moment.

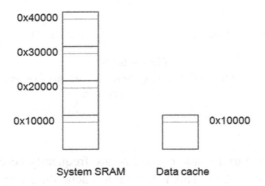

Figure 6.18
A cache line is mirrored to memory locations through the physical memory space.

The cache is mapped as a page of memory that is reflected up through the address space of the Cortex-M7 processor (Fig. 6.18). This means that each cache line has a number of matching memory locations evenly spaced through the system memory.

When the system memory is accessed, the value stored in the system memory is loaded into the CPU and also into the matching cache location. The cache tag will be set to decode the data address in the system memory, and the valid bit will be set. Once loaded into the cache, further accesses to this location will read the value from the cache rather than the system memory. Depending on the cache policy which is set by the application code, writes to a memory location will cause a write to the matching cache line. Again the cache tag will be updated with the decode address of the system memory location. The valid bit will be set to indicate that the cache holds the correct, current data value. The point at which the system memory is updated will depend on the cache policy which is set by the C bits in the cache tag. If the cache memory holds a value that has not yet been written to the system memory, the D bit will be set. As the application code executes, data and instructions will be moved into the caches. If data has been loaded into a cache line and an access is made to another location in system memory that is mapped to the same cache line, then the old data will be "evicted." The new data will be copied into the cache line, and the cache tag will be updated (Fig. 6.19).

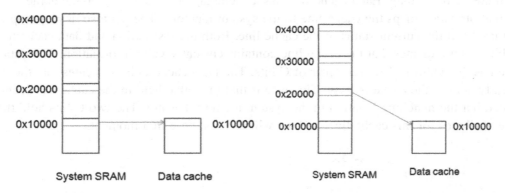

Figure 6.19
During execution data will be loaded into the cache. It can also be evicted if the mirror cache location is accessed.

If several locations mapped to a single cache line are frequently accessed, they will continually be evicting each other. This is known as cache thrashing, and a lot of thrashing will reduce the effectiveness of the cache.

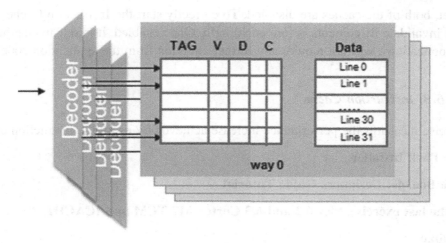

Figure 6.20
To improve cache efficiency the cache I arranged as a number of parallel pages. This provides multiple locations to store cached data.

To reduce the impact of thrashing, the cache is arranged as a set of parallel pages called "ways" rather than one contiguous block of memory (Fig. 6.20). The Instruction cache is arranged as two ways, and the data cache is arranged as four ways. Now the cache memory is mapped up through the system at the granularity of each cache way. This creates a number of parallel locations within the cache that a given memory location can be mapped into. This reduces the probability of cache thrashing and leads to more efficient use of the cache memory.

Instruction Cache

The instruction cache is controlled by a small group of functions provided by the CMSIS-Core specification as shown in Table 6.1. Functions are provided to enable and disable the Instruction cache. There is also a function to invalidate the cache contents. When the instruction cache is invalidated, all of the valid bits are cleared, effectively emptying the cache of any loaded instructions. As execution continues, fetches to the instruction address will start to reload the cache.

Table 6.1: CMSIS Core I-Cache management functions

CMSIS I-Cache Management Functions
void SCB_EnableICache(void)
void SCB_DisableICache(void)
void SCB_InvalidateICache(void)

After reset, both of the caches are disabled. To correctly start the Instruction Cache, you must first invalidate its contents before enabling it. Once enabled, it is self-managing and, in most applications, will not require any further attention from the application code.

Exercise 6.3: Instruction Cache

This exercise measures the performance increase achieved by using the instruction cache.

Open the Pack Installer.

Select the Boards::Designers Guide Tutorial.

Reopen the last exercise, "Ex 6.2 and 6.3 Cortex-M7 TCM and ICACHE."

Open main.c.

Remove the comments at lines 59 and 75.

This adds a new block of code to the last example. Here we enable the instruction cache and re-run the AXI loop functions to see the change in timing.

Open the view/serial windows/debug window.

Run the code and observe the difference in run time compared to the original example.

Measure Loop Timing in AXI FLASH with ICache Enabled

AXI loop timing with ICache and DTCM RAM = 62395

AXI loop timing with ICache and SDRAM = 76300

The instruction cache is a big improvement over executing from the FLASH but not as good as the Instruction Tightly Coupled Memory.

Data Cache

The Data Cache is also located on the AXI-M bus to cache any system RAM access. While this speeds up access to program data it can introduce unexpected side effects due to data coherency issues between the system memory and the Data Cache. The Data Cache is in effect, shadowing the system memory so that a program variable could be living in two places, within the cache and within the SRAM memory. This would be fine if the CPU were the only bus master in the system. In a typical Cortex-M7 microcontroller, there may be several other bus masters in the form of peripheral DMA units or even other Cortex-M processors. In such a multimaster system the cache may now introduce the problem of data coherency (Fig. 6.21).

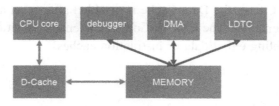

Figure 6.21
The D-Cache can introduce data coherency problems. The System memory can be updated by other bus masters. The D-Cache can "hide" these changes from the CPU.

Since our data may now be held is several locations within the memory hierarchy (CPU register, Data Cache, microcontroller SRAM) we must be careful not to introduce cache coherence problems. In particular the SRAM may be holding a historical value that is different to the current data located in the Data Cache. A DMA unit will only have access to the system memory, so it will use the older historical value rather than the current value held in the cache. Alternatively, a DMA update to the microcontroller SRAM may be masked from the CPU by an outdated value held in the data cache. This means that the Data Cache does require more management than the instruction cache. In the CMSIS-Core specification, there are eight Data Cache management functions as shown in Table 6.2.

Table 6.2: CMSIS Core D-Cache management functions

CMSIS Core Data Cache Functions
SCB_EnableDCache(void)
SCB_DisableDCache(void)
SCB_InvalidateDCache(void)
SCB_CleanDCache(void)
SCB_CleanInvalidateDCache(void)
SCB_InvalidateDCache_by_Addr(uint32_t*addr,int32_tdsize)
SCB_CleanDCache_by_Addr(uint32_t*addr,int32_tdsize)
SCB_CleanInvalidateDCache_by_Addr(uint32_t*addr,int32_tdsize)

Like the instruction cache, there are functions to enable, disable, and invalidate the data cache. In addition, we can clean the cache. This forces all the data held in the cache to be written to the system memory making the whole memory store coherent. It is also possible to clean and invalidate regions of the system memory which may be held in the cache.

Data Cache Invalidate—Flush what's in the cache memory and force a reload from system memory.

Data Cache Clean—Write the data held in cache memory to the system memory.

Data Cache Clean and Invalidate—Write cache memory to system memory and flush the cache memory.

As with the Instruction Cache, the Data Cache is disabled after reset and must first be invalidated before it is enabled. The CMSIS-Core cache enable functions performs a cache invalidation before enabling either data or instruction caches.

Memory Barriers

In Chapter 5, Advanced Architecture Features, we saw that the Cortex-M Thumb2 instruction set contains a group of memory barrier instructions. The memory barrier instructions are used to maintain data and instruction coherency within a Cortex-M microcontroller. These instructions are rarely used in most Cortex-M projects, and when they are, it is mainly for defensive programming. However, when working with the Cortex-M7 instruction and data caches, they are used to ensure that the cache operations have finished before allowing the CPU to continue processing. The CMSIS-Core cache functions include the necessary memory barrier instructions to ensure that all cache operations have been completed when the function returns.

```
__STATIC_INLINE void SCB_EnableICache (void)
{
   #if (__ICACHE_PRESENT = = 1U)
   __DSB();
   __ISB();
   SCB->ICIALLU = 0UL; /* invalidate I-Cache */
   SCB->CCR |= (uint32_t)SCB_CCR_IC_Msk; /* enable I-Cache */
   __DSB();
   __ISB();
   #endif
}
```

Example 6.4: Data Cache

This exercise measures the performance increase achieved by enabling the Data Cache.

Open the Pack Installer.

Select the Boards::Designers Guide Tutorial.

Reopen the last exercise "Ex 6.2−6.4 Cortex-M7 TCM and I-CACHE."

Open main.c.

Remove the comments at lines 59 and 75.

This adds a new block of code to the last example. Here we enable the Instruction Cache and re-run the AXI loop functions to see the change in timing.

Open the view/serial windows/debug window.

Run the code and observe the difference in run time compared to the original example.

Measure Loop Timing in AXI FLASH with ICache and DCache Enabled

AXI FLASH loop timing with ICache and DCache = 33305

So switching on the Data Cache gives another performance boost when data is located in the Internal or external RAM, but as expected the best level of performance comes from locating code and data into the Tightly Coupled Memories.

MPU and Cache Configuration

The MPU is common across all members of the Cortex-M processor family except for the Cortex-M0. As we saw in Chapter 5, Advanced Architecture Features, the Memory Protection Unit can be used to control access to different regions of the memory map within the Cortex-M0 + /M3 and M4. On the Cortex-M7, the MPU offers the same functionality, but in addition, it is used to configure the cache regions and cache policy.

Cache Policy

The cache policy for the different regions of memory within the microcontroller can be defined through the Memory Protection Unit regions with the cache configuration bits in the MPU "Attribute and Size" register (Fig. 6.22). The configuration options are shown in Table 6.3.

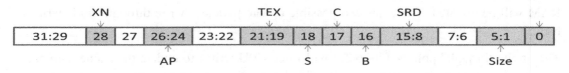

Figure 6.22
The MPU "Attribute and Size" register.

Table 6.3: The MPU D-Cache configuration options

Cache Configuration Field	Description
TEX	Type Extension
C	Cacheable
B	Bufferable
S	Sharable

The Sharable option in the MPU "Attribute and Size" register defines if a memory region can be shared between two bus masters. If a region is declared as sharable, its data will be kept coherent between the cache and the physical memory. This means that any bus master accessing the data will always get the correct, current value. However, in order to do this, the CPU will have to write through the Data Cache when it updates a RAM value. If the RAM is updated by another bus master, it must be loaded into the cache. So making the SRAM shareable removes coherency problems but reduces the overall Data Cache performance.

The cache policy also defines how data is written between the cache memory and the system memory. First, we can define when data is allocated into the cache. There are two options for the allocate policy. We can force data to be loaded into the cache when a RAM location is read, or we can select to load a RAM location into the cache when it is read or written to. This policy is set by configuring the Type Extension bits (TEX) as shown in Table 6.4.

Table 6.4: Cache allocate policy settings

TEX	Cache Allocate Policy
000	Read Allocate, load memory location into cache when it is read from
001	Read and Write Allocate, load memory location into cache when it is read from or written to

The remaining TEX values are there to support more complex memory systems, which may have a second outer (L2) cache.

Once a memory location has been loaded into the cache, we can define when the system RAM will be updated. There are two possible update policies "write through" and "write back."

The "write through" policy (Fig. 6.23) will force CPU writes to update the cache memory and the system memory simultaneously. This lowers the performance of the cache but maintains coherency for cache writes.

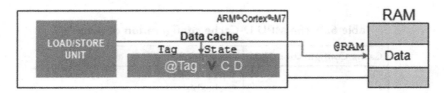

Figure 6.23
Data cache with Write through policy.

The write back policy (Fig. 6.24) will force writes to the cache memory only. The data will only be written to the system memory when the cache data is evicted, or a cache clean instruction is issued. This maximizes the performance of the cache for data writes at the expense of cache coherency.

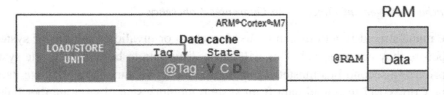

Figure 6.24
Data cache with write back policy.

The cache update policy is configured through the C and B bits in the MPU register as shown in Table 6.5.

Table 6.5: Cache update policy

C	B	Cache Update Policy
1	1	Write Back—Cacheable and Bufferable
1	0	Write Through—Cacheable but no Buffering

Managing the Data Cache

As we have seen the Instruction Cache will look after itself, and in most cases, we just need to enable it. However, we need to pay more attention to the Data Cache and the overall system coherency.

Switch off the Cache

The simplest thing to do is switch off the Data Cache. While this will remove any coherency problem, it will reduce the performance of the Cortex-M7. However, it can be a useful option at the beginning of a project in order to simplify the system.

Disable Caching Over a Region of System Memory

This is similar to the first option, except we can program an MPU region to prevent data caching on a region of the system memory. We can then locate the data that is common to the Cortex-M7 and other bus masters into this region. Since all access will be to the system memory, there will be no coherency problems but accesses to this memory region with the Cortex-M7 will be at their lowest.

Change the Cache Policy for a Region of System Memory

If the cache policy for a region is set to write through, the system memory will always be updated. So this is more limited option where the application code is writing into memory shared by bus masters.

Use the Cache Management Functions to Guarantee Coherency

The cache management functions can be used to clean or invalidate regions of system memory before they are accessed. This will cause dirty data to be written to the system memory prior to a second bus master accessing the data. We can also invalidate data causing the Cortex-M7 to reload data from system memory shared with another bus master.

Exercise 6.5: Data Cache Configuration

This exercise demonstrates the behavior of various cache configurations. If you are planning to use a Cortex-M7, you should experiment with this example to get a thorough understanding of the Data Cache.

```
Initilise_Device();
MPU_Config();
//SCB_EnableDCache();  //cache is disabled at the start of the project
Initilise_Arrays();     //load data into array_sdram1 and zero array_sdram2
//Cache_Clean();
Update_Source_Array();//Write new values to array_sdram1
//Cache_Clean();
DMA_Transfer();         //copy sdram1 to array_sdram2 using a DMA bus master
//Cache_Invalidate();
Test_Coherence();       //Test the coherency between array_sdram1 and array_sdram2
Test_Coherence();       //Test the coherency between array_sdram1 and array_sdram2
```

In this exercise, the MPU configures the Data Cache to cover the first 128 K bytes of external SDRAM which is populated with two data arrays. The Application code fills an array with some ordered data and writes zero into a second. Then a DMA unit (a bus master) is used to copy the data from the first array into the second. Finally, we test for data coherence between the contents of the two arrays. We can use this simple program to see the effect of different cache settings. The MPU code is contained in a single source fine and header file so it can be reused in other projects.

Open the Pack Installer.

Select the Boards::Designers Guide Tutorial.

Reopen the last exercise, "Ex 6.5 Data Cache."

Comment out line 90 so the data cache is not enabled.

Build the project.

Start the debugger.

Open the view/serialwindows/debug printf window.

Run the code.

During this first run, the Data Cache is disabled to give us a baseline for different cache configurations. The cycles taken for each operation are as follows:

```
->CPU cycles spent for arrays initialization: 5217
->CPU cycles spent re initilizing source array 1065
->CPU cycles spent on DMA Transfer: 11985
->CPU cycles spent for comparison: 13912
There were 0 different numbers
```

Open main.c and uncomment line 90 to enable the data cache

Open mpu.h and select the configuration wizard (Fig. 6.25)

This is a templated configuration file for the code in mpu.c. The initial configuration maps the Data Cache region at the start of the external SRAM at 0xC0000000 for 128Kbyte. In the Data Cache configuration options, we are enabling the region as sharable with other bus masters within the microcontroller.

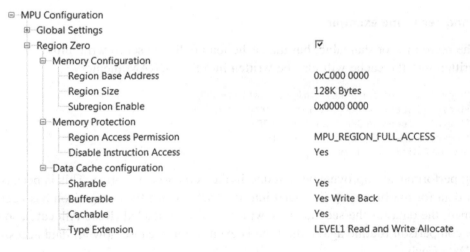

Figure 6.25
MPU configuration options.

Rebuild the code.

Start the debugger and run the code.

```
->CPU cycles spent for arrays initialization: 5184
->CPU cycles spent re initilizing source array 1064
->CPU cycles spent on DMA Transfer: 11985
->CPU cycles spent for comparison: 13910
There were 0 different numbers
```

While the Data Cache is now operating, it does not have a significant effect because it is set to write and read from the system memory in order to maintain coherency.

In mpu.h change the sharable option from Yes to NO (Fig. 6.26).

Figure 6.26
Set the D-Cache region to be nonsharable.

Build and rerun the example.

Now the region is not shareable, but the cache load policy is set to write through, so array data written into the cache will also be written into the SDRAM memory.

```
->CPU cycles spent for arrays initialization: 3388
->CPU cycles spent re initializing source array 1057
->CPU cycles spent on DMA Transfer: 11981
->CPU cycles spent for comparison: 4418
There were 0 different numbers
```

The big performance improvement is found in the comparison routine. This is because the current data for the first array is loaded into the cache. Once the comparison has been performed, the data for the second array will now also be loaded (Read Allocate), so if we run the comparison routine again, the time is even faster because now all data accesses are to the Data Cache.

```
->CPU cycles spent for comparison: 2083
There were 0 different numbers
```

Open mpu.h.

Change Bufferable to Yes Writeback.

Change Type extension to Level 1 Read and Write Allocate (Fig. 6.27).

```
⊟  Data Cache configuration
     Sharable                              No
     Bufferable                            Yes Write Back
     Cachable                              Yes
     Type Extension                        LEVEL1 Read and Write Allocate
```

Figure 6.27
Set the D-Cache policy to be "Write Back" and "Read and Write Allocate."

Build and run the code.

This time the code fails because the array data values are held in the Data Cache. The correct values are not copied by the DMA. Also, since we have selected read and Write, Allocate the array_sdram2[] s loaded into the Data Cache when it is initialized. This means that the DMA updates the SDRAM, but the Cortex-M7 will only see the old values stored in the cache.

Uncomment line 98 Invalidate_Cache();

Uncomment line 94 Clean_Cache();

Build and run the code.

Again the code fails but in a more interesting way. This time the Invalidate_Cache() function has loaded the values in the SDRAM into the cache, but they are the wrong values. We have initialized both arrays and then called the Clean_Cache() function. This places the initial values in the SDRAM. However, we then update the array_sdram1[] values which will only be changed in the cache.

Uncomment Line 96 Clean_Cache();

Build and run the code.

```
->CPU cycles spent for arrays initialization: 3051
->CPU cycles spent on cache clean: 1067
->CPU cycles spent re initilizing source array 1062
->CPU cycles spent on cache clean: 1095
->CPU cycles spent on DMA Transfer: 11999
->CPU cycles spent on cache invalidate: 371
->CPU cycles spent for comparison: 2696
There were 0 different numbers
->CPU cycles spent for comparison: 2081
There were 0 different numbers
```

This time the code runs correctly because we have maintained coherency between the Data Cache and the SDRAM.

Double Precision Floating Point Unit

Like the Cortex-M4, the Cortex-M7 has a Hardware Floating Point Unit. This unit is tightly integrated into the processor pipeline to allow very fast processing of both single and double-precision floating point values. The Cortex-M7 FPU is enabled through the CPARC register in the Cortex-M7 processor System Control Block. This is typically done by the standard system startup code. Floating-point calculations will now make use of the hardware unit rather than software libraries, provided the compiler options have been set correctly. Because the Cortex-M7 may be fitted with a single or double-precision floating-point unit, the CMSIS-Core specification provides a function to query the processor and return the type of FPU fitted (Fig. 6.28).

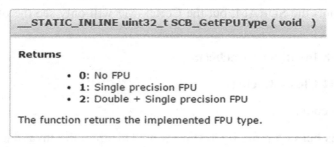

Figure 6.28
CMSIS Core get FPU type function.

We will look at using the FPU in Chapter 9 practical DSP for Cortex-M Microcontrollers.

Functional Safety

Many design sectors require compliance with some form of safety standard. Some obvious sectors are Medical, Automotive, and Aerospace. Increasingly other sectors such as Industrial Control, Robotics, and Consumer Electronics all require compliance to safety standards. As a rule of thumb, design processes required for functional safety are adopted by mainstream designers tomorrow. Developing a functional safety project is a lot more than writing and testing the code thoroughly. The underlying silicon device must be the product of an equally rigorous development process with available documentation to back this up. The microcontroller must also incorporate features (Error correction codes on bus interfaces, Built In Self Test) that allow the software to validate that the processor is running correctly and must be able to detect and correct faults. An IEC 61508−3 (SIL3) level system will often consist of dual processors running either in lockstep or with an application and supervisor arrangement. This allows the system to detect and manage processor faults.

The Cortex-M7 has been designed with both hardware and process/safety documentation that allows it to be used in high-integrity systems. At the time of writing, no microcontrollers using these features have yet been released.

Cortex-M7 Safety Features

The Cortex-M7 introduces a range of safety features that allow a Silicon Vendor to design a device suitable for a high-reliability system. The safety features shared with other Cortex-M processors are shown in Table 6.6.

Table 6.6: Common Cortex-M safety features

Memory protection Unit Instruction and Data Trace Fault exceptions

The Cortex-M7 extends the available safety features with the additional features shown in Table 6.7.

Table 6.7: Cortex-M7 safety features

Safety Feature	Description
Error Correction Codes	The cortex-M7 may be configured to detect and correct hard and soft errors in Cache RAMS using ECC. The TCM interfaces support interfacing to memory with ECC hardware
Dual Core Lock Step	The Cortex-M7 may be implemented as a dual core design where both processors are operating in Lock Step
MBIST	The Cortex-M7 may be synthesized with a Memory Built in self-test interface that supports memory validation during production and run time

Safety Documentation

To enable Silicon Vendors to design a new microcontroller that is suitable for safety use, Arm also provides extensive documentation in the form of a safety package which contains the documentation shown in Table 6.8.

Table 6.8: Cortex-M7 safety documentation

Document	Description
Safety Manual	A Safety Manual describing in detail the processor's fault detection and control features and information about integration aspects in Silicon Partner's device implementations
FMEA Manual	A Failure Modes and Effects Analysis with a qualitative analysis of failure modes within the processor logic, failure effects on the processor's behavior, and an example of quantitative hardware metrics
Development Interface Report	Report making clear how the Silicon Partner's engineers should manage the Arm deliverables and what to expect from them

These reports are only of interest to Silicon Designers and some companies specializing in safety testing and tools. As a Designer, you will be more interested in the software tools rather than the Silicon Design process. We will look at developing Functional Safety Software in Chapter 11, RTOS Techniques.

Conclusion

In this chapter, we have seen that the Cortex-M7 retains the Cortex-M programmers model allowing us to easily transition to this new processor. However, it does take a while to realize the big performance increase over the Cortex-M4. This will become more marked a Silicon Vendors moving the smaller process technologies and achieving even higher clock frequencies. The Cortex-M7 is also the first Cortex-M processor with a full suite of safety documentation which allows Silicon Vendors to design devices suitable for safety-critical applications.

CHAPTER 7

Armv8-M Architecture and Processors

Introduction

As we saw in the introduction, the current Cortex-M processors are based on the Armv6-M (Cortex-M0/M0 +) and Armv7-M (Cortex-M3\M4\M7) architectures. Toward the end of 2015, Arm announced the next generation of Cortex-M processors that are based on a new architectural standard, Armv8-M, and its extension Armv8.1. The Armv8-M architecture is a 32-bit architecture based on the existing Armv6-M and Armv7-M architectures and extends the current Cortex-M programmer's model. This allows your application code to be easily ported onto the new architecture with minimal changes. While Armv8-M is the latest architectural design from Arm, it does not make Armv7-M or Armv6-M devices obsolete, they will be around for a long time yet, but Armv8-M paves the way for highly scalable, cost-effective families of microcontroller and introduces a hardware security model that is a foundation for secure connected devices. The Armv8.1 architecture introduces vector processing support to the Cortex-M family, which provides a huge performance uplift for Machine Learning (ML) and Digital Signal Processing (DSP) algorithms. In this chapter we will first look at the Armv8-M architecture and then see the extensions provided by Armv8.1.

Armv8-M

The Armv8-M architecture consists of a Mainline profile and a Baseline subprofile (Fig. 7.1). While these two profiles are analogous to the original Armv7-M and Armv6-M split, the Armv8-M architecture creates a unified architecture for microcontroller processors with significant new features.

The Designer's Guide to the Cortex-M Processor Family.
DOI: https://doi.org/10.1016/B978-0-323-85494-8.00002-4

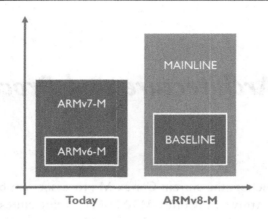

Figure 7.1
The Armv8-M architecture consists of two profiles, Mainline and Baseline, which are analogous to the existing Armv7-M and Armv6-M architectures.

The Armv8-M architecture is also designed to be highly modular for silicon designers; this allows more flexibility in processor feature selection within the two profiles. The Armv8-M architecture also introduces an important new hardware security extension called TrustZone, which is available in both the Mainline and Baseline profiles.

Common Architectural Enhancements

The two Armv8-M profiles have a common set of architectural enhancements (Table 7.1).

Table 7.1: Common Architectural Enhancements for the Mainline and Baseline Profiles

Load acquire store release instructions
New Instructions for Trust Zone support
New style Memory Protection Unit programmer's model
Better Debug capability

Armv8-M introduces a new group of "load acquire store and release" instructions which are required to provide hardware support for the latest C language standards. These instructions provide atomic variable handling that is required by the latest C and C++ standards. The Armv8-M Architecture includes the MPU as an optional unit in both the Mainline and Baseline profiles. In Armv8-M, the MPU has a new programmer's model called the "Protected Memory System Architecture version 8" (PMSAv8). The new PMSAv8 model allows greater flexibility in defining memory protection regions. In the Armv7 MPU (PMSAv7), a region has to have a start address which is a multiple of the region size, and the region size has to be a power of two. This restriction could often mean using two or more MPU regions to cover a single contiguous block of memory. In the PMSAv8 model, an MPU region is defined with a start address and an end address. The only limitation is that the start and end address have a 32-byte granularity. There are also enhancements to the data watch and breakpoint debug units along with the inclusion of comprehensive trace features.

Armv8 Baseline Enhancements

In addition, the baseline profile has the following enhancements (Table 7.2).

Table 7.2: Architectural Enhancements for the Baseline Profile

Hardware divide instruction
Compare and Branch and 32-Bit branch instructions
Exclusive access instruction
16-bit immediate data handling instructions
Support for more interrupts

The Armv6-M architecture uses the original 16-bit THUMB instruction set. In Armv8-M baseline, the THUMB instruction set is replaced with a subset of the THUMB2 32-bit instruction set. This will mean something like a 40% increase in performance over the current Cortex-M0 and M0 + . The Baseline THUMB2 instruction set now adds a number of useful instructions that Armv6-M was missing. Some of these instructions, such as hardware divide and the improved branch instruction, will give an obvious boost to the baseline performance. However, the inclusion of the 16-bit data handling instructions (MOVT and MOVW) is more subtle. These two instructions are used to load two 16-bit immediate values into the upper and lower half of a register to form a 32-bit word. This allows you to build large immediate values very efficiently, but more importantly, since the immediate value is part of the encoded instruction, it may be run from eXecute Only Memory (XOM). This allows both the Mainline and Baseline profiles to build application code that can run in the XOM as protected firmware that cannot be reverse-engineered through data or debug accesses. The Armv8-M architecture also adds the exclusive access instructions to the baseline profile. This is to increase support for multicore devices. There have already been a few multicore Cortex-M-based microcontrollers, notably the NXP LPC43xx and LPC15000 families. These are Asymmetric multiprocessor designs that consist of a Cortex-M4 and a Cortex-M0 (LPC43xx) or a Cortex-M4 and a Cortex-M0 + . In both cases, the developers at NXP had to implement their own interprocessor messaging system because the current M0 and M0 + do not have exclusive access instructions. Armv8-M-based devices will be much easier to integrate into a multicore system, be it a standard microcontroller or a custom System On Chip. Finally, the Armv8-M baseline will also extend the number of interrupt channels available on the NVIC. The Armv6-M-based processors (Cortex-M0 and

Cortex-M0 +) have a maximum of 32 interrupt channels, and on some devices, this is becoming a limitation.

Armv8-M Mainline Enhancements

The mainline profile has the following enhancements (Table 7.3).

Table 7.3: Architectural Enhancements for the Mainline Profile

Floating Point Extension Architecture v5 Optional addition of the DSP instructions

While the Mainline profile includes all of the common architectural improvements listed above, it also gets a new floating-point unit and the option to include the DSP instructions with or without the floating-point unit.

Coprocessor Interface

The Armv8 architecture also includes a coprocessor interface designed to allow the integration of hardware accelerators that extend the compute capabilities of the Cortex-M processor. The coprocessors are not peripherals in that they do not provide an interface to the external World like an ADC. Typically, the coprocessor will be an accelerator for specific algorithms such as a DSP or cryptographic accelerator. Each coprocessor is tightly coupled to the Cortex-M processor to provide a low latency interface. Dedicated single cycle instructions are used to move 32- or 64-bit data to and from the Cortex-M processor registers. The Armv8-M architecture provides control and data channels for up to eight separate coprocessors with an addressing scheme shown in Table 7.4

Table 7.4: Coprocessor Support

Coprocessor	Description
CP0−CP7	Available for vendor implementation
CP8−CP9	Reserved
CP10−CP11	Reserved for FPU
CP12−CP15	Reserved

When an Armv8-M-based microcontroller is fitted with a coprocessor, the silicon vendor will typically provide a supporting library of functions that are used to access the coprocessor features. For example, the NXP lpc55s69 is a Cortex-M33-based microcontroller that includes a cryptographic coprocessor and a DSP coprocessor. Both are supported by vendor libraries so you don't need to do any low-level code development.

Trust Zone

A key technique used to create secure online devices is to place sensitive algorithms and data into a "security island" which is isolated from the application code. The security island may be a second microcontroller which acts as a secure element that provides security services to the application microcontroller. If the application microcontroller is compromised by a network or software attack, the sensitive data will remain protected in the secure element. This dual microcontroller approach is often too expensive and it is often preferable to use a single microcontroller and create a security island as a software partition within the microcontroller.

With an Armv6-M- or Armv7-M-based device, it is perfectly possible to design such a secure software architecture. However, Armv8-M provides hardware support that allows you to partition the CPU resources into secure and nonsecure partitions (Fig. 7.2). The TrustZone security peripheral is an established technology on Cortex-A processors, and with Armv8-M it is now available to Cortex-M microcontrollers for the first time. TrustZone is a set of hardware extensions that allow you to create a boundary between Secure and Nonsecure resources with a minimum of application code. TrustZone has a negligible impact on the performance of an Armv8-M processor. This all means simple, efficient code that is easier to develop, debug and validate.

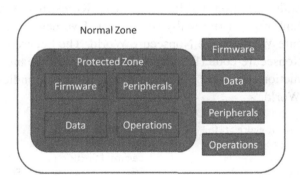

Figure 7.2
TrustZone creates a protected zone within the microcontroller system.

TrustZone extends the programmer's model to include Secure and Nonsecure "worlds," each with its own privilege modes (Fig. 7.3). The TrustZone peripheral is similar to the MPU in that it defines memory regions which are then assigned to the Secure or Nonsecure worlds. When the processor is running in Secure State it can access Secure and Nonsecure memory areas. When it is running in the Nonsecure state, it can only access Nonsecure memory areas. In addition, each security "world" has its own set of stack pointers, Memory

Protection Unit, SysTick timer, and System Control Block. Thus the Secure World becomes a security island that is used to hold sensitive code and data that cannot be snooped on by code in the Nonsecure World.

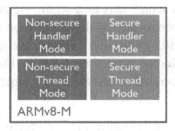

Figure 7.3
TrustZone creates two new states, secure state and nonsecure state.

We can build code to run in the Nonsecure state, call functions, access Nonsecure peripherals, and service Nonsecure interrupts as normal. Nonsecure code can also make function calls to access code in the Secure State. When Nonsecure code makes a call to Secure code, it must make an entry to Secure code through a valid entry point (Fig. 7.4). These valid entry points are a new instruction called the Secure Gateway (SG) instruction. If Nonsecure code tries to enter the Secure World by accessing anything other than an SG instruction a Secure Fault exception will be raised. The SG instruction also clears the least significant bit of the Link register (normally set to one). Two new instructions are provided to return from the Secure World to the Nonsecure World. The Secure code can use a Branch Exchange to Nonsecure code (BXNS) or Branch Link Exchange to Nonsecure code (BLXNS). These instructions use the LSB of the return address to indicate a transition from Secure to Nonsecure World.

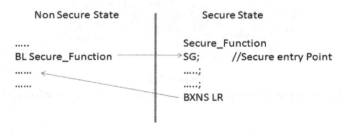

Figure 7.4
Nonsecure code can call secure code but it must enter secure state via an SG instruction.

It is also possible for Secure code to call Nonsecure code (Fig. 7.5). When this happens, the Secure world return address is stored on the Secure stack, and a special code FNC_RETURN is stored in the link register. When the Nonsecure function ends, it will branch on the Link register. This branch instruction will see the FNC_RETURN code. This triggers the unstacking of the true return address from the Secure stack and a return to the Secure function.

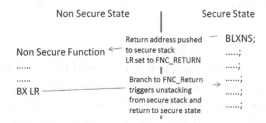

Figure 7.5
Secure code can call Nonsecure code. Its return address is hidden from Nonsecure code.

At the heart of the TrustZone technology are the Security Attribution Unit (SAU) and an Implementation Definition Attribution Unit that provide a memory "marking" to define default regions of secure and nonsecure resources. The SAU is then used to customize the default memory marking. An Armv8-M processor with TrustZone will always boot in Secure mode, and the Secure software can then define the Secure and Nonsecure partitions by programming the SAU regions. The SAU memory region may be defined with one of three security attributes. We can define a memory region as Secure or Nonsecure but there is also an additional level called "Secure Nonsecure Callable" (NSC) which can be read as "located in the secure region and callable from the nonsecure region"(Fig. 7.6).

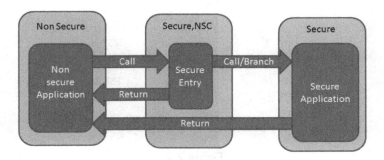

Figure 7.6
The security model implements three zones: Secure, Nonsecure, and Nonsecure callable.

In a real application, it is possible that constant data may inadvertently have the same binary pattern as an SG instruction. This would create a false entry point into the Secure World that could then be exploited by an attacker. The SAU allows us to define an NSC region that will contain all the SG instructions and can be more easily policed for false entry points. Any SG instruction pattern outside of the NSC region will raise a security exception if accessed. In practice, the NSC region is populated by a set of branch tables to the Secure code. When a secure project is built, the compiler and linker can generate a library of entry points which is used by the nonsecure code. This allows us to build application code that can access secure functions without any visibility of their internal data or instructions.

TrustZone also includes an additional instruction that allows you to check the bounds of a data object. The Test Target (TT) instruction can be used to return the SAU region number for a selected address. We can use the TT instruction to request the SAU region number for the start and end address of a memory object to ensure it does not span across Secure and Nonsecure regions. This allows us to easily create bounds-checking code for any objects with dynamic ranges such as pointers.

The Secure state adds an additional pair of stack pointers (MSP_S and PSP_S) for the secure stacks and a pair of stack limit registers in both the Secure and Nonsecure states (Fig. 7.7). The limit registers are used to define the size of each stack space, and a security fault will be generated if any stack space is exceeded.

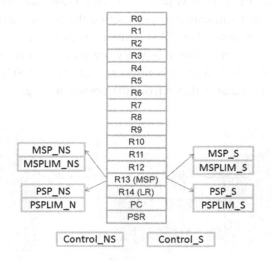

Figure 7.7
The CPU register file is extended with secure and Nonsecure stack pointers and CONTROL registers. Stack limit registers are provided for each of the four stacks.

The Armv8-M processor peripherals are also extended with a Secure and Nonsecure MPU, SysTick, and System Control Blocks (Fig. 7.8).

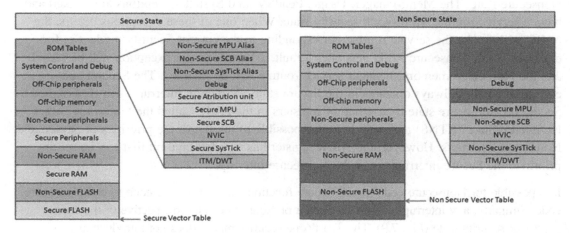

Figure 7.8

The secure state has access to all nonsecure resources plus a set of alias registers to access the nonsecure system control block. The nonsecure state only has access to its own resources.

When the processor is in the Secure state, it can access the Nonsecure peripherals through a set of alias registers.

Interrupts and Exceptions

The Armv8-M architecture retains the NVIC as the interrupt processing unit and provides new extensions to support the TrustZone Secure and Nonsecure states. The processor exception table is the same as for Armv7-M with the addition of a Security Fault exception (Table 7.5).

Table 7.5: Armv8-M Processor Exception Table

Exception	Processing State
NMI	Default: Secure State Can be routed to Nonsecure state
Hard Fault	Default: Secure State Can be routed to Nonsecure state
MemManager	Banked
Bus Fault	Default: Secure State Can be routed to Nonsecure state
Usage Fault	Banked
SecureFault	Secure State
DebugMonitor	Can target Secure or Nonsecure state (Defined by hardware)
PendSV	Banked
SysTick	Banked
Peripheral Interrupts	Programmable through new NVIC "Interrupt Target Nonsecure State" registers

A new SecureFault exception is added.

The behavior of each exception is defined for each security state. The NMI, Hard Fault, and Bus Fault exceptions default to the Secure state but may be configured to execute in the Nonsecure state. The MemManager, Usage, PendSV, and Systick exceptions are banked and have a service routine within each security state. When one of these exceptions occurs, the NVIC will invoke the service routine corresponding to the current Security mode. So, for example, if the Nonsecure MPU generates a fault, a MemManager exception will be raised, and the Nonsecure memory manager service routine will be invoked. The SecureFault exception handler always operates in the Secure state. Peripheral interrupts can be assigned to Secure on Nonsecure state by a group of registers in the NVIC called the "Interrupt Target Nonsecure State" (ITNS) registers. It is also possible to interleave the priorities of Secure and Nonsecure interrupts. However, the AICR register has an additional bit that can be set to prioritize the Secure interrupts over the Nonsecure interrupts.

It is possible for Nonsecure code to call Secure functions and for Secure code to call Nonsecure code. Similarly, any interrupt can run in Secure or Nonsecure state irrespective of the current processor security state (Fig. 7.9). The TrustZone security model does not introduce any additional hardware latency except when the processor is running in Secure state and has to service a Nonsecure interrupt. In this case, the processor pushes all the CPU registers to the Secure stack and then writes zero to the CPU registers to hide the Secure world data from the Nonsecure code. This gives the processor more work to do and introduces a "slightly longer" interrupt latency.

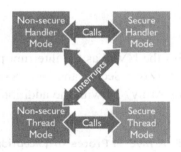

Figure 7.9
Interrupts can be configured as secure or Nonsecure and may be served irrespective of the current processor state.

If we have a mixture of Secure and Nonsecure interrupts and exceptions, it follows that we must isolate the Secure interrupt vectors, so they are not visible from the Nonsecure code. In order to do this TrustZone creates two interrupt vector tables, a Secure vector table and a Nonsecure vector table. The location of each vector table in the processor memory map is controlled by a pair of vector table offset registers, one for Nonsecure state (VTOR_NS) and one for the Secure state (VTOR_S). The Secure vector table will be located at the start of the memory map so the processor can boot into Secure mode (Fig. 7.8). The Secure startup code can then program the SAU to define the Secure and Nonsecure regions, and then we can locate the Nonsecure vector table at the start of the Nonsecure code space (Fig. 7.8). Each peripheral interrupt can then be routed to the secure or nonsecure NVIC through the ITNS registers.

CMSIS Trust Zone Support

The CMSIS core specification provides extensive support for TrustZone. Table 7.6 shows the range of support that is provided.

Table 7.6: CMSIS Support for TrustZone

CMSIS TrustZone support	Description
Core register access functions	Supports access to NonSecure processor registers from Secure state
NVIC Functions	Supports access to the Nonsecure NVIC registers when in secure state
Systick Functions	Supports access to the Nonsecure Systick timer when in secure state
Stack Sealing Functions	Functions to support the stack sealing technique to protect against potential stack underflow attacks
RTOS Context Management	Supports RTOS Thread management for calling secure modules from Nonsecure state

Platform Security Architecture

Alongside the introduction of the Armv8-M architecture and TrustZone Arm have also developed an open-source security project called the Platform Security Architecture (PSA). The PSA project defines a robust security model for IoT or any other connected device against a network software attack. The PSA also provides a reference Trusted Firmware for Cortex-M, which is designed to reside within the "Secure Processing Environment" (SPE) (Fig. 7.10).

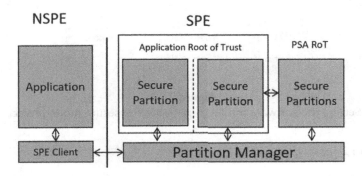

Figure 7.10
PSA Trusted firmware provides secure services in isolated partitions within the Secure Processing Environment. The Partition manager provides a point of entry from the Nonsecure processing Environment.

The Trusted Firmware provides a set of security services which are accessed from the "Nonsecure Processing Environment" through a partition manager, which a single point of entry to the SPE. While this is a complicated system once installed, the Secure Services are accessed from the Nonsecure application code using a set of standard API calls. So if you intend to develop an IoT device, you should implement it using the PSA. If it is a commercial device, the PSA also provides a route to certification that meets and exceeds all current global legal requirements.

Exercise 7.1: TrustZone Configuration

In this example, we will examine how to design an application to configure the TrustZone security peripheral and make calls from the nonsecure application code to the secure functions.

In the pack installer select Exercise 7.1 and press the copy button.

This is a multiproject workspace that includes a secure and nonsecure project (Fig. 7.11).

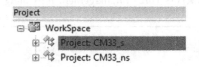

Figure 7.11
Workspace with Secure and nonsecure projects.

Open the CM33_s project/options for Target (Fig. 7.12).

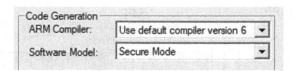

Figure 7.12
Setting the software model to generate the secure mode project.

This project will be built in secure mode.

Expand the CM33_S project and open the device/partition.h header file (Fig. 7.13).

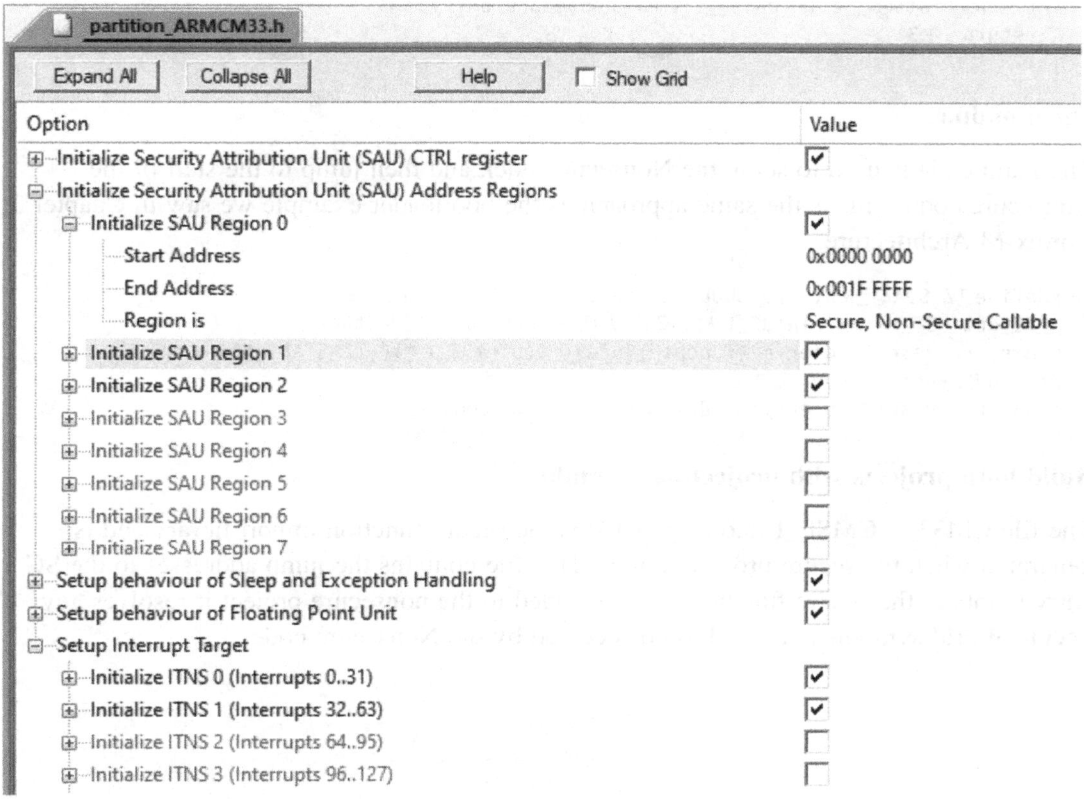

Figure 7.13
The TrustZone SAU configuration is defined in partition.h.

This is a templated file that allows us to define all the values required to configure the TrustZone peripheral at startup of the secure code. Within this header file, it is possible to define each of the SAU regions and also route the peripheral interrupts to the Secure or Nonsecure NVIC.

Partition.h also includes the function TZ_SAU_Setup() that is called from the system_ARMCM33.c code at startup.

Open Interface/interface.c.

This module provides some test functions that can be called from the nonsecure code.

Any function that will be callable from the nonsecure code must be defined with the extension __attribute__((cmse_nonsecure_entry)). This will create a table of entry points in an import library module that can be used by the Nonsecure code to call the secure functions.

```
int func1(int x) __attribute__((cmse_nonsecure_entry)) {
  return x + 3;
}
```

Open main.c.

The main code is used to setup the Nonsecure stack and then jump to the start of the Nonsecure code. This is the same approach as the bootloader example we saw in Chapter 3, Cortex-M Architecture.

```
#define TZ_START_NS (0x200000U) // Start on the Non Secure code
__TZ_set_MSP_NS(*((uint32_t *)(TZ_START_NS))); // set NS stack
NonSecure_ResetHandler = (funcptr_void)(*((uint32_t *)((TZ_START_NS) + 4U)));
//read NS reset addres
NonSecure_ResetHandler(); // Jump to start of NS code
```

Build both projects with project/batch build.

The file CM33_s_CMSE_Lib.o (Fig. 7.14) is the secure function import library and is generated when the secure project is built. This file contains the jump addresses to the SG entry points to the secure functions. When added to the nonsecure project it resolves any Secure World symbols that need to be accessed by the Nonsecure code.

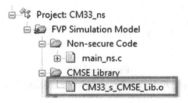

Figure 7.14
The secure project create an interface library that is used by the nonsecure code.

This project is also built in Nonsecure mode (Fig. 7.15) and is linked to start at 0x200000. Its system_ARMCM33.c startup file will move its interrupt table to this address by programming the Nonsecure VTOR register.

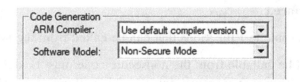

Figure 7.15
The application code is built to run in the NonSecure world.

Start the debugger so that it runs to main() in the secure code.

View the register window and the peripherals/core peripherals/SAU (Fig. 7.16).

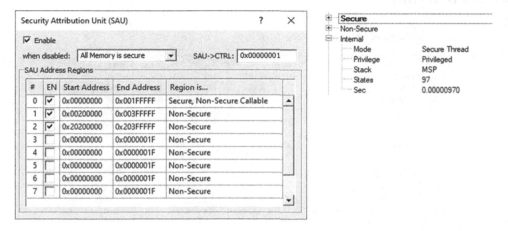

Figure 7.16
The register window provides details of the current operating mode while the peripheral/SAU window shows the TrustZone configuration.

Here, we can see the security mode in the register window and how the TrustZone regions are configured in the SAU peripheral view.

Set a breakpoint at the start of interface.c/Func1 in the Secure project.

Open main_ns.c and set a breakpoint at main();

Run the code.

Execution will leave the Secure code and jump to the Nonsecure code (Fig. 7.17).

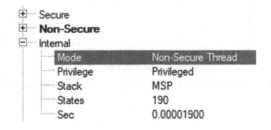

Figure 7.17
Now we have returned to the nonsecure world.

Open the peripheral/core peripheral/NVIC.

Here, we can see the routing of each interrupt source and also the location of each hardware vector table (Fig. 7.18).

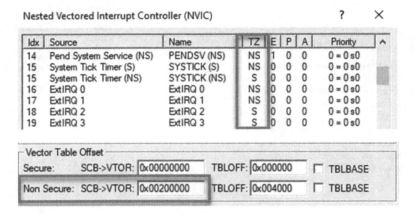

Figure 7.18
The Peripheral/NVIC window shows the interrupt routing and location of the hardware vector tables.

Now run the code again.

The nonsecure code will call func1() which is in the secure partition.

Now we are back in the secure partition and can run any sensitive code in isolation from the application code (Fig. 7.19). Once we return from func1() execution will return to the nonsecure World.

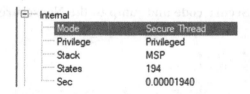

Figure 7.19
When a secure function is called the processor swaps to secure world processing.

Armv8.1-M

Currently, the Armv8.1-M architecture consists of two new processors. The Cortex-M55 that was announced in 2020, and at the time of writing the first microcontrollers based on Cortex-M55 are just being released. The second processor is Cortex-M85 which has just been announced by Arm, and there is generally a 2-year lag before an actual microcontroller is available. However, it should be noted that tools support already exists in the form of enhancements to the Arm C compiler along with simulation models of each new processor. Alongside the latest processors, Arm has also developed a range of Neural processing units, including a microNPU which is intended to be a companion to a Cortex-M processor. The Ethos U55 microNPU is designed to provide a massive acceleration to ML and other algorithms with computationally demanding workloads. In this section, we will examine the Armv8.1 processors.

The Armv8.1-M architecture adds several significant extensions to the existing Cortex-M processors. The key feature extensions are shown in Table 7.7.

Table 7.7: Armv8.1 Feature Extensions

Feature	Description
CPU extension	Additional instructions to support the new hardware features and low overhead looping support.
Helium vector extension	Vector processing for fixed and floating-point calculations
Floating point extension	Provides half, single, and double precision
Debug enhancements	Break and watchpoint improvements. Additional unprivileged debug extension
ECC and RAS	Addition of memory error correction code and error reporting through a Reliability Availability and Serviceability RAS extension
Performance monitoring unit	Monitoring of performance events within the CPU

Helium Vector Extension

Helium is an optional extension to the CPU that provides a Cortex-M processor with a vector processing as an extension (MVE). The Vector extension any be added in a range of different configurations as shown in Table 7.8.

Table 7.8: Helium MVE Configuration Options

Configuration	Notes
Scalar FPU ()	Single or double precision
Vector Integer	8, 16, and 32-bit fixed-point Q numbers
Vector Integer and scaler FPU	As above
Vector Integer, Vector Floating point and Scalar Floating point	As above plus support for half-precision float

Helium also adds over 150 instructions to the processor instruction set that provide a range of very efficient data handling techniques in addition to the vector processing capabilities.

FPU Register Organization

The Helium Vector processing extension reuses the FPU register bank to create an array of input vectors (Fig. 7.20). We will look at the Floating point unit in more detail in Chapter 9, Practical DSP for Cortex-M Microcontrollers, but for now, when used for scalar operations, the Floating-point register bank is composed of 32 registers which are each 32 bits wide. When used as part of the MVE these registers are reused as eight vector registers, each 128 bits wide.

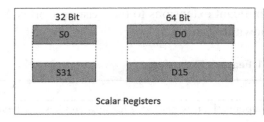 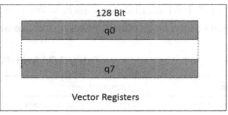

Figure 7.20
The FPU registers may be used as scalar (32 and 64 bits) registers or an array of 128-bit vector registers.

Each of these vector registers can be used to support a wide range of data types (Fig. 7.21) which include a vector of 8, 16, or 32-bit fixed-point numbers known as Q numbers. We will take a closer look at Q numbers in Chapter 9, Practical DSP for Cortex-M Microcontrollers. Helium can also process floating-point vectors consisting of half-precision float and single-precision floating-point. The MVE also supports complex number arithmetic for fixed-point and single-precision floating-point values.

```
8 bit (q7)
16 bit (q15)
32 bit (q31,float32)
```

Figure 7.21
MVE fixed-point and floating-point formats.

Within the array of vectors, each set of parallel data values is referred to as a lane. So, for example, in an 8-bit vector, there are 16 lanes, while in a single-precision floating-point value, there are four lanes.

Lane Prediction

Many of the new Helium instructions support lane prediction. This is a form of conditional execution for a lane within the vector. So, for example, is it possible to perform a conditional compare and then only process the data values that actually matched the compare condition. When a vector operation is performed, the condition codes for each lane are updated in a new register called the Vector Predication Status and Control register. We can now construct an if-then block of up to four instructions by using the "Vector Predicate Then" and "Vector Predicate Set" Instructions. An example is shown in Table 7.9.

Table 7.9: Conditional Lane Processing

Instruction	Pseudo code	Note

MOVS R0, #1
MOVS R1, #3
VIDUP.U32 Q0, R0, #1
VPTE.S32 GE, Q0, R1
VDUPT.32 Q0, R1
VMULE.S32 Q0, Q0, R1
	R0 = 1
	R1 = 3
	Q0 = [1 2 3 4]
	If Q0[i] > = R1
	Then Q0[i] = R1
	Then Q0 = [3 6 3 3]
		For(I = 0;i < 4;i + +) {Q0[i] = (I + 1)};

As we will see below, the Helium extension also supports loop prediction, which is a form of low overhead looping that can efficiently process a data set as a collection of vectors, and we will look at these instructions later in this chapter.

Big Integer Support

Finally, the vector processing extension also provides big integer support. This allows the 128-bit vectors to be chained together and processed as a single big integer. This is very useful for cryptographic "Big Number" libraries used for asymmetric (public key) encryption.

Data Load and Store Instructions

The helium extension provides a range of instructions that provide different methods to load and unload the MVE vector register.

Vector Load and Store

The most basic way to load data into the vector registers is to load each lane in sequence, starting from a base address specified in a CPU register. A vector store will unload each lane in turn to an incrementing base address.

Data Interleaving and Deinterleaving

A deinterleaving load will take data from consecutive memory locations and separate the data into vector registers. For example, if you have an audio codec that has separate registers for left and right channel data a deinterleaving load can be used to load the right channel data into the lanes of one vector register while the left channel data is loaded into a second vector register. An interleaving store is used to take the data from two vector registers and combine them into a single data stream.

Scatter Load and Unload

A gather load is used to gather data from noncontiguous locations into a vector register. Here, a second array contains the address location for each data value allowing the data to be assembled directly into a vector register. A matching set of scatter store instructions can place the individual vector elements into a set of noncontiguous memory locations.

Helium Data Throughput

The MVE processing of each vector is defined as a quarter-beat system. This means that a quarter of the input vector can be processed per beat. In the case of 8-bit data this would mean processing four bytes per beat. When implemented in the Cortex-M55, the MVE is a dual-beat system. This means that half of the vector may be processed per CPU cycle. In the case of our 8-bit vector the MVE would process 8 bytes per CPU cycle.

Developing Applications with Helium

You can develop an application to use the Helium extension in three different ways.

CMSIS-Enabled Helium Libraries

Both the CMSIS DSP and CMSIS NN libraries have build options that are optimized to take advantage of the Helium extension while providing the API that is consistent across all the Cortex-M processors. This allows you to move existing application code from earlier Cortex-M processors to a processor with a Helium MVE extension.

Helium Code Development

It is possible to write hand-optimized assembly code using the Helium instructions. A set of Helium intrinsics are also available in the header arm_mve.h that allows you to generate C code to access the Helium extension.

Auto Vectorizing Compiler

Writing assembly code or low-level C code that provides optimal performance is time-consuming and requires experience of the low-level workings of the helium extension. The Arm compiler version 6 is an auto vectorizing compiler that will generate code to take advantage of the MVE extensions where possible. The two main optimizations are shown in Table 7.10.

Table 7.10: Compiler Auto Vectorizing Optimizations

Optimization	Description
Loop vectorization	Unrolling loops to reduce the number of iterations
SuperWord-level parallelism	Concatenating scalar operations to use full-width Helium instructions

The advantages of using auto-vectorization include ease of use and code portability. When you define a project, you just need to enable the Helium extension and then let the compiler get on with it. The source code remains as easily portable C code free of inline assembler or intrinsic functions. This makes the project easier to maintain as it does not need any specialist knowledge to understand it. Therefore it is best to start by using the auto vectorizing support in the compiler as this will meet the requirements of most projects.

CPU Extension

The CPU extension expands the CPU instruction set to support the new processor hardware features along with some new instructions that improve the overall efficiency of the Armv8.1 processor. As we saw with the Cortex-M7, one of the limitations of a general-purpose CPU compared to a DSP is execution overhead in a program loop because the Thumb-2 instruction set requires a conditional branch to test for the end of the loop. The Cortex-M7 reduces this impact with the introduction of a Branch Target Address Cache. This issue is re-addressed in the Armv8.1-M architecture with the introduction of a low overhead branching extension.

Low Overhead Branch Extension

The low overhead branch extension is used to increase the efficiency of for(;;), while() and do while() loops by introducing a dedicated set of instructions that remove the need for a conditional branch (Table 7.11).

Table 7.11: Low Overhead Looping Instructions

Instruction	Description
WLS	While loop start
LE	Loop end
DLS	Do Loop Start
WLST	While loop start with Tail prediction
DLST	Do Loop start with Tail prediction
LET	Loop end with Tail prediction

The while loop start instruction WLS is used in conjunction with a loop end Loop end (LE) instruction. The WLS instruction loads the link register LR with a loop iteration count and the end address of the loop. The end of loop address is cached in an internal CPU register. The LE instruction is used to cache the startof loop address in an internal CPU register.

In the case of a simple for() loop, the WLS instruction loads the loop count into the Link register and also sets the end of loop address. Once executed we enter the loop code which is terminated with a LE Instruction. When executed, the LE instruction sets the StartOf Loop address. Once the loop conditions are set up the loop instructions will now execute for the required number of iterations without any further execution of the WLS or LE instruction. On each iteration the count in the Link Register will be decremented. Once the required number of iterations has been completed and the count in the link register has reached zero, the code will jump to the stored EndOf Loop address.

```
For( index = 10;index>0; index--)
movS RO,#10
WLS LR,RO,EndOfLoop
{
StartOfLoop
..........   // loop code
LE LR,StartOfLoop
}
EndOfLoop
```

If an interrupt occurs during the execution of a low overhead loop the loop address Cache is cleared and the LE instruction is executed once to restore the StartOfLoop address.

The low overhead loop instructions can also be used to efficiently process vector data. In this case we use the WLST instruction to set the number of elements to be processed (not the number of loops). The LE Prediction instruction is used at the end of the loop to again set the loop start address. We can now process arrays of data with the Helium vector instructions. If the data set is not a multiple of an input vector (128 bits), the last vector will be partially filled. In this case, the Tail prediction will ensure only the active lanes are processed.

Exercise 7.2: Armv8.1 Performance

In this exercise we will compare the performance of a Cortex-M4, Cortex-M33, and a Cortex-M55 by measuring the time taken to execute a set of matrix processing examples.

In the pack installer select Exercise 7.2 and press the copy button.

Use the dropdown in the toolbar to select Cortex-M55_FP_MVE (Fig. 7.22).

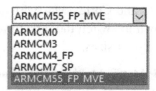

Figure 7.22
Select the Cortex-M55 build and simulation model.

Check that the Helium vector extension is enabled in the options for target dialog
(Fig. 7.23).

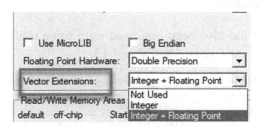

Figure 7.23
Enabling the Vector Extension.

Build the code.

Start the debugger and run to the first breakpoint.

In the bottom right corner of the debugger reset the stopwatch timer (Fig. 7.24).

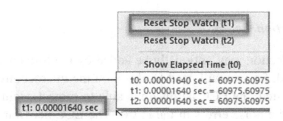

Figure 7.24
Reset the stopwatch timer.

Run the code until it hits the second breakpoint.

Record the time taken to execute the code.

Repeat the exercise for Cortex-M33 and Cortex-M4 (with and without the floating-point unit enabled) by using the dropdown to switch the target processor. Table 7.12 shows the expected execution time, code, and data sizes.

Table 7.12: CMSIS DSP Matrix Multiplication Examples Execution Time

CPU	Time (uS)	Code Size	RW-Data size	ZI data
M4	1967	7264 + 1152	4	3892
M4-FP	556	5920 + 1152	4	3892
M7-FP	543	5904 + 1152	4	3892
M55-MVP	167	3928 + 2176	4	3892

Coresight Debug Enhancements

The CoreSight Debug unit has also been improved with some useful refinements to the breakpoint and data watch units shown in Table 7.13.

Table 7.13: Coresight Debug Enhancements

Feature	Description
Data Watchpoint bit mask	Used for value matching
Breakpoint with counter	Trigger the BP when a count value is reached

While these features may have been offered in a debugger for earlier Cortex-M processors, the feature was implemented in firmware within the debug adapter. This would result in the debugger becoming intrusive and adds overhead to the processor code execution. The new break and watchpoint features are part of the internal CoreSight debug architecture and are much more efficient and run nonintrusively within the processor.

Memory Error Correction Codes

The ECC memory protection may optionally be added by the Silicon vendor when the processor is synthesized. So you will need to check on the microcontroller datasheet to see if it is present. When fitted, the ECC hardware is able to protect against data errors, address errors and transient white noise errors in the RAM. The ECC hardware is designed to support two error detection schemes: Single Error correct, Double Error Detect; and also Double error Detect.

Poison Signaling

The ECC error detection hardware includes a poison signal that is used to mark corrupted data. This is typically used to detect errors in the data cache but can also be applied to the data TCM. In the case of the data cache, the ECC may detect a corrupted data line when data is being evicted from the Cache. If more than one bit of error is detected, a poison signal within the AXI bus will be asserted, and the data will be stored in the RAM but marked as being in a poisoned state. When the data is next accessed by the CPU or another bus master, the poison state is signaled to the CPU, and a fault exception is raised. While this may seem a bit involved, the poison signal has two important benefits. First, it prevents any bus master from using the poisoned data. Second, the fault exception is raised synchronously with the use of the erroneous data, which makes it much easier to deal with. If the fault was raised when the data was evicted from the Cache the code currently executing would not be associated with the data making it very difficult to manage the error.

Reliability Availability and Serviceability

The RAS extension is added to the Armv8.1-M architecture to manage the dependability and reliability of a microcontroller system. The RAS extension was originally developed for Cortex-A processors to ensure high reliability for server platforms. The introduction of the RAS extension is an important new addition to Cortex-M processors that are used in functional safety applications. The RAS provides a standard interface that is used to record and report errors that have occurred in memory regions protected by an Error Correcting Code scheme. The memory regions supported include the L1 Cache plus the data and instruction TCM. The RAS extends the Thumb-2 instruction set with two new memory barrier instructions.

Error Synchronization Barrier

A new Error synchronization barrier (ESB) instruction is provided. When this instruction is executed all outstanding transactions in the memory system are completed, and any BusFault exceptions are raised before the ESB instruction completes. It is also possible to enable an Implicit Error Synchronization Barrier by setting the AIRCR.IESB bit in the system control space. When enabled an IESB synchronization point will be inserted at the end of exception entry and exit register stacking and unstacking operations. An IESB instruction is also inserted for floating-point lazy stacking operations. The IESB instruction allows the application code to spot errors during context switches. However, it will potentially increase the overall interrupt latency because all memory transactions must complete before the code can continue.

RAS Error Event

When an error is detected, the RAS will store a record of the event. Currently, the Armv8.1-M processors can log a single event that provides the record shown in Table 7.14.

Table 7.14: RAS Record

Error record Field	Size	Description
Address Valid	Bit	Address is valid for TCM ECC error or a precise bus fault
Status Valid	Bit	Set for a valid RAS event
Uncorrected Errors	Bit	Set for at least one uncorrected error
Error Reported	Bit	Set for BusFault caused by RAS event
Overflow	Bit	Set when at least two RAS events have occurred
Miscellaneous registers valid	Bit	Set for
Corrected errors	2 Bits	0x00 No corrected errors 0x01 at least one corrected error has been detected
Deferred Errors	Bit	Set for
POISON	Bit	Set for
Uncorrectable Error Type	2 Bits	
Architecturally defined primary error code	Byte	0 No error. 2 TCM ECC error. 6 L1 data cache or instruction cache data RAM ECC error. 7 L1 data cache or instruction cache tag RAM ECC error. 21 Poison BusFault

Logging and monitoring of RAS events allow the application code to determine if there is any degradation in the microcontroller memory and system components. This allows a safety-critical system a means of ensuring its functional reliability and availability.

Performance Monitoring

Within the Armv8.1-M a performance monitoring unit is used to provide a deep view into the internal run-time activity of the processor. The PMU can be used to monitor over 180 different event types within the processor. Monitoring of event activity is done by a set of event counters. There are up to eight 16-bit event counters and one 32-bit cycle counter. Each event counter can be configured to monitor a single event source. It is also possible to cascade a pair of event counters to create 32-bit counters. The PMU can be configured through a set of CMSIS core functions, as shown in Table 7.15.

Table 7.15: CMSIS Core PMU Functions

CMSIS Core PMU function	Description
Arm_PMU_CNTR_Disable	Disable selected event counters and cycle counter
Arm_PMU_CNTR_Enable	Enable selected event counters and cycle counter
ARM_PMU_Set_EVTYPER	Select event to count
ARM_PMU_CYCCNT_Reset	Reset the debug cycle counter
ARM_PMU_Disable	Disable the PMU
ARM_PMU_Enable	Enable the PMU
ARM_PMU_EVCNTR_ALL_Reset	Reset all event counters
ARM_PMU_Get_CCNTR	Read the cycle counter
ARM_PMU_Get_CNTR_OVS	Read counter overflow status
ARM_PMU_Get_EVCNTR	Read the selected event counter
ARM_PMU_Set_CNTR_IRQ_Disable	Disable counter overflow interrupt request
ARM_PMU_Set_CNTR_IRQ_Enable	Enable counter overflow interrupt request
ARM_PMU_Set_CNTR_OVS	Clear counter overflow status
ARM_PMU_CNTR_Increment	Increment one or more selected counters

The PMU event counters can also raise a debug monitor exception to trigger a local debug agent that transfers the current event count to an external debugger.

Security

The Armv8.1 Architecture also adds an optional security feature that can prevent common styles of software attacks. There are also security improvements to the TrustZone handling of the FPU and Debug support.

Pointer Authentication and Branch Target Identification Extension

A group of software attacks can be derived from a simple buffer overflow attack. As its name implies, vulnerable code allows an attacker to input data to a device that goes beyond the range of the input array and overwrites other data values stored in memory. Sophisticated attacks known as "stack smashing" can be devised if the input buffer is located on the stack. In such a case an attacker can overwrite local variables, which may contain pointer addresses and also the current function return addresses. This type vulnerability allows an attacker to build successful exploits using techniques such as Return Orientated Programming and Jump Orientated programming. Compromising pointer addresses and return addresses is an attacker's stock in trade.

The Armv8.1-M architecture includes an optional security extension designed to prevent this style of exploitation by providing Pointer Authentication and Branch Target Identification (PACBTI).

Pointer Authentication

Pointer Authentication provides a hardware-based method of validating that a pointer address has not been manipulated by an attacker. When a cortex-M processor is fitted with the PACBTI extension, a Pointer Authentication Code (PAC) will be created and stored in a CPU general-purpose register. The PAC can then be regenerated to validate the address before the pointer is used to prove that the address has not been tampered with. A PAC is generated and validated on the execution of a specific range of PACBTI instructions shown in Table 7.16.

Table 7.16: PAC Code Generating and Authenticating Instructions

Instruction	Description
PAC	Sign LR using SP as the modifier and the PAC is stored in R12
PACBTI	Sign LR using SP as the modifier and the PAC is stored in R12
PACG	Sign a GPR Rn using Rm as the modifier and the PAC is stored in Rd
AUT	Authenticate LR using SP as the modifier and PAC in R12
BXAUT	Indirect branch with pointer authentication. Branch to GPR Rn using Rm as the modifier and the PAC in Rd
AUTG	Authenticate a GPR Rn using Rm as the modifier and the PAC is in Rd

When a supporting instruction is executed, the code is generated by extending the pointer address and the current value of the stack pointer to 64-bits. The two values and a 128-bit key are then passed through a lightweight block cipher, and the resulting 32-Bit code is stored in a general-purpose register (Fig. 7.25).

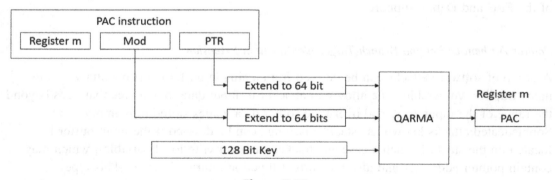

Figure 7.25
PAC generation.

The authentication instructions are used to validate the PAC code before the pointer address value is used. If the validation fails an INVSTATE Usage Fault exception is raised. The PACBTI extension is enabled separately for each privilege and security state. Each privilege and security state has its own key. Each key is stored in supporting CPU system registers.

The encryption system used may be customized by the silicon vendor when the microcontroller is designed but currently defaults to a lightweight block cipher called QARMA, which has been designed to be used as a keyed Hash function to generate runtime tags such as the PAC.

Branch Target Identification

While Pointer Authentication may be used to validate directly address memory, we need a second technique to detect tampering of indirectly addressed memory.

When the PACBTI extension is fitted the Execution Program Status Register (EPSR) includes an additional BTI bit with the label EPSR.B. The BTI instructions are used to set and clear this bit at the beginning and end of each indirect branch. The BTI instructions are divided into a group of setting instructions (Table 7.17) and a set of clearing instructions (Table 7.18). At the start of the jump, the EPSR.B bit is set. The jump must target a bit clearing instruction as a branch landing instruction to unset the bit. A Usage Fault exception will be raised if any other instruction is executed when EPSR.B is set.

Table 7.17: BTI Setting Instructions

Instruction	Description
BX, BXNS:	Only when LR is not used
BLX,BLXNS,BFX	Only when LR is not used, updates CPU register LO_BRANCH_INFO.BTI
BFLX	Updates CPU register LO_BRANCH_INFO.BTI
LDR (register)	Only when PC is updated by the instruction
LDR (literal)	Only when PC is updated by the instruction
LDR (immediate)	Only when PC is updated and the base address register is either not the SP or the SP and write-back of the SP does not occur
LDM, LDMIA, LDMFD	Only when PC is updated and the base address register is either not the SP or the SP and write-back of the SP does not occur
LDMDB, LDMEA	Only when PC is updated and the base address register is either not the SP or the SP and write-back of the SP does not occur

Table 7.18: BTI Clearing Instructions

Instruction	Description
BTI	Branch Target Identification
SG	Secure Gateway
PACBTI	Pointer Authentication code for link register with BTI clearing using key

PACBTI Compiler Support

The Arm compiler version 6 provides support for the PACBTI extension with the following compiler switches shown in Table 7.19.

Table 7.19: Compiler PACBTI Options

Compiler option	Description
-mbranch-protection	Adds PACBTI checking to branches
—library_security = protection	Selects a security hardened version of the compiler libraries

These options are currently available for Cortex-A devices. You will need to check the compiler manual to find the latest support for Armv8.1-M-based processors.

TrustZone Support

The Armv8.1 architecture also improves support for the TrustZone security peripheral. New instructions have been added to help manage the Floating-Point Unit registers during a context switch between security states. A new "privileged execute never" attribute has also been added to the Memory Protection Unit. This is designed to prevent an elevation of privilege attack by a hacker who can compromise the stack and attempt to execute nonprivileged thread code at a privileged level.

Unprivileged Debug Extension

The debugger also provides an unprivileged debug extension as a security enhancement. In a system fitted with the TrustZone security peripheral, it is possible to disable Secure debug, restricting debug access to the Nonsecure code. However, in Armv8.1-M, secure code running in privileged mode can selectively enable debug support through the Debug Authentication Control register. This allows the Secure World entry code to enable debug support for certain components while disabling debug for others (Fig. 7.26).

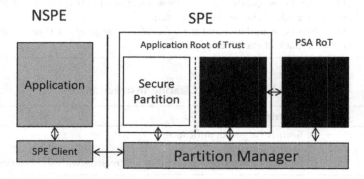

Figure 7.26
Unprivileged debug allows the secure partition to enable unprivileged debug for selected software components.

In the Arm Platform Security Architecture, the Secure World is entered through a partition manager which routes a request to the relevant secure service. In Armv8.1-M the partition manager can be programmed to allow debug support for a selected secure service. This allows multiple secure service libraries to coexist in the Secure World and their vendors can decide if they wish to provide debug support.

CPU Comparison

Table 7.20 provides a comparison between the features implemented in the different Cortex-M architectural variants.

Table 7.20: Comparison of Cortex-M Architectural Features

Feature	Cortex-M4	Cortex-M33	Cortex-M55	Cortex-m85
Arm Architectural version	Armv7-m	Armv8.0-m	Armv8.1-m	Armv8.1-m
Coremark/MHz[a]	3.42	4.02	4.2	TBA[b] > 5.01
Performance Monitoring Unit			Y	Y
Reliability Availability and Serviceability			Y	Y
Arm Custom instruction		Y	Y	Y
Coprocessor interface		Y	Y	Y
Coresignt Debug and trace	Y	Y	Y	Y
PACBTI				Y
Unprivileged debug extension			Y	Y
Stack Limit checking		Y	Y	Y
TrustZone		Y	Y	Y
MPU	PMSAv7	PMSAv8	PMSAv8	PMSAv8
Helium MVE			Y	Y
FPU	fp32—PFv4	fp-32—FPv5	fp16, fp32, fp64—FPv5	fp16, fp32, fp64—FPv5
Architecture	Armv7-M	Armv8-M Main Extension	Armv8.1-M Main Extension	Armv8.1-M Main Extension

[a]Via Wikichip https://en.wikichip.org/wiki/coremark-mhz.
[b]The coremark benchmark for Cortex-M85 has not been publicly announced but it will be the highest performing Cortex-M processor.

Conclusion

The Armv8-M architecture is more evolution than revolution. It provides an easy migration to the next generation of Cortex-M microcontrollers. The introduction of TrustZone into the Cortex-M profile provides an easy-to-use security model that is vitally important as the devices become more connected and the Internet of Things starts to become a reality. The Armv8.1 architecture continues this evolution by adding the Helium Vector extension to provide the compute power required by ML algorithms.

Debugging With CoreSight

Introduction

Many developers who start work with a Cortex-M microcontroller assume that its debug system is a form of "JTAG" interface. In fact, a Cortex-M processor has a debug architecture called "CoreSight," which is considerably more powerful. In addition to the run control and memory access features provided by "JTAG," the "CoreSight" debug system provides a number of real-time trace units that provide you with a detailed debug view of the processor as it runs. In this chapter, we will see what features are available and how to configure them.

Going back to the dawn of modern times, microcontroller development tools were quite primitive. The application code was written in assembler and tested on an EPROM (Fig. 8.1) version of the target microcontroller. Each EPROM had to be erased by UV light before it could be reprogrammed for the next test run.

Figure 8.1
The Electrically Erasable Programmable Read Only Memory was the for runner of today's FLASH memory.

To help debugging the code was "instrumented" by adding additional lines of code to write debug information out of the UART or to toggle an IO pin. Monitor debugger programs were developed to run on the target microcontroller and control execution of the application code. While monitor debuggers were a big step forward, they consumed resources on the

The Designer's Guide to the Cortex-M Processor Family.
DOI: https://doi.org/10.1016/B978-0-323-85494-8.00013-9
© 2023 Elsevier Ltd. All rights reserved.

microcontroller and any bug that was likely to crash the application code would also crash the monitor program just at the point you needed it.

If you had the money an alternative was to use an In-Circuit Emulator (Fig. 8.2). This was a sophisticated piece of hardware that replaced the target microcontroller and allowed full control of the program execution without any intrusion on the CPU runtime or resources. In the late 1990s, the more advanced microcontrollers began to feature various forms of on-chip debug units. One of the most popular on-chip debug units was specified by the "Joint Test Action Group" and is known by the initials JTAG. The JTAG debug interface provides a basic debug connection between the microcontroller CPU and the PC debugger via a low-cost debug adapter (Fig. 8.3).

Figure 8.2
An In Circuit Emulator provides nonintrusive real time debug for older embedded microcontrollers.

Figure 8.3
Today a low-cost debug adapter allows you to control devices fitted with on-chip debug hardware.

JTAG allows you to start and stop the CPU running. It also allows you to read and write to memory locations and insert instructions into the CPU. This allows the debugger designer to halt the CPU, save the state of the processor, run a series of debug commands and then restore the state of the CPU and restart execution of the application program. While this process is transparent to the user, it means that the PC debugger program has run control of the CPU (reset, run, halt, and set a breakpoint) and memory access (read/write to user memory and peripherals). The key advantage of JTAG is that it provides a core set of debug features with the reliability of an emulator at a much lower cost. The disadvantage of JTAG is that you have to add a hardware socket to the development board, and the JTAG interface uses some of the microcontroller pins. Typically, the JTAG interface requires 5 GPIO pins which may also be multiplexed with other peripherals. More importantly, the JTAG interface needs to halt the CPU before any debug information can be provided to the PC debugger. This run/stop style of debugging becomes very limited when you are dealing with a real-time system such as a communication protocol or motor control. While the JTAG interface was used on ARM7/9-based microcontrollers, a new debug architecture called CoreSight was introduced by ARM for all the Cortex M/R- and A-based processors.

CoreSight Hardware

When you first look at the datasheet of a Cortex-M-based microcontroller, it is easy to miss the debug features available or assume it has a form of JTAG interface. However, the CoreSight architecture provides a very powerful set of debugging features that go way beyond what can be offered by JTAG. First of all, on the practical side, a basic CoreSight debug connection only requires two pins, Serial IO and serial Clock. An optional third debug pin, Serial Wire Out (SWO), provides real-time streaming data from the internal data-trace units.

The JTAG hardware socket is a 20-pin berg connector (Fig. 8.4) that often has a bigger footprint on the PCB than the microcontroller that is being debugged. CoreSight specifies two connectors a 10-pin connector for the standard debug features and a 20-pin connector for the standard debug features and instruction trace. The recommended connectors are shown in Table 8.1. We will talk about trace options later, but if your microcontroller supports instruction trace, then it is recommended to fit the larger 20-pin socket, so you have access to the trace unit even if you don't initially intend to use it. A complex bug can be sorted out in hours with a trace tool, whereas with basic run/stop debugging, it could take weeks.

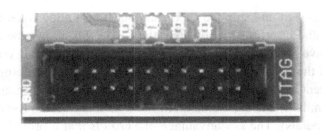

Figure 8.4
The coresight debug architecture replaces the JTAG berg connector with two styles of subminiature connector.

Table 8.1: CoreSight debug sockets

Socket	Samtec	Don Connex
10 pin standard debug	FTSH-105−01-L-DV-K	C42−10-B-G-1
20 pin Standard + ETM Trace	FTSH-110−01-L-DV-K	C42−20-B-G-1

This standard debug system uses the 10-pin socket (Fig. 8.5). This requires a minimum of two microcontroller pins, serial wire IO (SWIO) and serial wire clock (SWCLK), plus the target Vcc, Ground, and reset. As we will see below the Cortex-M3 and Cortex-M4 are fitted with two trace units that require the extra SWO pin. Some Cortex-M3 and M4 devices are fitted with an additional instruction trace unit. This is supported by the 20-pin connector and uses an additional four-processor pin for the instruction trace pipe (Fig. 8.5).

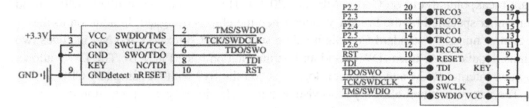

Figure 8.5
The two debug connectors require a minimum number of processor pins for hardware debug.

Debugger Hardware

Once you have your board fitted with a suitable CoreSight socket, you will need a debug adapter unit that plugs into the socket and is connected to the PC, usually through a USB or Ethernet connection. An increasing number of low-cost development boards also incorporate a USB debug interface based on the CMSIS Debug Access Port (CMSIS-DAP) specification.

CoreSight Debug Architecture

There are several levels of debug support provided over the Cortex-M family. In all cases the debug system is independent of the CPU and does not use processor resources or runtime.

The minimal debug system available on the Cortex-M3 and Cortex-M4 consists of the serial wire interface connected to a debug control system which consists of a run control unit, breakpoint unit, and a memory access unit (Fig. 8.6). The breakpoint unit supports up to eight

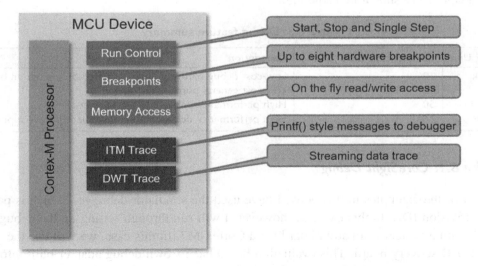

Figure 8.6
In addition to run control the Cortex-M3 and M4 basic debug system includes two trace units. A data trace and an instrumentation trace.

hardware breakpoints (Four on Cortex-M0 and M0 +). The total number available will depend on the number specified by the Silicon Vendor when the chip is designed. In addition to the debug control units, the standard CoreSight debug support for Cortex-M3 and M4 includes two trace units, a data watch trace (DWT) and an Instrumentation trace(Fig. 8.6). The DWT allows you to view internal RAM and peripheral locations "on the fly" without using any CPU cycles. The data watch unit allows you to visualize the behavior of your application data.

Debug Adapters

In order to connect to the Coresight debug hardware within the Cortex-M processor, you will need an external debug adapter. This is a small unit that connects to the Coresight debug port and, in turn, connects to the development PC, normally using a USB connection. The debug adapter software interfaces to the Coresight debug hardware using either a UART method or a much faster Manchester encoded streaming method (Table 8.2).

Table 8.2: Debug feature summary

Trace Port Interface	Maximum Bit Rate (Mb/s)
Serial Wire USART	10
Serial Wire Manchester encoding	100

While this isn't too important in the early stages of a project, as you make increased use of the more advanced on-chip streaming trace units, the additional bandwidth allows you to capture much more detailed real-time debug information. The connection rates of common debug adapters are shown in Table 8.3.

Table 8.3: Debug feature summary

Debug Unit	Coresight Bit Rate (Mb/s)	Description
ST Link	10	Low-cost debug hardware often included on evaluation boards
J Link	10	Low-cost general purpose debug unit
Ulink Plus	50	High performance debug unit with power analysis
Ulink Pro	100	High performance debug unit with instruction trace support

Exercise 8.1: CoreSight Debug

For most of the examples in this book, I have used the simulator debugger, which is part of the Microvision IDE. In this exercise, however, I will run through setting up the debugger to work with a typical evaluation board for a Cortex-M7. In this case, we will use the STM32F7 Discovery board. This evaluation board has its own debug adapter built into the board, so we just need a USB cable rather than a stand-alone debug adapter. While there are a plethora of evaluation boards available for different Cortex devices, the configuration of the hardware debug interface is essentially the same for all boards.

Hardware Configuration

If you are developing your own hardware and software, an external debug adapter is connected to the development board through the 10-pin CoreSight debug socket. The debug adapter is in turn connected to the PC via a USB cable (Fig. 8.7).

Figure 8.7
A typical evaluation board provides a JTAG/CoreSight connector. The evaluation board must also have its own power supply. Often this is provided via a USB socket.

It is important to note here that when you migrate to a custom board it has its own power supply. The debug adapter will sink some current into the development board, enough to switch on LEDs on the board and give the appearance that the hardware is working. However, not enough current is provided to allow the board to run reliably, always ensure that the development board is powered by its usual power supply.

In this example, we will configure the hardware debugger built into the STM32F7 Discovery board (Fig. 8.8). This board is connected to the development PC by using a USB

Fig 8.8
STM32F7xx discovery board with built in ST Link debug hardware.

cable. On this board, the USB cable provides power to the discovery board and also connects to the built-in debug interface.

Software Configuration

Open the Pack Installer.

Select the Boards::Designers Guide Tutorial.

Select the example tab and Copy "Ex 8.1 Hardware Debug."

This is a version of the Blinky example for the ST Discovery board

Now open the options for Target/Debug tab (Fig. 8.9).

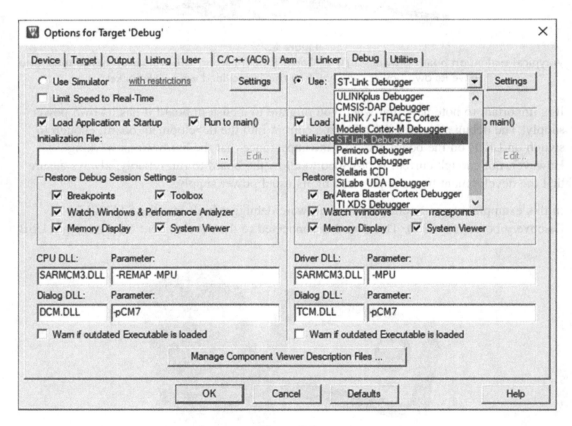

Figure 8.9
The debug tab allows you to configure the debugger for hardware debug and select the debug adapter.

It is possible to switch from using the simulator to the CoreSight Debugger by clicking on the "Use" radio button on the right-hand side of the menu. In the drop-down box, make sure ST-Link Debugger is selected.

Now press the settings button (Fig. 8.10).

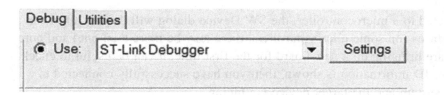

Debug | Utilities |

◉ Use: | ST-Link Debugger ▼ | Settings

Figure 8.10
The driver setup menu will initially show you that the debug adapter is connected and working. Successfully reading the coreID proves the microcontroller is running.

This will open the debug adapter configuration window. The two main pains in this dialog show the connection state of the debug adapter to the PC and to the microcontroller. The debug adapter USB pain shows the debug adapter serial number and firmware version. It also allows you to select which style of debug interface to use when connecting to the microcontroller. For Cortex-M microcontrollers, you should normally always use Serial Wire (SW), but JTAG is also available should you need it.

When you are connected to a microcontroller, the SW Device dialog will display all of the available debug interfaces (on some microcontrollers, there may be more than one) and core IDcodes. When you are bringing up a new board for the first time, it can be useful to check this screen. If the core ID information is shown, then you have successfully connected to the CoreSight debug system and the target device is running.

The debug dialog (Fig. 8.11) in the setting dialog allows you to control how the debugger connects to the hardware. The connect option defines if a reset is applied to the microcontroller when the debugger connects, you also have to option to hold the microcontroller in reset. Remember that when you program code into the flash and reset the processor, the code will run for many cycles before you connect the debugger. If the application code does something to disturb the debug connection, then the debugger may fail to connect. Being able to reset and halt the code can be a way around such problems. The reset method may also be controlled. This can be a hardware reset applied to the whole microcontroller or a software reset caused by writing to the "SYSRESET" or "VECTRESET" registers in the NVIC. In the case of the "SYSRESET" option, it is possible to do a "warm" reset that is resetting the Cortex-M CPU without resetting the microcontroller peripherals. The cache options affect when the physical memory is read and displayed. If the code is cached, then the debugger will not read the physical memory but hold a cached version of the program image in the PC memory. If you are writing self-modifying code, you should uncheck this option.

Figure 8.11
The debug section allows you to define the connection and reset method used by the debugger.

Now click on the Trace tab (Fig. 8.12).

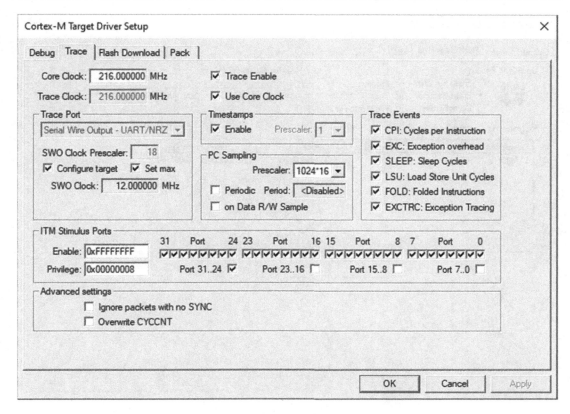

Figure 8.12

The trace tab is used to configure the Instruction, data, and Instrumentation trace units and also enable the performance counters.

This dialog allows us to configure the internal trace units of the Cortex-M device. To ensure accurate timing, the core clock frequency must be set to the Cortex processor CPU clock frequency. The trace port options are configured for the serial wire interface using the UART communication protocol. In this menu, we can also enable and configure the DWT and enable the various trace event counters. This dialog also configures the instrumentation trace (ITM), and we will have a look at this later in this chapter.

Now click the flash download tab (Fig. 8.13).

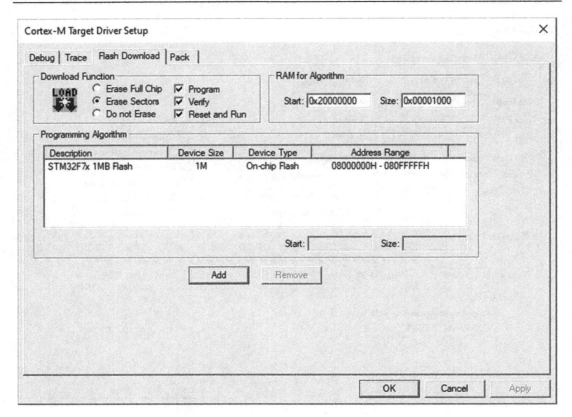

Figure 8.13
The programming algorithm for the internal microcontroller FLASH memory.

This dialog allows you to set the flash download algorithm for the microcontroller flash memory. This will normally be configured automatically when the project is defined. This menu allows you to update or add algorithms to support additional external parallel or serial flash memory.

Now click the Packs Tab (Fig. 8.14).

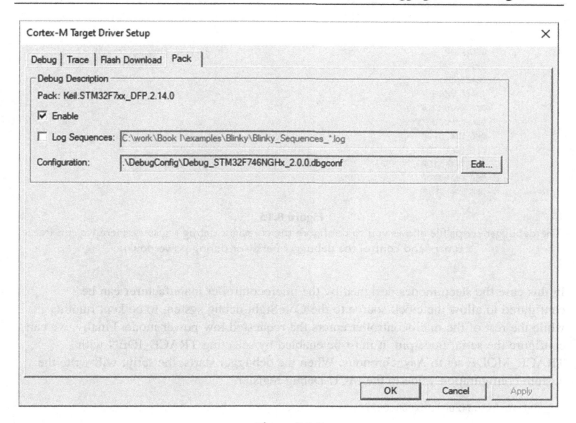

Figure 8.14
The CMSIS packs tab is used to control debugger startup scripts for custom support.

The Packs tab is used to define a script file (debug_<microcontroller>.dbgconf) that is executed when the debugger is started. This allows any custom debug support to be added to the debugger by the pack system. In the case of the STM32F7 the manufacturer has implemented options that allow some of the user peripherals to be halted when the Cortex-M CPU is halted by the debug system. The user timers, watchdog's and CAN module may be frozen when the CPU is halted (Fig. 8.15). The script also allows us to enable debug support when the cortex CPU is placed into low-power modes.

```
⊟ Debug MCU Configuration
    ─ DBG_SLEEP                              ☑
    ─ DBG_STOP                               ☑
    ─ DBG_STANDBY                            ☑
    ─ TRACE_IOEN                             ☑
    ─ TRACE_MODE                             Asynchronous
    ─ DBG_IWDG_STOP                          ☐
    ─ DBG_WWDG_STOP                          ☐
    ─ DBG_TIM1_STOP                          ☐
    ─ DBG_TIM2_STOP                          ☐
    ─ DBG_TIM3_STOP                          ☐
    ─ DBG_TIM4_STOP                          ☐
    └ DBG_CAN_STOP                           ☐
```

Figure 8.15
The debugger script file allows you to configure the coresight debug registers. Here we can freeze timers and control the debugger behavior during powerdown.

In this case the sleep modes designed by the microcontroller manufacturer can be configured to allow the clock source to the CoreSight debug system to be kept running while the rest of the microcontroller enters the requested low-power mode. Finally, we can configure the serial trace pin. It must be enabled by selecting TRACE_IOEN with TRACE_MODE set to Asynchronous. When the debugger starts, the script will write the custom configuration value to the MCU Debug register.

```
_WDWORD(0xE0042004, 0x00000027);
```

Once configured, you can start the debugger as normal, and it will be connected to the hardware in place of the simulation model. This allows you to exercise the code on the real hardware through the same debugger interface we have been using for the simulator. With the DWT enabled, you can see the current state of your variables in the watch and memory windows without having to halt the application code. It is also possible to add global variables to the system analyzer to record and graphically display the values of a variable over time. This can be incredibly useful when working with real-time data.

The DWT windows give some high-level information about the run time performance of the application code. The DWT windows provide a raw high-level trace exception and data access (Fig. 8.16). An exception trace is also available, which provides detailed information of exception and interrupt behavior (Fig. 8.17).

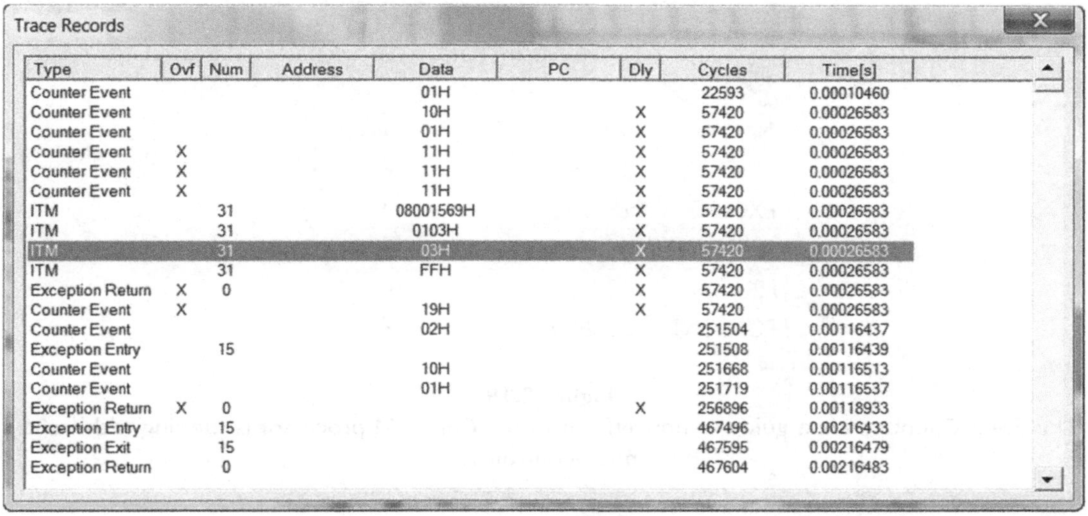

Type	Ovf	Num	Address	Data	PC	Dly	Cycles	Time[s]
Counter Event				01H			22593	0.00010460
Counter Event				10H		X	57420	0.00026583
Counter Event				01H		X	57420	0.00026583
Counter Event	X			11H		X	57420	0.00026583
Counter Event	X			11H		X	57420	0.00026583
Counter Event	X			11H		X	57420	0.00026583
ITM		31		08001569H		X	57420	0.00026583
ITM		31		0103H		X	57420	0.00026583
ITM		31		03H		X	57420	0.00026583
ITM		31		FFH		X	57420	0.00026583
Exception Return	X	0				X	57420	0.00026583
Counter Event	X			19H		X	57420	0.00026583
Counter Event				02H			251504	0.00116437
Exception Entry		15					251508	0.00116439
Counter Event				10H			251668	0.00116513
Counter Event				01H			251719	0.00116537
Exception Return	X	0				X	256896	0.00118933
Exception Entry		15					467496	0.00216433
Exception Exit		15					467595	0.00216479
Exception Return		0					467604	0.00216483

Figure 8.16
A log of execution events can be viewed in the trace records window.

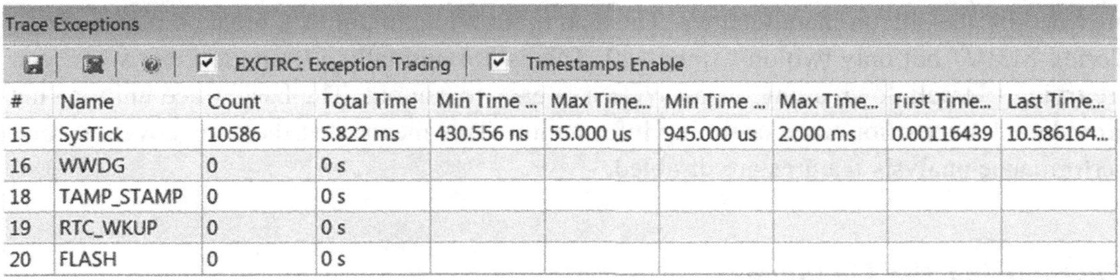

Trace Exceptions

☑ EXCTRC: Exception Tracing   ☑ Timestamps Enable

#	Name	Count	Total Time	Min Time ...	Max Time...	Min Time ...	Max Time...	First Time...	Last Time...
15	SysTick	10586	5.822 ms	430.556 ns	55.000 us	945.000 us	2.000 ms	0.00116439	10.586164...
16	WWDG	0	0 s						
18	TAMP_STAMP	0	0 s						
19	RTC_WKUP	0	0 s						
20	FLASH	0	0 s						

Figure 8.17
The Exception tracing and Event counters provide useful code execution metrics.

The CoreSight debug architecture also contains a number of counters that show the performance of the Cortex-M processor (Fig. 8.18). The extra cycles per instruction count is the number of wait states that the processor has encountered waiting for the instructions to be delivered from the FLASH memory. This is a good indication of how efficiently the processor is running. These are interesting numbers if you are comparing different microcontrollers but don't have much practical use in day-to-day debugging.

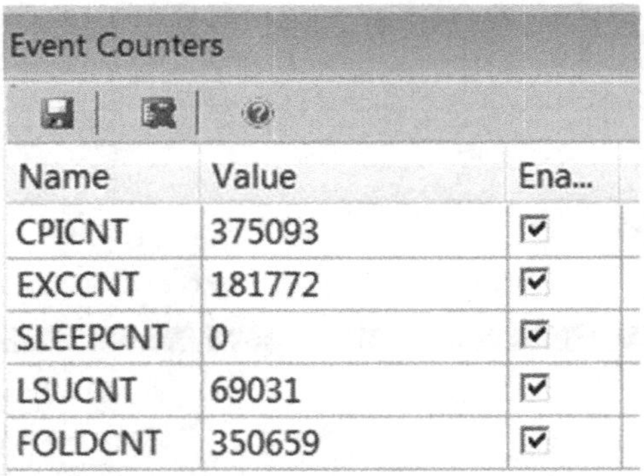

Event Counters		
Name	Value	Ena...
CPICNT	375093	☑
EXCCNT	181772	☑
SLEEPCNT	0	☑
LSUCNT	69031	☑
FOLDCNT	350659	☑

Figure 8.18
The Event Counters are a guide to how efficiently the Cortex-M processor is running within the microcontroller.

Debug Limitations

When the PC debugger is connected to the real hardware, there are some limitations compared to the simulator. First, you are limited to the number of hardware breakpoints provided by the silicon manufacturer. This can up to a maximum of 8 breakpoints on Cortex-M3/4/7 but only two on Cortex-M0. This is not normally a limitation, but when you are trying to track down a bug or test code it is easy to run out. The basic trace units do not provide any instruction trace or timing information. This means that the code coverage and performance analysis features are disabled.

Customizing the Debugger

Instrumentation Trace

In addition to the DWT, the basic debug structure on Cortex-M3, and Cortex-M4 includes a second trace unit called the Instrumentation Trace (ITM). You can think of the Instrumentation Trace as serial port which is connected to the debugger. You can then add code to your application that writes custom debug messages to the ITM which are then displayed within the debugger console window. By instrumenting your code this way, you can send complex debug information to the debugger. This can be used to help locate obscure bugs, but it is also especially useful for software testing.

The Instrumentation Trace (ITM) is slightly different from the other two trace units in that it is intrusive on the CPU in that it does use a few CPU cycles. To use it you need to instrument your code by adding simple send and receive hooks. These hooks are part of the CMSIS standard and are automatically defined in the standard microcontroller header file. The hooks consist of three functions and one variable.

```
static __INLINE uint32_t ITM_SendChar (uint32_t ch);    //Send a character to the ITM
volatile int ITM_RxBuffer = ITM_RXBUFFER_EMPTY;         //Receive buffer for the ITM
static __INLINE int ITM_CheckChar (void);               //Check to see if a character has
                                                        been sent //from the debugger
static __INLINE int ITM_ReceiveChar (void);             //Read a character from the ITM
```

The ITM is actually a bit more complicated in that it has 32 separate channels. Currently, channel 31 is used by the RTOS kernel to send messages to the debugger for the kernel aware debug windows. Channel 0 is the user channel that can be used by your application code to send printf() style messages to a console window within the debugger.

Exercise 8.2: Setting up the ITM

In this exercise we will look at configuring the Instrumentation trace to send and receive messages between the microcontroller and the PC debugger.

Open the Pack Installer.

Select the Boards::Designers Guide Tutorial.

Select the example tab and Copy "Ex 8.1 Hardware Debug."

We will configure the Blinky project so that it is possible to send messages through the ITM from the application code. First, we need to configure the debugger to enable the application ITM channel.

Open the "Options for target\debug" dialog.

Press the ST-Link settings button and select the trace tab (Fig. 8.19).

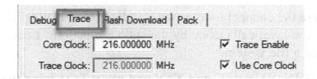

Figure 8.19
The core clock frequency must be set correctly for the ITM to work successfully.

To enable the ITM, the core clock must be set to the correct frequency as discussed in the last exercise, and the trace must be enabled.

In the ITM stimulus ports menu port 31 will be enabled by default (Fig. 8.20). To enable the application port we must enable port 0. We want to be able to use the Application ITM form both privileged and unprivileged mode. This can be done by ensuring that the Privileged port 8.0.0 box is unchecked.

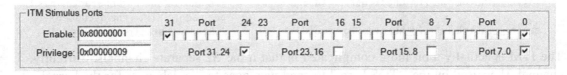

Figure 8.20
The ITM has 32 channels. Currently, channel 31 and 0 are used for the RTOS event view and the ITM user debug channel.

Once the trace and ITM settings are configured click OK and return back to the editor.

Open the Run Time Environment (Green diamond icon).

Select the Compiler::IO section (Fig. 8.21).

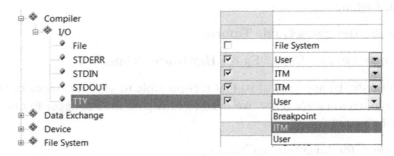

Figure 8.21
The Compiler::I/O options allow you to select the low-level STDIO channel.

Here, we can configure the channel used as the standard IO (STDIO). The File option provides support for low-level calls made by the stdio.h functions. Leave this unchecked unless you have added a file system.

Enable STDERR, STDIN, STDOUT, and TTY and select ITM from the dropdown menu.

This adds a "retarget_io.c" file to the project (Fig. 8.22). This file contains the low-level STDIO driver functions which read and write a single character to the ITM.

Figure 8.22
Updated project with the ITM support for STDIO in retarget_IO.c.

Open Blinky.c.

Add stdio.h to the list of include files.

```
#include "RTE_Components.h"
#include <stdio.h>
```

Add a printf statement before the kernelStart() API call.

```
printf("RTX Started\n");
osKernelStart();
while (1) {
```

Start the debugger.

Open the view\serial windows\Debug(printf) viewer (Fig. 8.23).

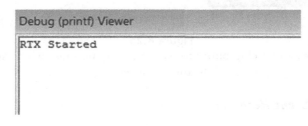

Figure 8.23
Debug message sent from the application code to the debugger console window.

Run the code.

The printf() message is sent to the STDIO channel which is the ITM. This message is read and displayed in the debugger console window.

We will see how to use the ITM to send runtime diagnostic messages in this chapter, and also how to use it as part of a software testing scheme in Chapter 13, Test-Driven Development.

Event Recorder

The ITM provides an ideal debug channel for diagnostic messages. However, it is not available on the cortex M0 and M0 + . The Microvision debugger provides an additional debug serial channel called the Event Recorder. This channel works on all Cortex-M processors at the expense of using some additional on-chip RAM as a buffer. When the event recorder is configured you must specify a small region of on-chip SRAM. This region is used as a circular buffer and a set of macros is used to write diagnostic messages into this buffer. The Cortex M debugger can then read the buffer on the fly and display the messages in the console window of the debugger (Fig. 8.24).

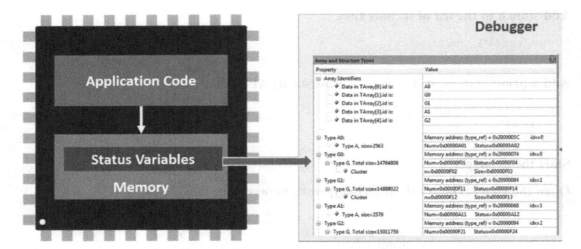

Figure 8.24

The Event recorder uses an on-chip buffer to record debug messages which are in turn uploaded to the debugger.

Exercise 8.3: Basic Event Recorder

In this exercise, we will configure the event recorder to send user debug messages to a dedicated window in the Microvision debugger. The event recorder can also be used as a channel for STDOUT in place of the ITM.

In the pack installer select Exercise 8.3 and press the copy button.

Open the RTE and check that the compiler:: Event Recorder option is selected (Fig. 8.25).

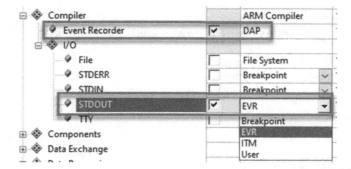

Figure 8.25
The event recorder is enabled in the RTE. It can also be used as the STDOUT channel to send printf( ) messages to the debugger console.

The stdout channel must be set to EVR to select the event recorder.

Click on the OK button to close the RTE.

In order to work correctly, the event recorder buffer must be located in uninitialized RAM. Typically, this will mean dedicating a small region of user RAM for use by the event recorder.

Open the options for target dialog and select the target menu (Fig. 8.26).

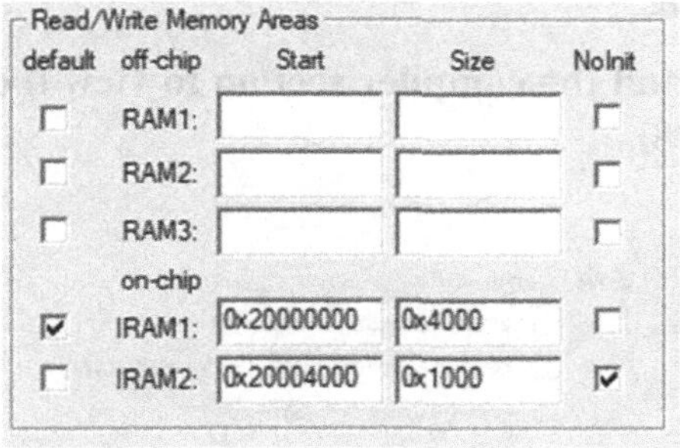

Figure 8.26
The Event recorder buffer requires an uninitialized RAM region.

Here, we must divide the internal RAM into two regions IRAM1 is used by the application by default. IRAM2 is uninitialized and will be dedicated for use by the event recorder.

Select the compiler branch and open its local options (Fig. 8.27).

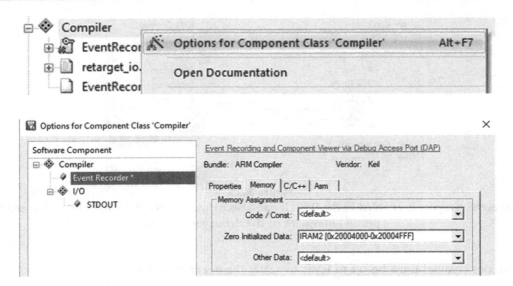

Figure 8.27
The event recorder buffer can be located in uninitialized RAM using its local memory options.

Highlight the Event recorder component and select its memory dialog.

Select IRAM2 for the "Zero Initialized Data."

Click OK to close the dialog.

In the project window expand the compiler section to view the event recorder configuration files (Fig. 8.28).

Figure 8.28
The event recorder and retarget support files.

We will view these in more detail in the next exercise. The event recorder code is contained in a single file with a configuration header file. The retarget.c file contains different implementations of putchar() and getchar() to support different IO channels.

Open main.c.

Support for the event recorder functions is added with a header file.

```
#include "EventRecorder.h"
```

At the beginning of the main() function we must start the event recorder agent with the following line of code.

```
EventRecorderInitialize(EventRecordAll, 1);
```

Now when the printf() function is called its output will be written the event recorder buffer. The debugger will then upload the message in the buffer and display it in its console window.

Build and run the project in the debugger.

Observe the messages in the debugger console window (view\serial\printf viewer).

Event Recorder Annotations

The event recorder is much more than a replacement for the ITM. In addition to replacing the printf() diagnostic channel, the event recorder provides a set of annotation macros that can send formatted messages to the debugger. These annotation messages are in turn, supported by a set of dedicated windows in the debugger.

The macros are defined as part of the event recorder header file that we added in the last example.

The annotations define two types of messages. The first is used to upload data values to an event recorder window in the debugger, while the second is used to provide execution timing information.

When added to a project, the event viewer provides a set of macros that are used to write formatted messages into the event viewer buffer (Table 8.4).

Table 8.4: Event recorder message macros

Function	Description
EventRecord2()	Record an event with two 32-bit data values
EventRecord4()	Record an event with four 32-bit data values
EventData()	Record an event with variable data size

In addition to uploading diagnostic messages and data to the debugger the event recorder also provides a set of macros used for statistical timing analysis (Table 8.5). These can be used to measure min, max, and average runtime between points in the application code. Results from the timing annotations can then be viewed in a timing statistics window.

Table 8.5: Event recorder timing macros

Function	Description
EventStartx(slot)	Generate a start event for group x Slot n
EventStartx (slot, v_1, v_2)	Generate a start event for group x Slot n and records integer values v_1 v_2
EventStop(slot)	Generate a stop event for group x Slot n
EventStopx(slot,v_1, v_2)	Generate a stop event for group x Slot n and records integer values v_1 v_2

Where $x = A-D$ and slot $= 0-15$.

The event recorder timing annotations are split into four slots $A-D$ each slot is further subdivided into 16 entries. Each slot + entry has a start and stop function, and this provides stopwatch style timing measurement between different points in your code.

Exercise 8.4: Event Viewer Annotations

In this exercise we will expand the use of the event viewer to use the annotation macros and explore the supporting windows in the debugger.

Exercise 8.4 and press the copy button.

This project has the event viewer installed using the steps given in the last example.

Open Compiler:EventRecorderConf.h (Fig. 8.29).

Figure 8.29
The event recorder configuration file controls the size of the buffer. You must also set the timestamp clock for accurate timing values.

The configuration file is used to define the size of the event recorder buffer, its timestamp source, and the CPU frequency.

The size of the event recorder buffer is defined by the number of records it can hold. The overall buffer size can be calculated as follows

$$\text{Buffer size in bytes} = 164 + (\text{number of Records} \times 16)$$

The event recorder also needs a hardware timer to provide a timestamp for each event recorder record. A range of timers are supported as shown in Table 8.6.

Table 8.6: Event recorder timing sources

Time Stamp	Cortex-M3/M4/M7/M33/M55	Cortex-M0/M0 + /M23
DWT Cycle counter	Y	N
Systick	Y	Y
CMSIS-RTOS2	Y	Y
User Timer	Y Requires user config	Y Requires user config
User Timer	Y Requires user config	Y Requires user config

The DWT cycle counter is part of the debug system and is not available on the smaller Cortex-M processors. It is also not part of the simulator but can be supported with a script file.

The event recorder will configure the selected hardware timer so you must also provide the timestamp clock frequency. Generally, this will be the CPU frequency, however, you may need to account for any clock dividers. This particularly applies to the SysTick input frequency, which may be divided down from the CPU core clock. In the case of a user timer provide all the necessary configuration code for the selected hardware timer.

Open main.c.

The main loop now has event recorder macros added to provide timing analysis and also to upload diagnostic data to the debugger.

```
unsigned int counter = 1;
unsigned int numLED;
EventRecorderInitialize(EventRecordAll, 1);
LED_Initialize();
numLED = LED_GetCount();
while(1)
{
EventRecord2 (1,counter, 0);
EventStartA(0);
LED_SetOut (counter);
delay(counter);
counter++;
if (counter > numLED) counter = 1;
EventStopA(0);
}
```

Build the project and start the debugger.

Start the code running.

Open the event recorder window view\analysis\event recorder (Fig. 8.30).

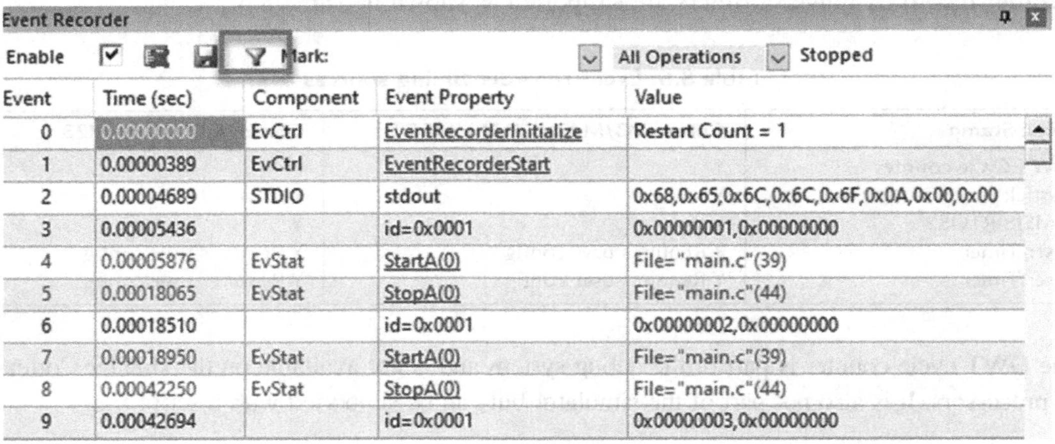

Event Recorder									
Enable	☑	▦	🖫	▽	Mark:		All Operations	Stopped	

Event	Time (sec)	Component	Event Property	Value
0	0.00000000	EvCtrl	EventRecorderInitialize	Restart Count = 1
1	0.00000389	EvCtrl	EventRecorderStart	
2	0.00004689	STDIO	stdout	0x68,0x65,0x6C,0x6C,0x6F,0x0A,0x00,0x00
3	0.00005436		id=0x0001	0x00000001,0x00000000
4	0.00005876	EvStat	StartA(0)	File="main.c"(39)
5	0.00018065	EvStat	StopA(0)	File="main.c"(44)
6	0.00018510		id=0x0001	0x00000002,0x00000000
7	0.00018950	EvStat	StartA(0)	File="main.c"(39)
8	0.00042250	EvStat	StopA(0)	File="main.c"(44)
9	0.00042694		id=0x0001	0x00000003,0x00000000

Figure 8.30
Event recorder messages are displayed in a dedicated trace window.

This provides a raw view of all the event recorder annotations (printf, data, and timing). The filter icon at the top of the screen can be used to manage the annotations displayed in the view.

Open the event statistics window view\analysis\statistics (Fig. 8.31).

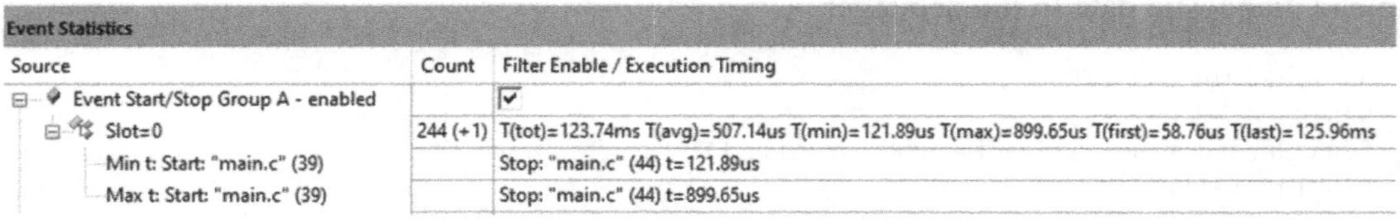

Event Statistics		
Source	Count	Filter Enable / Execution Timing
⊟ ◆ Event Start/Stop Group A - enabled		☑
⊟ ⏱ Slot=0	244 (+1)	T(tot)=123.74ms T(avg)=507.14us T(min)=121.89us T(max)=899.65us T(first)=58.76us T(last)=125.96ms
Min t: Start: "main.c" (39)		Stop: "main.c" (44) t=121.89us
Max t: Start: "main.c" (39)		Stop: "main.c" (44) t=899.65us

Figure 8.31
The event statistics window displays min, max, and average values for event recorder timing macros.

The statistics window displays the max, min, average, and total time for each pair of timing annotations as well as the first and last time executed.

While the event recorder can be used as a replacement for the ITM, it also supports a set of dedicated macros that can be viewed as structured messages in a dedicated debugger window.

When the event recorder messages are received by the debugger the raw binary information can be decoded and displayed as meaningful text strings. This decoding is preconfigured for standard software components such as the RTX RTOS that we will meet in Chapter 10, Using a Real-Time Operating System. It is also possible to extend this decoding by creating a custom XML file that describes the annotations' structure and meaning.

Exercise 8.5: Customizing the Debugger I

In this exercise we will extend the event recorder view to decode our diagnostic messages to meaningful text.

In the pack installer select Exercise 8.5 and press the copy button.

In Microvision open the options for Target\debug window.

Click on the "Manage Component Viewer Debug files."

The default event recorder message file is already loaded but we can add to this by defining our own custom messages.

In the project directory, select <project>\component\customRecorder.xml.

Restart the debugger and run the code.

Examine the output in the event recorder window.

Open the customRecorder.xml file.

This file describes the diagnostic message data structure and decodes the data into meaningful text strings which are then displayed in the event recorder window.

When creating the event recorder XML file, we can structure each message ID to take full advantage of the supporting event recorder window.

The event recorder allows an event ID to be constructed from three fields, an event level, the originating software component ID, and the specific message number. The 32-bit ID is subdivided into these three fields as shown in Table 8.7.

Table 8.7: Event recorder message structure

Macro	Description
EventID(Level,Component Number, Message Number)	((level & 0x30000U) | ((comp_no & 0xFFU) << 8) | (msg_no & 0xFFU))

The event-level field allows us to provide some additional information about the origin of the message. The event level #defines are declared in eventRecorder.h as shown in Table 8.8.

Table 8.8: Event recorder ID fields

Event Level	Description
Error	Runtime errors in the component
API	Info about API function calls.
OP	Relates to an internal component operation
Detail	Provides additional detailed information about internal component events

In the custom XML file we can now create an event lookup table for each message ID that provides extended debug information which is overlaid on the raw event recorder message.

```
<events>
<group>
<component name = "MyComponent" brief = "MyCo" no = "0x0A" prefix = "EvrMyCo_" info = "My
Component - Demo example"/>
</group>
<event id = "0xA0B" level = "Op" property = "SendComplete" value = "size = %d[val1]"
info = "Event on MyComp_send - completed"/>
<event id = "0xA0C" level = "Error" property = "SendFailed" value = "" info = "Event on
MyComp_send - send failed"/>
</events>
```

This file defines a component name for a group of messages. This allows you to relate a specific message back to a region of code and also easily manage filters for the event recorder view. Once the group name is defined, we can create definitions for each message within the component group. Additional text can be displayed alongside any event message data. The event recorder data may also be displayed and formatted to present a meaningful value.

Within our application code the event recorder file provides an EventID() macro, which can be used to manage the construction of the overall event message ID number.

```
#define EvtMyCo_No 0x0A /// < Number of the component with short name 'EvtMyCo_No'
// Event id list for "MyComponent"
#define EvtMyCo_InitEntry EventID (EventLevelAPI, EvtMyCo_No, 0x00)
#define EvtMyCo_InitStatus EventID (EventLevel, EvtMyCo_No, 0x01)
```

We will look at these macros in more detail in Chapter 14, Software Components, when we look at the design of Software components.

Component Viewer

The event recorder xml file can also be extended to create a custom "component viewer" window. These windows are a bit like the peripheral system viewer in that it displays a fixed group of values, but in the component viewer, we can display the current values of variables within a component module. Unlike the system viewer, the component viewer window is linked to the variable symbols rather than absolute addresses, so as you rebuild code, the component window will always align with the current address of each variable.

Exercise 8.6: Customizing the Debugger II

In this exercise, we will look at how to create a custom component viewer window to display data values held within application variables.

In the pack installer select Exercise 8.6 and press the copy button.

Build the project and start the debugger.

Open view watch\ and then each of the component view options (Fig. 8.32).

Add\view\watch

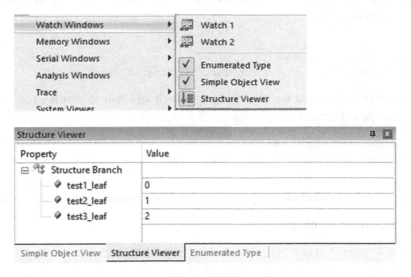

Figure 8.32
Custom component viewer windows.

This is a custom component window that displays the current value of several variables that are within the application code. The component view creates three branches that demonstrate displaying a simple variable, a variable with typedefs, and a formatted C structure.

Run the code and observe the data update in the component viewer window.

Exit the Debugger.

In the project directory open customComponent.xml.

This file defines three custom component viewer objects and is loaded in the debug menu in the same way as the customRecorder file in the last example.

The first lines of the file load the XML schema.

```
<?xml version = "1.0" encoding = "utf-8"?>
<component_viewer schemaVersion = "0.1" xmlns:xs = "http://www.w3.org/2001/XMLSchema-
instance" xs:noNamespaceSchemaLocation = "Component_Viewer.xsd">
```

We can then add a custom component.

```
<component name = "custom Component" version = "0.0.1"/>
```

The component consists of a set of objects which are used to display the variables of interest.

Here, we create the object container and our first simple object that will display the current value of a 32-bit unsigned int. The object definition is declared in two parts: first we create the read name. This defines a local name for the variable, its C type and its true symbol name in the underlying code.

```
<objects>
<object name = "SimpleObject">
<read name = "counter" type = "uint32_t" symbol = "counter" const = "0"/>
```

Once the read name is declared we can create a branch and leaf to display the value. The display value can be formatted using a range of % options.

```
<item property = "Simple Object Branch">
<item property = "Counter Leaf" value = "%d[counter]" info = "uint32_t demo"/>
</item>
```

The component viewer also includes the use of typedefs to support more complex data types.

Here, we can declare an enumerated type.

```
<typedef name = "state" size = "4" info = "State variable">
<member name = "val" type = "uint32_t" offset = "0">
<enum name = "OFF" value = "0" info = "LED OFF" />
<enum name = "ON" value = "1" info = "LED ON" />
</member>
</typedef>
```

And then a new branch to the object. By using the typedef we can display the enumerated name in place of the raw value.

```
<item property = "Enumerated Branch">
<item property = "Stages Leaf" value = "%E[state.val]" info = "LED State"/>
</item>
```

Although creating these support files is additional work over the lifetime of a project they are incredibly useful and it is highly recommended to spend the extra time creating them. It is also extra beneficial if your code is reused in another project or passed on to a third party.

There are also a number of examples and a tutorial available in the following pack directory.

```
C:\Keil\ARM\PACK\Keil\ARM_Compiler\<version>\SCVD_Examples
```

System Control Block Debug Support

The CoreSight debugger interface allows you to control the execution of your application code and examine values in the memory and peripheral registers. Combined with the various trace units this provides you with a powerful debug system for normal program development. However, as we saw in Chapter 3, Cortex-M Architecture, the Cortex-M processors have up to four fault exceptions (Table 8.9) which will be triggered if the application code makes incorrect use of the Cortex-M processor or the microcontroller hardware.

Table 8.9: Fault exceptions

Fault Exception	Priority	Cortex Processor
Hard Fault	-1	Cortex-M0, Cortex-M0 + Cortex-M3 Cortex-M4
Bus Fault	Programmable	Cortex-M3 Cortex-M4
Usage fault	Programmable	Cortex-M3 Cortex-M4
Memory Manager Fault	Programmable	Cortex-M3 Cortex-M4 (Optional)

When this happens, your program will be trapped on the default fault handler in the startup code. If this happens it can be very hard to work out how you got there. If you have an instruction trace tool you can work this out in seconds. If you don't have access to instruction trace then resolving a run time crash can take a long long time. In this section, we will look at configuring the fault exceptions and then looking at how to track back to the source of a fault exception.

The behavior of the fault exceptions can be configured by registers in the system control block. The key registers are listed in Table 8.10.

Table 8.10: Fault exception configuration registers

Register	Processor	Description
Configuration and control	M3, M4	Enable additional fault exception features
System Handler control and state	M3, M4	Enable and pending bits for fault exceptions
Configurable fault status register	M3, M4	Detailed fault status bits
Hard Fault Status	M3, M4	Reports a hard fault or fault escalation
Memory manager fault address	M3, M4	Address of location that caused the Mem manager fault
Bus Fault Address	M3, M4	Address of location that caused the bus fault

When the Cortex-M processor comes out of reset, only the Hard Fault Handler is enabled. If a Usage, Bus, or Memory Manager Fault is raised and the exception handler for these faults is not enabled, then the fault will "escalate" to a hard fault. The hard fault status register provides two status bits that indicate the source of the Hard Fault (Table 8.11).

Table 8.11: Hard fault status register

Name	Bit	Use
FORCED	30	Reached the hard fault due to fault escalation
VECTTBL	1	Reached the hard fault due to a faulty read of the vector table

The "System Handler Control and State" Register contains enable, pending and active bits for the Bus, Usage, and Memory Manager exception handlers. We can also configure the behavior of the fault exceptions with the "Configuration and Control" Register as shown in Table 8.12.

Table 8.12: Configuration and control register

Name	Bit	Use
STKALIGN	9	Configures 4- or 8-byte stack alignment
BFHFMIGN	8	Disables data bus faults caused by load and store instructions
DIV_0_TRP	4	Enables usage fault for divide by zero
UNALIGNTRP	3	Enables a usage fault for unaligned memory access

The "divide by zero" exception can be a useful trap to enable, particularly during development. The remaining exceptions should be left disabled unless you have a good reason to switch them on. When a Memory Manager Fault exception occurs, the address of the instruction that attempted to access a prohibited memory region will be stored in the "Memory Fault Address Register," similarly when a Bus Fault is raised, the address of the instruction that caused the fault will be stored in the "Bus Fault Address" Register. However, under some conditions, it is not always possible to write the fault addresses to these registers. The configurable "Fault Status" Register contains an extensive set of flags that report the Cortex-M processor error conditions that help you track down the cause of a Fault exception.

Tracking Faults

If you have arrived at the Hard Fault handler, first check the "Hard Fault Status" Register. This will tell you if you have reached the Hard Fault due to fault escalation or a vector table read error. If there is a fault escalation, next check the "System Handler Control and State" Register to see which other fault exception is active. The next port of call is the "Configurable Fault Status" Register. This has a wide range of flags that report processor error conditions (Table 8.13).

Table 8.13: Configurable fault status register

Configurable Fault Status Register		
Name	Bit	Use
DIVBYZERO	25	Divide by zero error
UNALIGNED	24	Unaligned memory access
NOCP	19	No Coprocessor present
INVPC	18	Invalid PC load
INVSTATE	17	Illegal access the the execution program status register EPSR
UNDEFINSTR	16	Attempted execution of an undefined instruction
BFARVALID	15	Address in bus fault address register is valid
STKERR	12	Bus fault on exception entry stacking
UNSTKERR	11	Bus fault on exception exit unstacking
IMPRECISERR	10	Data bus error. Error address not stacked
PRECISERR	9	Data bus error. Error address stacked
IBUSERR	8	Instruction bus error
MMARVALID	7	Address in the memory manager fault address register is valid
MSTKERR	4	Stacking on exception entry caused a memory manager fault
MUNSTKERR	3	Stacking on exception exit caused a memory manager fault
DACCVIOL	1	Data access violation flag
IACCVIOL	0	Instruction access violation flag

When the processor fault exception is entered a stack frame will normally be pushed onto the stack. In some cases, the Stack frame will not be valid, and this will be indicated by the flags in the "Configurable Fault Status" Register. When a valid stack frame is pushed, it will contain the PC address of the instruction that generated the fault. By decoding the stack frame, you can retrieve this address and locate the problem instruction. The System Control Block provides a memory and bus fault address register which depending on the cause of the error, may hold the address of the instruction that caused the error exception.

Exercise 8.7: Processor Fault Exceptions

In this project we will generate a fault exception and look at how it is handled by the NVIC. Once the fault exception has occurred, we can interrogate the stack to find the instruction which caused the error exception.

Open the Pack Installer.

Select the example tab and Copy "Ex 8.7 Fault Tracking."

```
volatile uint32_t op1;
int main (void)
{
int op2 = 0x1234, op3 = 0;
        SCB->CCR = 0x0000010;    //Enable divide by zero usage fault
        op1 = op2/op3;           //perform a divide by zero to generate an usage exception
while(1);
}
```

The code first enables the "divide by zero usage fault" and then performs a divide by zero to cause the exception.

Build the code and start the debugger.

Set a breakpoint on the line of code that contains the divide instruction (Fig. 8.33).

```
10      SCB->SHCSR = 0x00060000;
11      SCB->CCR = 0x0000010;
●12     op1 = op2/op3;
```

Figure 8.33
Set a breakpoint on the divide statement.

Run the code until it hits this breakpoint.

Open the Peripherals\Core Peripherals\System Control and Configuration window and check that the Divide by Zero trap has been enabled (Fig. 8.34).

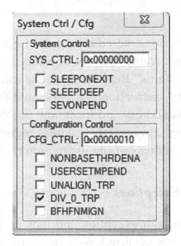

Figure 8.34
Check the Divide by Zero trap is enabled by using the "System Configuration and Control" peripheral view.

Single step the divide instruction.

A Usage Fault exception will be raised. We have not enabled the Usage Fault exception vector so the fault will elevate to a Hard Fault (Fig. 8.35).

```
127                    ENDP
128   HardFault_Handler\
129                    PROC
130                    EXPORT  HardFault_Handler          [WEAK]
⇒ 131         |        B       .
132                    ENDP
```

Figure 8.35
An Error exception will cause you to hit the Hard Fault handler if the other Fault Exceptions have not been enabled.

Open the Peripherals\Core Peripherals\Fault Reports window (Fig. 8.36).

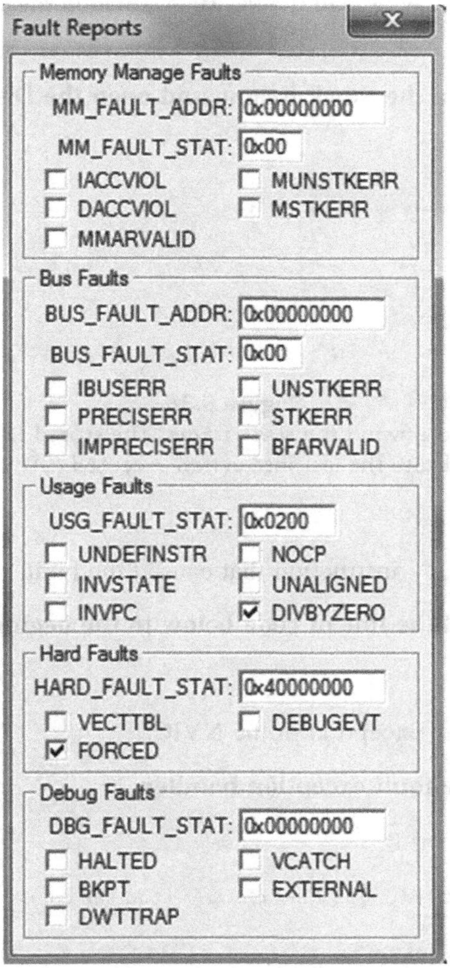

Figure 8.36
If a fault occurs you can view a digest of the fault diagnostic registers in the peripheral fault reports window.

This window shows that the Hard Fault has been forced by another fault exception. Also, the divide by zero flag has been set in the Usage Fault Status Register.

In the register window read the contents of R13 the Main Stack Pointer and open a Memory window at this location (Fig. 8.37).

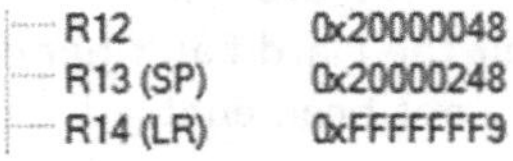

```
R12        0x20000048
R13 (SP)   0x20000248
R14 (LR)   0xFFFFFFF9
```

Figure 8.37
Use the register window the read the current address stored in the stack pointer. Then use the memory window to read the PC value stored in the stack frame.

Read the PC value saved in the Stack Frame and open the Disassembly Window at this location (Fig. 8.38).

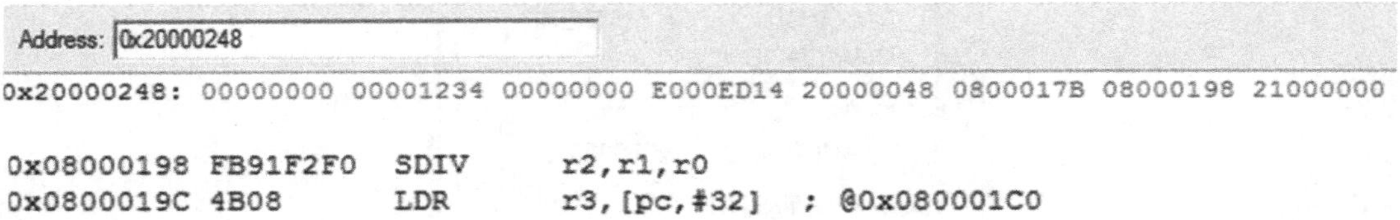

```
Address: 0x20000248

0x20000248:  00000000 00001234 00000000 E000ED14 20000048 0800017B 08000198 21000000


0x08000198 FB91F2F0  SDIV    r2,r1,r0
0x0800019C 4B08      LDR     r3,[pc,#32]   ; @0x080001C0
```

Figure 8.38
Use the memory window to unwind the stack frame. The stored PC (0x08000198) value will indicate the last instruction executed (SDIV).

This takes us back to the SDIV instruction that caused the fault.

Exit the debugger and add the line of code below to the beginning of the program.

```
SCB->SHCSR = 0x00060000;
```

This enables the Usage Fault Exception in the NVIC.

Now add a "C" level usage fault exception handler.

```
void UsageFault_Handler (void)
{
error_address = (uint32_t *)(__get_MSP());   // load the current base address of the stack
pointer
error_address = error_address + 6;            // Locate the PC value in the last stack frame
while(1);
}
```

Build the project and start the debugger.

Set a breakpoint on the while loop in the exception function (Fig. 8.39).

```
19    void UsageFault_Handler (void)
20  ⊟{
21       error_address = (uint32_t *)(__get_MSP());
22       error_address = error_address + 6;
23       while(1);
24  }
```

Figure 8.39
The Usage Fault Exception routine can be used to read the PC value from the stack.

Run the code until the exception is raised and the breakpoint is reached.

When a Usage Fault occurs this exception routine will be triggered. It reads the value stored in the Stack Pointer and extracts the value of the PC stored in the Stack Frame.

Power Analysis

Some of the latest debug adapters include support for power analysis. This feature allows you to visualize and optimize the power consumption of your target hardware as the application code executes.

The Ulink Plus (Fig. 8.40) provides an additional interface that may be connected to a power domain on the target hardware via a shunt resistor. As the code executes, the target current and voltage values are logged and can then be displayed as real-time graphical data in the system analyzer window (Fig. 8.41) while a separate window provides power statistics details.

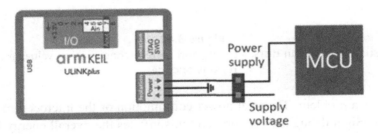

Figure 8.40
The Ulink Plus includes additional support for power analysis during a debug session.

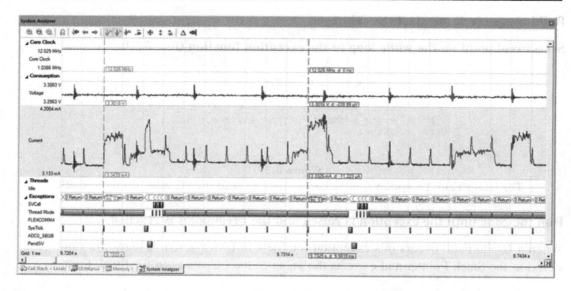

Figure 8.41
The system analyzer window can display voltage and current values against code execution (RTOS Threads and interrupts).

The power analysis features can also be combined with the event Statistics to provide a detailed analysis of different regions of code (Fig. 8.42).

Event Statistics			
Source	Count	Execution Timing	
Event Start/Stop Group A			
Slot=0 (Errors=29)	723	T(tot)=2.08s Q(tot)=108.23mAs T(avg)=2.88ms T(min)=1.71ms T(max)=9.10ms T(first)=2.43ms T(last)=3.59s	
Min t: Start: v1=501 v2=0		Stop: "main.c" (49) t=1.71ms Q=92.67µAs I=54.09mA U=3.30V	
Max t: Start: v1=792 v2=0		Stop: "main.c" (45) t=9.10ms Q=471.56µAs I=51.84mA U=3.31V	
Min Q: Start: v1=501 v2=0		Stop: "main.c" (49) t=1.71ms Q=88.85µAs I=51.85mA U=3.31V	
Max Q: Start: v1=792 v2=0		Stop: "main.c" (45) t=9.10ms Q=471.56µAs I=51.84mA U=3.31V	

Figure 8.42
The event statistics window can display timing and current consumption values for user-selected regions of code.

However, there is a problem. The total power consumption of the microcontroller also includes the CoreSight debug architecture, and this inflates the overall energy figures. Within Microvision it is possible to use the power analysis features independently of an active debug session. This switches off the debug units and provides an accurate power analysis (Fig. 8.43).

Figure 8.43
Power analysis is available with and without an active debug session.

Needless to say, if you are developing a low-power or battery-based application, this level of detailed information is invaluable.

Instruction Trace With the Embedded Trace Macro Cell

The Cortex-M3, Cortex-M4, and Cortex-M7 may have an optional debug module that may be fitted by the silicon vendor when the microcontroller is designed. The Embedded Trace Macro Cell (ETM) is a third trace unit that provides an Instruction Trace as the application code is executed on the Cortex-M processor (Fig. 8.44). Because the ETM is an additional cost to the silicon vendor it is normally only fitted to higher end microcontrollers. When selecting a device, it will be listed as a feature of the microcontroller in the datasheet.

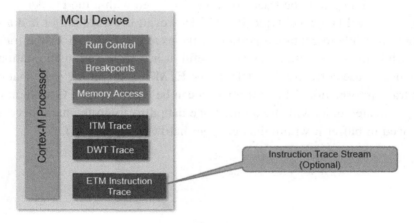

Figure 8.44
The Cortex-M3 and Cortex-M4 may optionally be fitted with a third trace unit. The Embedded trace macrocell (ETM) supports Instruction trace. This allows you to quickly find complex bugs. The ETM also enables code coverage and performance analysis tools which are essential for software validation.

The ETM trace pipe requires four additional pins which are brought out to a larger 20-pin socket that incorporates the serial wire debug pins and the ETM trace pins. Standard JTAG/ CoreSight debug tools do not support the ETM Trace channel, so you will need a more sophisticated debug unit (Fig. 8.45).

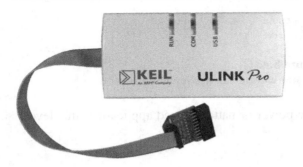
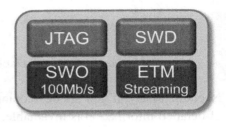

Figure 8.45

An ETM trace unit provides all the features of the standard hardware debugger plus instruction trace.

At the beginning of this chapter, we looked at various debug methods that have been used historically with small microcontrollers. For a long time, the only solution that would provide any kind of instruction trace was an In Circuit Emulator. The Emulator hardware would capture each instruction executed by the microcontroller and store it in an internal trace buffer. When requested, the trace could be displayed within the PC debugger as assembly or high-level language, typically "C". However, the trace buffer had a finite size, and it was only possible to capture a portion of the executed code before the trace buffer was full. So while the trace buffer was very useful, it had some serious limitations and took some experience to use correctly. In contrast, the ETM is a streaming trace that outputs compressed trace information. This information can be captured by a CoreSight trace tool, the more sophisticated units will stream the trace data directly to the hard drive of the PC without the need to buffer it within the debugger hardware (Fig. 8.46).

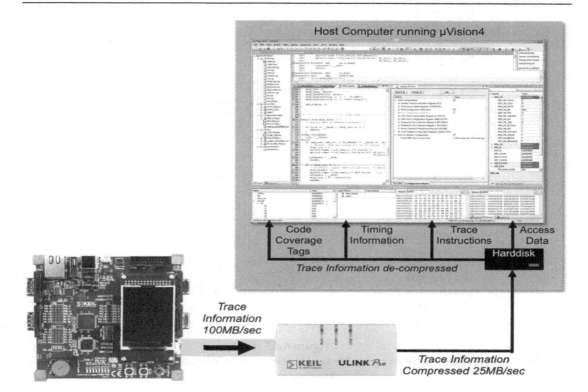

Figure 8.46
A "Streaming" trace unit is capable of recording every instruction directly onto the hard drive of the PC. The size of the trace buffer is only limited by the size of your hard disk.

This streaming trace allows the debugger software to display 100% of the instructions executed along with execution times. This also means that the trace buffer is only limited by the size of the PC hard disk. It is also possible to analyze the trace information to provide accurate code coverage and performance analysis information.

Exercise 8.8: Using the ETM Trace

In this exercise, we will look at how to configure the Ulink Pro debug adapter for streaming trace and then review the additional debug features available through the ETM.

In the Pack Installer select the example tab and Copy "Ex 8.6 Instruction Trace."

Open options for target\debug\settings\pack tab and click the Configuration edit button (Fig. 8.47).

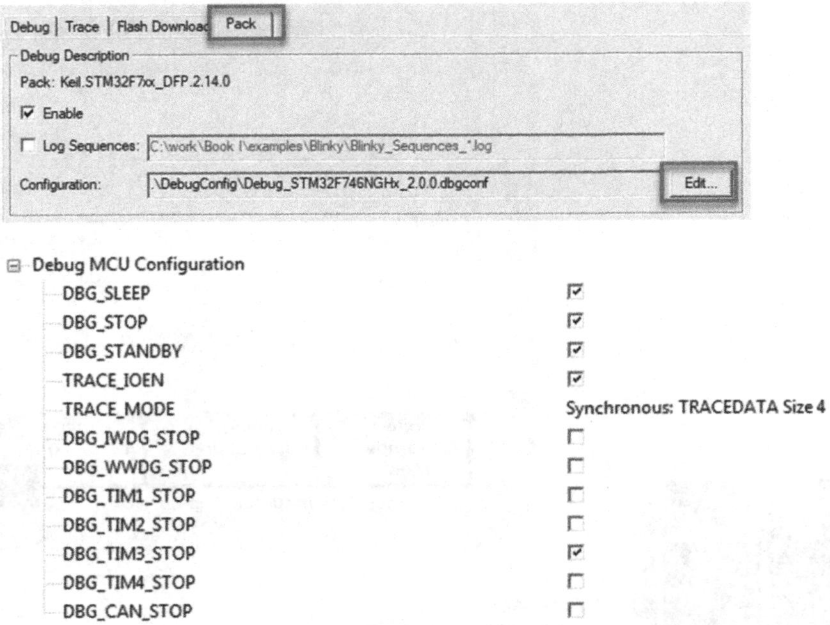

Debug | Trace | Flash Download | Pack

Debug Description

Pack: Keil.STM32F7xx_DFP.2.14.0

☑ Enable

☐ Log Sequences: C:\work\Book I\examples\Blinky\Blinky_Sequences_*.log

Configuration: .\DebugConfig\Debug_STM32F746NGHx_2.0.0.dbgconf   Edit...

⊟ Debug MCU Configuration

DBG_SLEEP	☑
DBG_STOP	☑
DBG_STANDBY	☑
TRACE_IOEN	☑
TRACE_MODE	Synchronous: TRACEDATA Size 4
DBG_IWDG_STOP	☐
DBG_WWDG_STOP	☐
DBG_TIM1_STOP	☐
DBG_TIM2_STOP	☐
DBG_TIM3_STOP	☑
DBG_TIM4_STOP	☐
DBG_CAN_STOP	☐

Figure 8.47
The debug script file is used to enable the additional four instruction trace pins when the debugger starts.

This is the same script file that was used with the standard debug adapter. This time the TRACE_MODE has been set for 4-bit synchronous trace data. This will enable both the ETM trace pipe and switch the external microcontroller pins from GPIO to debug pins.

Now press the ULINK-PRO settings button (Fig. 8.48).

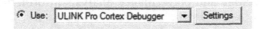

⊙ Use: ULINK Pro Cortex Debugger ▼ Settings

Figure 8.48
Select a debug adapter that is capable of capturing data from the ETM.

Select the Trace tab (Fig. 8.49).

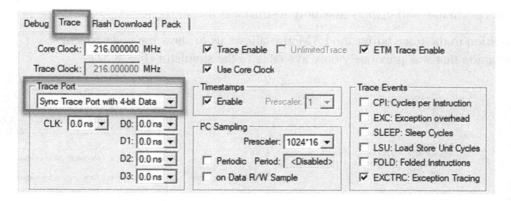

Figure 8.49
The Trace port can now be configured to access the ETM.

When the ULINK Pro Trace tool is connected you have the option to enable the ETM trace. The unlimited trace option allows you to stream every instruction executed to a file on your PC hard disk. The trace buffer is then only limited by the size of the PC hard disk. This makes it possible to trace the executed instructions for days if necessary, yes days.

Click OK to quit back to the μVision editor.

Start the Debugger.

Select the View\Trace\Trace Data window (Fig. 8.50).

Nr.	Time	Address	Opcode	Instruction		Src Code	
32,739	0.038 093 547 s	0x00000684	B168	CBZ	r0,0x000006A2		
32,740	0.038 093 577 s	0x000006A2	EA950006	EORS	r0,r5,r6	if (ad_val ^ ad_val_) {	/...
32,741	0.038 093 587 s	0x000006A6	D005	BEQ	0x000006B4		
32,742	0.038 093 617 s	0x000006B4	480D	LDR	r0,[pc,#52] ; @0x000006EC	if (clock_1s) {	
32,743	0.038 093 637 s	0x000006B6	7800	LDRB	r0,[r0,#0x00]		
32,744	0.038 093 657 s	0x000006B8	B130	CBZ	r0,0x000006C8		
32,745	0.038 093 687 s	0x000006C8	E7DA	B	0x00000680	while (1) {	/* Lo...
32,746	0.038 093 717 s	0x00000680	4815	LDR	r0,[pc,#84] ; @0x000006D8	if (AD_done) {	/* L...
32,747	0.038 093 737 s	0x00000682	7800	LDRB	r0,[r0,#0x00]		
32,748	0.038 093 757 s	0x00000684	B168	CBZ	r0,0x000006A2		
32,749	0.038 093 787 s	0x000006A2	EA950006	EORS	r0,r5,r6	if (ad_val ^ ad_val_) {	/...
32,750	0.038 093 797 s	0x000006A6	D005	BEQ	0x000006B4		
32,751	0.038 093 827 s	0x000006B4	480D	LDR	r0,[pc,#52] ; @0x000006EC	if (clock_1s) {	
32,752	0.038 093 847 s	0x000006B6	7800	LDRB	r0,[r0,#0x00]		
32,753	0.038 093 867 s	0x000006B8	B130	CBZ	r0,0x000006C8		

Figure 8.50
The debugger trace window can now display all of the instructions executed by the Cortex-M processor (streaming trace).

The trace window will display assembly instructions and alongside "C" code.

In addition to the trace buffer the ETM also allows us to show the code coverage information that was previously only available in the simulator (Fig. 8.51).

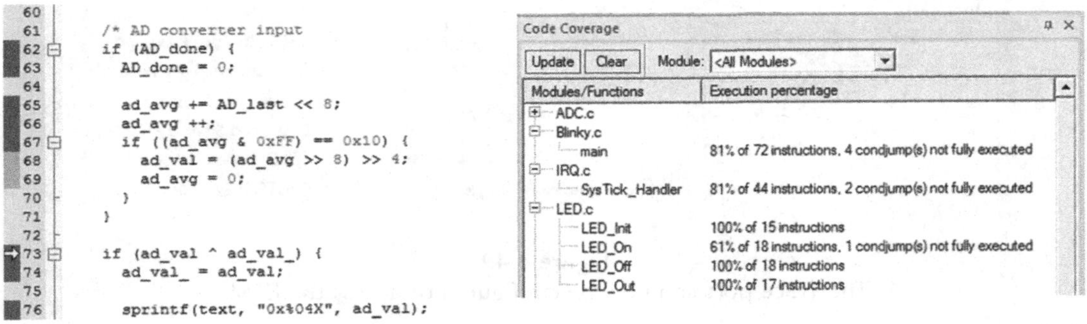

```
60
61        /* AD converter input
62 ⊟    if (AD_done) {
63        AD_done = 0;
64
65        ad_avg += AD_last << 8;
66        ad_avg ++;
67 ⊟    if ((ad_avg & 0xFF) == 0x10) {
68        ad_val = (ad_avg >> 8) >> 4;
69        ad_avg = 0;
70    }
71    }
72
73 ⊟    if (ad_val ^ ad_val_) {
74        ad_val_ = ad_val;
75
76        sprintf(text, "0x%04X", ad_val);
```

Code Coverage		ㅁ ×
Update Clear Module: \<All Modules\> ▼		
Modules/Functions	Execution percentage	▲
⊞ ADC.c		
⊟ Blinky.c		
main	81% of 72 instructions, 4 condjump(s) not fully executed	
⊟ IRQ.c		
SysTick_Handler	81% of 44 instructions, 2 condjump(s) not fully executed	
⊟ LED.c		
LED_Init	100% of 15 instructions	
LED_On	61% of 18 instructions, 1 condjump(s) not fully executed	
LED_Off	100% of 18 instructions	
LED_Out	100% of 17 instructions	

Figure 8.51
A streaming trace allows you to display accurate code coverage information.

Similarly, timing information can be captured and displayed alongside the C code or as a performance analysis report (Fig. 8.52).

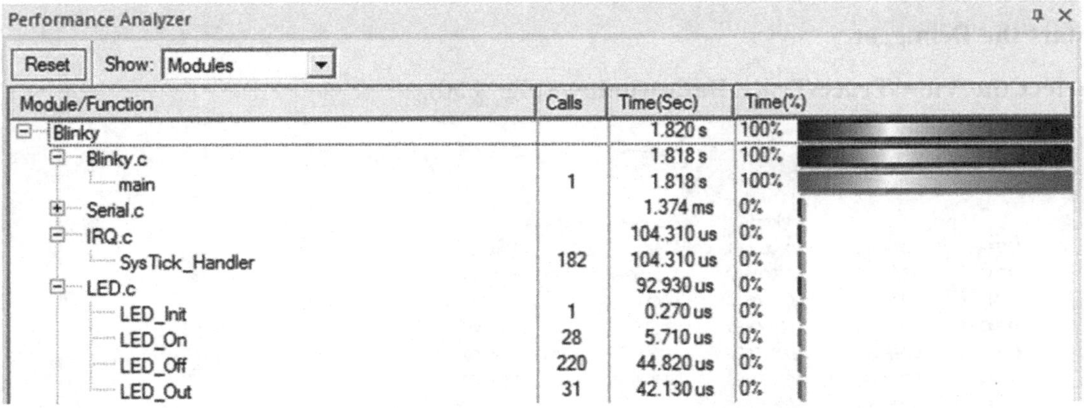

Performance Analyzer				ㅁ ×
Reset Show: Modules ▼				
Module/Function	Calls	Time(Sec)	Time(%)	
⊟ Blinky		1.820 s	100%	
⊟ Blinky.c		1.818 s	100%	
main	1	1.818 s	100%	
⊞ Serial.c		1.374 ms	0%	
⊟ IRQ.c		104.310 us	0%	
SysTick_Handler	182	104.310 us	0%	
⊟ LED.c		92.930 us	0%	
LED_Init	1	0.270 us	0%	
LED_On	28	5.710 us	0%	
LED_Off	220	44.820 us	0%	
LED_Out	31	42.130 us	0%	

Figure 8.52
The performance analyzer shows the cumulative execution time for each function.

CMSIS-DAP

The CMSIS-DAP specification defines the interface protocol between the CoreSight debugger hardware and the PC debugger software (Fig. 8.53). This creates a new level of interoperability between different vendors' software and hardware debuggers. The CMSIS-DAP firmware is designed to operate on very low-cost microcontrollers that have some GPIO and a USB interface (Fig. 8.54). The CMSIS-DAP firmware converts a

CMSIS-DAP

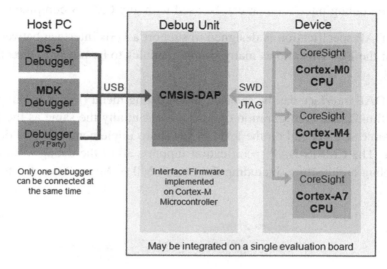

Figure 8.53
The CMSIS DAP specification is designed to support interoperability between different debugger hardware and debugger software.

Figure 8.54
The MDED module is the first to support the CMSIS DAP specification.

microcontroller to become a debug adapter. It can then be added to an evaluation board alongside the target microcontroller. This allows even the most basic evaluation modules to host a common debug interface that can be used with any CMSIS-compliant tool chain.

The CMSIS-DAP specification is designed to support a USB interface between the target hardware and the PC. This allows many simple modules to be powered directly from the PC USB port.

The CMSIS-DAP interface can be selected in the debug menu (Fig. 8.55) in place of the proprietary Ulink2. The configuration options are essentially the same as the ulink2 but the options available will depend on the level of firmware implemented by the device manufacturer. The CMSIS-DAP specification supports all of the debug features found in the CoreSight debug architecture, including the Cortex-M0 + Micro Trace Buffer (MTB).

Figure 8.55
The CMSIS DAP driver must be selected in the debugger menu.

Cortex-M0 + Micro Trace Buffer

While the ETM is available for the Cortex-M3 and Cortex-M4, no form of instruction trace is currently available for the Cortex-M0. However, the Cortex-M0 + has a simple form of instruction trace buffer called the MTB. The MTB uses a region of internal SRAM which is allocated by the developer. When the application code is running, a trace of executed instructions is recorded into this region. When the code is halted, the debugger can read the MTB trace data and display the executed instructions. The MTB trace SRAM can be configured as a circular buffer or a one-shot recording. While this is a very limited trace, the circular buffer will allow you to see "what just happened" before the code halted. While the one-shot mode can be triggered by the hardware breakpoints to start and stop allowing you to track down more elusive bugs.

Exercise 8.9: Micro Trace Buffer

This exercise is based on the Freescale Freedom board for the MKL25Z microcontroller. This was the first microcontroller available to use the Cortex-M0 + .

Connect the freedom board via its USB cable to the PC.

Open the Pack Installer.

Select the example tab and Copy "Ex 8.9 Micro Trace Buffer."

Open the options for target\debug menu (Fig. 8.56).

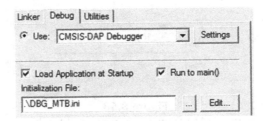

Figure 8.56
Select the CMSIS-DAP debug adapter and the micro trace buffer script file.

Here, the CMSIS DAP interface is selected along with an initializing file for the microtrace buffer. With newer devices this script file may also be located in the packs tab.

The initializing script file has a wizard that allows you to configure the size and configuration of the micro trace buffer (Fig. 8.57).

Expand All	Collapse All		Help	☐ Show Grid
Option			Value	
⊟ Trace: MTB (Micro Trace Buffer)			☑	
Buffer Size			4kB	
Stop Trace when buffer is full			☐	
Stop Target when buffer is full			☐	

Figure 8.57
The script file I used to configure the micro trace buffer.

Here, we can select the amount of internal SRAM that is to be used for the trace. It is also possible to configure different debugger actions when the trace is full. Either halt trace recording or halt execution on the target.

By default, the MTB is located at the start of the internal SRAM (0x20000000). So it is necessary to offset the start of user SRAM by the size of memory allocated to the MTB. This is done in the Options For Target\Target dialog (Fig. 8.58).

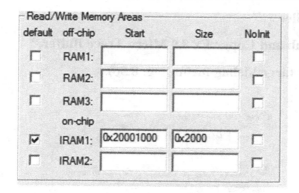

Figure 8.58
You must offset the user RAM from the region used by the Micro Trace Buffer.

Now start the debugger and execute the code.

Halt the debugger and open the trace window (Fig. 8.59).

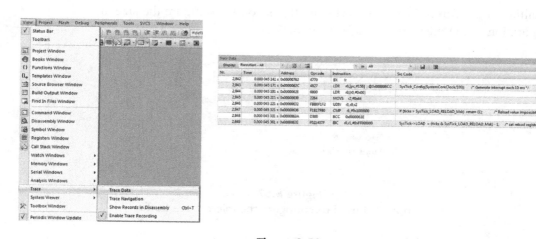

Figure 8.59
The contents of the Micro Trace buffer are downloaded to the PC and displayed in the debugger trace window.

While this is a limited trace buffer, it can be used by very low-cost tools and provides a means of tracking down run time bugs which would be time-consuming to find any other way.

System Viewer

The CMSIS system viewer description format is designed to provide silicon manufacturers with a method of creating a description of the peripheral registers in their microcontroller. The System Viewer Description files are then passed to third-party tool manufacturers so that compiler includes files and debugger peripheral view windows can be automatically created from the XML (Fig. 8.60). This means that there will be no lag in software development support when a new family of devices is released. As a developer, you will not normally need to work with these files, but it is useful to understand how the process works so that you can fix any errors that may inevitably occur. It is also possible to create your own additional peripheral debug windows. This would allow you to create a view of an external memory-mapped peripheral or provide a debug view of a complex memory object.

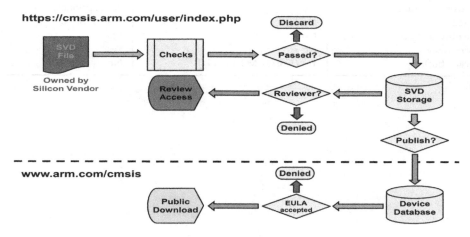

Figure 8.60

ARMs have created a repository of "system viewer description" files. This enables tools' suppliers to have support for new devices as they are released.

When the Silicon Vendor develops a new microcontroller, they also create an XML description of the microcontroller registers. A conversion utility is then used to create a binary version of the file that is used by the debugger to automatically create the peripheral debug windows.

Alongside the XML description definition, Arm has introduced a submission and publishing system for new SVD files. When a Silicon Vendor designs a new microcontroller, its system description file is submitted via the CMSIS website. Once it has been reviewed and validated, it is then published for public download on the main Arm website.

Conclusion

In this chapter, we have had a look through the advanced debug features available within the Coresight debug architecture for Cortex-M. Table 8.14 summarizes the features available on the different Cortex-M processors.

Table 8.14: Debug feature summary

Feature	Cortex-M0 Cortex-M0 +	Cortex-M3\M4\M7
Debug interface	Legacy JTAG or Serial Wire	Legacy JTAG or Serial Wire
"on the fly" memory access	Yes	Yes
Hardware breakpoint	4	6 Instruction + 2 Literal
Data watchpoint	2	4
Software breakpoint	Yes	Yes
ETM instruction trace	No	Yes (optional)
Data trace	No	Yes (optional)
Instrumentation trace	No	Yes
Serial wire viewer	No	Yes
Micro trace buffer	Yes (Cortex-M0 + only)	No

CHAPTER 9

Practical DSP for Cortex-M Microcontrollers

Introduction

From a developer's perspective, the Cortex-M4 and Cortex-M7 are versions of the Cortex-M processor that have additional features to support Digital Signal Processing (DSP). The key enhancements over the Cortex-M3 are the addition of "Single Instruction Multiple Data" or SIMD instructions, and an improved MAC unit for integer maths, plus the optional addition of a hardware "Floating Point Unit" (FPU). In the case of the Cortex-M4, this is a single-precision FPU. While the Cortex-M7 has the option of either a single- or double-precision FPU. These enhancements give the Cortex-M4 the ability to run DSP algorithms at high enough levels of performance to compete with dedicated 16-bit DSP processors (Fig. 9.1). As we saw in Chapter 6, Cortex-M7 Processor, the Cortex-M7 processor has a more advanced pipeline and the Branch Cache Unit. Both of these features dramatically improve its DSP capability. In this chapter, we will look at using the Cortex-M4/M7 to process real-world signals.

Figure 9.1
The Cortex-M4 and Cortex-M7 extend the Cortex-M3 with the addition of DSP instructions and fast maths capabilities. This creates a microcontroller capable of supporting real-time DSP algorithms, a digital signal controller.

Hardware Floating Point Unit

One of the major features of the Cortex-M4 and Cortex-M7 processors is the addition of a hardware FPU. The FPU supports floating-point arithmetic operations to the IEEE 754

The Designer's Guide to the Cortex-M Processor Family.
DOI: https://doi.org/10.1016/B978-0-323-85494-8.00010-3

standard. Initially, the FPU can be thought of as a coprocessor that is accessed by dedicated instructions to perform most floating-point arithmetic operations in a few cycles (Table 9.1).

Table 9.1: Floating point unit performance

Operation	Cycle Count
Add/Subtract	1
Divide	14
Multiply	1
Multiply Accumulate (MAC)	3
Fused MAC	3
Square Root	14

The FPU consists of a group of control and status registers and 31 single-precision scalar registers. The scalar registers can also be viewed as 16 double-word registers (Fig. 9.2).

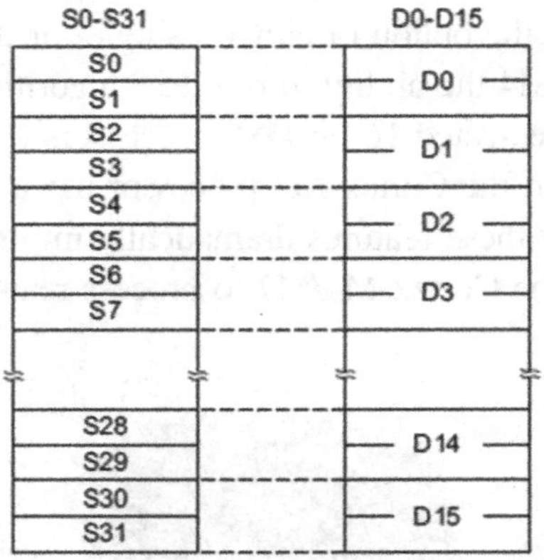

Figure 9.2
The FPU 32-bit scalar registers may also be viewed as 64-bit double word registers. This supports very efficient casting between C types.

While the FPU is designed for floating-point operations it is possible to load and store fixed point and integer values. It is also possible to convert between floating-point to fixed-point and integer values. This means "C" casting between floating-point and integer values can be done in a single cycle.

FPU Integration

While it is possible to consider the FPU as a coprocessor adjacent to the Cortex-M4 and M7 processors this is not really true. The FPU is an integral part of the Cortex-M processor, the floating-point instructions are executed within the FPU in a parallel pipeline to the Cortex-M processor instructions (Fig. 9.3). While this increases the FPU performance, it is "invisible" to the application code and does not introduce any strange side effects.

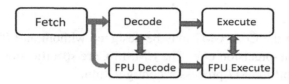

Figure 9.3
The FPU is described as a CoProcessor in the register documentation. In reality it is very tightly coupled to the main instruction pipeline.

FPU Registers

In addition to the scalar registers, the FPU has a block of control and status registers (Table 9.2).

Table 9.2: FPU control registers

Register	Description
CoProcessor Access Control	Controls the privilege access level to the FPU.
Floating-point context control	Configures stacking and lazy stacking options.
Floating-point Context Address	Holds the address of the unpopulated FPU stack space.
Floating-point default status control	Holds the FPU condition codes and FPU configuration options.
Floating-point status control	Holds the default status control values.

All of the FPU registers are memory-mapped except the Floating Point Status Control Register which is a CPU register accessed by the MRS and MSR instructions. Access to this function is supported by a CMSIS core function.

```
uint32_t  __get_FPSCR(void);
void      __set_FPSCR (uint32_t fpscr)
```

The FPSCR register contains three groups of bits. The top four bits contain condition code flags N, Z, C, V that match the condition code flags in the xPSR. These flags are set and cleared in a similar manner by the results of floating-point operations. The next groups of bits contain configuration options for the FPU. These bits allow you to change the operation of the FPU from the IEEE 754 standard. Unless you have a strong reason to do this, it is recommended to leave them alone. The final group of bits is status flags for the FPU exceptions. If the FPU encounters an error during execution an exception will be raised and the matching status flag will be set. The exception line is permanently enabled in the FPU and just needs to be enabled in the NVIC to become active. When the exception is raised, you will need to interrogate these flags to work out the cause of the error. Before returning from the FPU exception the status flags must be cleared. How this is done depends on the FPU exception stacking method which is covered in the next section.

Cortex-M7 FPU

Microcontrollers using the Cortex-M7 may be designed without an FPU or may be fitted with a single- or double-precision unit. The CMSIS-Core specification has been extended with a function to report the M7 processor configuration.

```
uint32_t SBC_GetFPUType(void)

0 = No FPU
1 = Single-Precision
2 = Double Precision
```

This function reads a configuration register in the System Control Block, the "Media and VFP feature register," and then returns the configuration of the Cortex-M7 processor. This function can only be used with a Cortex-M7.

Enabling the FPU

When the Cortex-M4 or Cortex-M7 leaves the reset vector the FPU is disabled. The FPU may be enabled by setting the coprocessor 10 and 11 bits in the CPARC register. It is necessary to use the data barrier instruction to ensure that the write is made before the code continues. The instruction barrier command is also used to ensure the pipeline is flushed before the code continues.

```
SCB->CPACR |= ((3UL << 10*2) | (3UL << 11*2));   // Set CP10 & CP11 Full Access
__DSB();                                          //Data barrier
__ISB();                                          //Instruction barrier
```

In order to write to the CPARC register, the processor must be in privileged mode. Once enabled the FPU may be used in privileged and unprivileged modes.

Exceptions and the FPU

When the FPU is enabled, an extended stack frame will be pushed when an exception is raised. In addition to the standard stack frame, the Cortex-M processor also pushes the first 16 FPU scalar registers and the FPSCR. This extends the stack frame from 32 to 100 bytes. Clearly pushing this amount of data onto the stack, every interrupt would increase the interrupt latency significantly. To keep the 12-cycle interrupt latency, the Cortex-M processor uses a technique called "lazy stacking." When an interrupt is raised, the normal stack frame is pushed onto the stack and the stack pointer is incremented to leave space for the FPU registers, but their values are not pushed onto the stack. This leaves a void space in the stack. The start address of this void space is automatically stored in the "Floating Point Context Address Register" (FPCAR). If the interrupt routine uses floating-point calculations, the FPU registers will be pushed into this space using the address stored in the FPCAR as a base address. The "Floating Point Context Control Register" is used to select the stacking method used (Table 9.3). Lazy stacking is enabled by default when the FPU is first enabled. The stacking method is controlled by the most significant 2 bits in the "Floating Point Context Control Register"; these are the Automatic State Preservation Enable (ASPEN) and Lazy State Preservation Enable (LSPEN).

Table 9.3: Lazy stacking options

LSPEN	ASPEN	Configuration
0	0	No automatic state preservation. Only use when the interrupts do not use floating-point
0	1	Lazy stacking disabled
1	0	Lazy stacking enabled
1	1	Invalid Configuration

Using the FPU

Once you have enabled the FPU, the compiler will start to use hardware floating-point calculations in place of software libraries. The exception is the square root instruction sqrt (), which is part of the math.h library. If you have enabled the FPU, the Arm compiler provides an intrinsic instruction to use the FPU square root instruction.

```
float __sqrtf(float x);
```

Note: The intrinsic square root function differs from the ANSI sqrt() library function in that it takes and returns a float rather than a double.

Exercise 9.1: Floating Point Unit

This exercise performs a few simple floating-point calculations using the Cortex-M4 processor so that we can compare the performance of the software and hardware floating-point execution times.

Open the Pack Installer.

Select the Boards::Designers Guide Tutorial.

Select the example tab and Copy "Ex 9.1 Floating Point Unit."

The code in the main loop is a mixture of maths operations to exercise the FPU.

```
#include <math.h>
float a,b,c,d,e;
int f,g = 100;

while(1){
        a = 10.1234;
        b = 100.2222;
        c = a*b;
        d = c-a;
        e = d+b;
        f = (int)a;
        f = f*g;
        a1 = (unsigned int) a;
        a = __sqrtf(e);
        //a = sqrt(e);
        a = c/f;
        e = a/0;
        }
}
```

Before we can build this project, we need to make sure that the compiler will build code to use the FPU.

Open the options for target and select the target menu (Fig. 9.4).

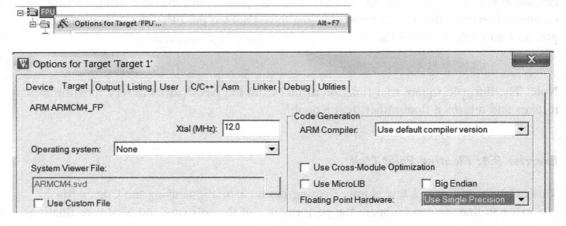

Figure 9.4
Hardware floating point support is enabled in the Project Target options.

We can enable floating-point support by selecting "Use FPU" in the floating-point hardware box. This will enable the necessary compiler options and load the correct simulator model.

Close the Options for the target menu and return to the editor.

In addition to our source code, the project includes the CMSIS startup and system files for the Cortex-M4 (Fig. 9.5).

Figure 9.5
This example project is using a simulation of the Cortex-M4 processor.

Now build the project and note down the build size (Fig. 9.6).

```
Build target 'FPU'
compiling system_ARMCM4.c...
assembling startup_ARMCM4.s...
compiling main.c...
linking...
Program Size: Code=224 RO-data=224 RW-data=4 ZI-data=1028
"cortexM4.axf" - 0 Error(s), 0 Warning(s).
```

Figure 9.6
Note the code size for a project using the Hardware FPU.

Start the debugger

When the simulator runs the code to main, it will hit a breakpoint that has been preset in the system_ARMCM4.c file (Fig. 9.7).

```
61   void SystemInit (void)
62  {
63    #if (__FPU_USED == 1)
64      SCB->CPACR |= ((3UL << 10*2) | (3UL << 11*2)  );
65    #endif
66
67    SystemCoreClock = __SYSTEM_CLOCK;
68
69  }
```

Figure 9.7
The CMSIS-Core SystemInit() function will enable the FPU on startup.

The standard microcontroller include file will define the feature set of the Cortex-M processor and will define the availability of the FPU. In our case, we are using a simulation model of the Cortex-M4 only and are using the minimal ARMCM4_FP.h include file provided as part of the CMSIS core specification. If the FPU is present on the microcontroller the SystemInit() function will make sure it is switched on before you reach your application's main() function.

Open ARMCM4_FP.H and locate the processor definitions.

```
/* = = = = = = = = = = = = = = = = = = = = = = = = = = = = = = = = = = = = =*/
/* = = = = = = = = = = = = = =    Processor and Core Peripheral Section    = = = =*/
/* = = = = = = = = = = = = = = = = = = = = = = = = = = = = = = = = = = = = =*/
/* -------- Configuration of the Cortex-M4 Processor and Core Peripherals -------       */
#define __CM4_REV          0x0001  /*!< Core revision r0p1                */
#define __MPU_PRESENT         1    /*!< MPU present or not                */
#define __NVIC_PRIO_BITS      3    /*!< Number of Bits used for Priority Levels  */
#define __Vendor_SysTickConfig 0   /*!< Set to 1 if different SysTick Config is used */
#define __FPU_PRESENT         1    /*!< FPU present or not                */


Then __FPU_USED is defined in core_cm4.h

#if (__FPU_PRESENT == 1U)
    #define __FPU_USED 1U
```

Open the main.c module and run the code to the main while() loop (Fig. 9.8).

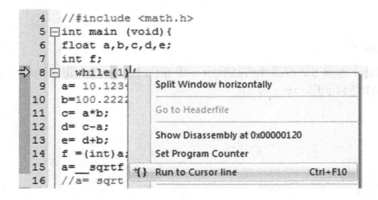

Figure 9.8
Use the local debugger options to run to the main() while(1) loop.

Now open the disassembly window (Fig. 9.9).

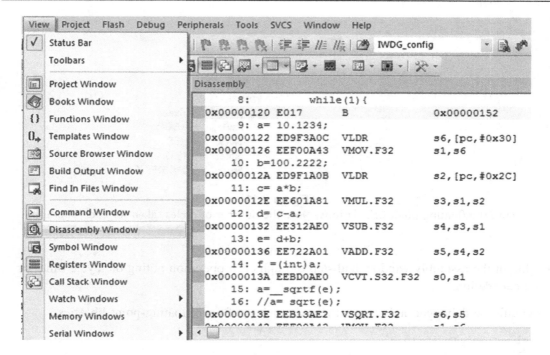

Figure 9.9
The disassembler window will show you the calls to the hardware FPU.

This will show the C source code interleaved with the Cortex-M4 assembly instructions.

In the project window select the registers window (Fig. 9.10).

⊟── FPU			⊟── FPU Reg	Float Format
├── S0	0x00000000		├── S0	0.000000
├── S1	0x00000000		├── S1	0.000000
├── S2	0x00000000		├── S2	0.000000
├── S3	0x00000000		├── S3	0.000000
└── S4	0x00000000		└── S4	0.000000

Figure 9.10
The Registers window allows you to view the FPU registers.

The register window now shows the 31 scalar registers in their raw format and the IEE754 Format.

The contents of the FPSC register are also shown (Fig. 9.11). In this exercise, we will also be using the states (cycle count) value which is also shown in the register window.

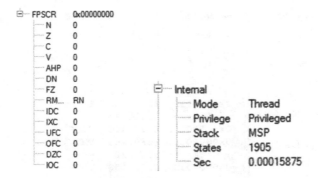

Figure 9.11
Execute the floating-point calculations. Note the number of cycles taken for each operation.

Highlight the assembly window and step through each operation noting the cycle time for each calculation.

Now quit the debugger and change the code to use software floating-point libraries.

```
  a1 = (unsigned int) a;
//a =__sqrtf(e);
  a = sqrt(e);
```

Comment out the __sqrtf() intrinsic and replace it with the ANSI C sqrt() function.

In the options for target\target settings, we also need to remove the FPU support (Fig. 9.12).

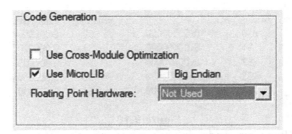

Figure 9.12
Disable the FPU support in the project target menu.

Now change the processor header file for the Cortex-M4 without the Floating point unit.

```
#include <ARMCM4.h>
```

Rebuild the code and compare the build size to the original version (Fig. 9.13).

```
Build target 'FPU'
compiling main.c...
linking...
Program Size: Code=1440 RO-data=224 RW-data=8 ZI-data=1024
"cortexM4.axf" - 0 Error(s), 0 Warning(s).
```

Figure 9.13
Rebuild the project and compare the code size the original project.

Restart the debugger run and run the code to main.

In the disassembly window step through the code and compare the number of cycles used for each operation to the number of cycles used by the FPU.

By the end of this exercise, you can clearly see not only the vast performance improvement provided by the FPU but also its impact on project code size. The only downside is the additional cost to use a microcontroller fitted with the FPU and the additional power consumption when it is running.

Cortex-M4/M7 DSP and SIMD Instructions

The Thumb2 instruction set has a number of additional instructions that are useful in DSP algorithms (Table 9.4).

Table 9.4: Thumb2 DSP instructions

Instruction	Description
CLZ	Count leading zeros
REV, REV16, REVSH, and RBIT	Reverse instructions
BFI	Fit field insert
BFC	Bit field clear
UDIV and SDIV	Hardware divide
SXT and UXT	Sign and Zero extend

The Cortex-M4 and M7 instruction set includes a new group of instructions that can perform multiple arithmetic calculations in a single cycle. The SIMD instructions allow 16-bit or 8-bit data to be packed into two 32-bit words to be operated on in parallel (Fig. 9.14). So, for example, you can perform two 16-bit multiplies and a 32-bit or 64-bit accumulate or a quad 8-bit addition in one processor cycle. Since many DSP algorithms work on a pipeline of data the SIMD instructions can be used to dramatically boost performance.

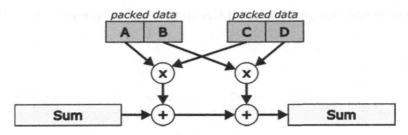

Figure 9.14
The SIMD instructions support multiple arithmetic operations in a single cycle. The operand data must be packed into 32-bit words.

The SIMD instructions have an additional field in the xPSR register (Fig. 9.15). The "Greater than or Equal" (GE) field contains four bits that correspond to the four bytes in the SIMD instruction result operand. If the result operand byte is greater than or equal to zero then the matching GE flag will be set.

Figure 9.15
The cortex M4 xPSR register has an additional Greater than or equal field. Each of the four GE bits is updated when a SIMD instruction is executed.

The SIMD instructions can be considered as three distinct groups. Add and Subtract operations, Multiply operations, and supporting instructions. The Add and Subtract operations can be performed on 8- or 16-bit signed and unsigned quantities. A signed and unsigned halving instruction is also provided; this instruction Adds or Subtracts the 8- or 16-bit quantities and then halves the result as shown in Table 9.5.

Table 9.5: SIMD add halving and subtract halving instructions

Instruction	Description	Operation
UHSUB16	Unsigned halving 16-bit subtract	$Res[15:0] = (Op1[15:0] - Op2[15:0])/2$ $Res[31:16] = (Op1[31:16] - Op2[31:16])/2$
UHADD16	Unsigned halving 16-bit add	$Res[15:0] = (Op1[15:0] + Op2[15:0])/2$ $Res[31:16] = (Op1[31:16] + Op2[31:16])/2$

The SIMD instructions also include an Add and Subtract with Exchange (ASX) and a Subtract and Add with Exchange (SAX). These instructions perform and Add and Subtract on the two half words and store the results in the upper and lower half words of the destination register (Table 9.6).

Table 9.6: SIMD add exchange and subtract exchange instructions

Instruction	Description	Operation
USAX	Unsigned 16-bit Subtract and add with exchange	Res[15:0] = Op1[15:0] + Op2[31:16] Res[31:16] = Op1[31:16] - Op2[15:0]
UASX	Unsigned 16-bit Add and subtract with exchange	Res[15:0] = Op1[15:0] + Op2[31:16] Res[31:16] = Op1[31:16] - Op2[15:0]

A further group of instructions combine these two operations in a Subtract and Add or Add and Subtract with exchange halving instruction. This gives quite a few possible permutations. A summary of the Add and Subtract SIMD instructions is shown in Table 9.7.

Table 9.7: Permutations of the SIMD add, subtract, halving, and saturating instructions

Instruction	Signed	Signed Saturating	Signed Halving	Unsigned	Unsigned Saturating	Unsigned Halving
ADD8	SADD8	QADD8	SHADD8	UADD8	UQADD8	UHADD8
SUB8	SSUB8	QSUB8	SHSUB8	USUB8	UQSUB8	UHSUB8
ADD16	SADD16	QADD16	SHADD16	UADD16	UQADD16	UHADD16
SUB16	SSUB16	QSUB16	SHSUB16	USUB16	UQSUB16	UHSUB16
ASX	SASX	QASX	SHASX	UASX	UQASX	UHASX
SAX	SSAX	QSAX	SHSAX	USAX	UQSAX	UHSAX

The SIMD instructions also include a group of Multiply instructions that operate on packed 16-bit signed values. Like the Add and Subtract instructions the Multiply instructions also support saturated values. As well as Multiply and Multiply Accumulate the SIMD Multiply instructions support Multiply Subtract and Multiply Add as shown in Table 9.8.

Table 9.8: SIMD multiply instructions

Instruction	Description	Operation
SMLAD	Q setting dual 16-bit signed multiply with single 32-bit accumulator	X = X + (AxB) + (CxD)
SMLALD	Dual 16-bit signed multiply with single 64-bit accumulator.	X = X + (AxB) + (CxD)
SMLSD	Q setting dual 16-bit signed multiply subtract with 32-bit accumulate.	X = X + (AxB)-(BxC)
SMLSLD	Q setting dual 16-bit signed multiply subtract with 64-bit accumulate.	X = X + (AxB)-(BxC)
SMUAD	Q setting sum of dual 16-bit signed multiply.	X = (AxB) + (CxD)
SMUSD	Dual 16-bit signed multiply returning difference.	X = (AxB)-(CxD)

To make the SIMD instructions more efficient a group of supporting pack and unpack instructions have also been added to the instruction set. The pack/unpack instructions can be used to extract 8 and 16-bit values from a register and move them to a destination register (Fig. 9.16). The unused bits in the 32-bit word can be set to zero (unsigned) or one (signed). The pack instructions can also take two 16-bit quantities and load them into the upper and lower half words of a destination register. The full range of supporting instructions is shown in Table 9.9.

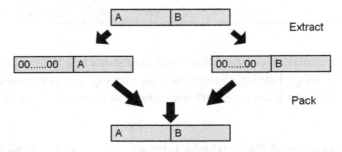

Figure 9.16
The SIMD instruction group includes support instructions to pack 32-bit words with 8- and 16-bit quantities.

Table 9.9: SIMD supporting instructions

Mnemonic	Description
PKH	Pack Halfword
SXTAB	Extend 8-bit signed value to 32 bits and add
SXTAB16	Dual extend 8-bit signed value to 16 bits and add
SXTAH	Extend 16-bit signed value to 32 bits and add
SXTB	Sign extend a byte
SXTB16	Dual extend 8-bit signed values to 16 and add
SXTH	Sign extend a half word
UXTAB	Extend 8-bit signed value to 32 bits and add
UXTAB16	Dual extend 8 bits to 16 and add
UXTAH	Extend a 16-bit value and add
UXTB	Zero extend a byte
UXTB16	Dual zero extend 8 bits to 16 and add
UXTH	Zero extend a half word

When a SIMD instruction is executed, it will set or clear the xPSR GE bits depending on the values in the resulting bytes or halfwords. An addition select (SEL) instruction is provided to access these bits. The SEL instruction is used to select bytes or halfwords from two input operands depending on the condition of the GE flags (Table 9.10).

Table 9.10: xPSR "Greater than or equal" bit field results

GE bit[3:0]	GE bit = 1	GE bit = 0
0	Res[7:0] = OP1[7:0]	Res[7:0] = OP2[7:0]
1	Res[15:8] = OP1[15:8]	Res[15:8] = OP2[15:8]
2	Res[23:16] = OP1[23:16]	Res[23:16] = OP2[23:16]
3	Res[31:24] = OP1[31:24]	Res[31:24] = OP2[31:24]

Exercise 9.2: SIMD Instructions

In this exercise, we will have a first look at using the Cortex-M4/M7 SIMD instructions. In this exercise, we will simply Multiply and Accumulate two 16-bit arrays first using a SIMD instruction and then using the standard Add instruction.

Open the CMSIS core documentation and the SIMD Signed multiply-accumulate intrinsic __SMLAD (Fig. 9.17).

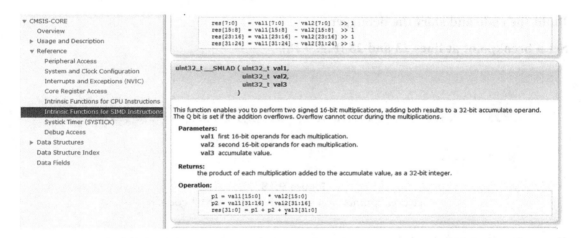

Figure 9.17
CMSIS-DSP documentation for the __SMLAD() intrinsic.

Open the Pack Installer.

Select the Boards::Designers Guide Tutorial.

Select the example tab and Copy "Ex 9.2 SIMD."

The application code defines two sets of arrays as a union of 16-bit and 32-bit quantities.

```
union _test{
int16_t    Arry_halfword[100];
int32_t    Arry_word[50];
};
```

The code first initializes the arrays with the values 0 to 100.

```
for(n=0;n<100;n++){
opl.Arry_halfword[n] = op2.Arry_halfword[n] = n;  }
```

Then multiply accumulates first using the SIMD instruction then the standard multiply accumulate.

```
for(n=0;n<50;n++){
Result = __SMLAD(opl.Arry_word[n],op2.Arry_word[n],Result);}
```

Result is then reset and the calculation is repeated without using the SIMD instruction.

```
Result = 0;
for(n=0;n<100;n++){
Result = Result + (opl.Arry_halfword[n]* op2.Arry_halfword[n]);}
```

Build the code and start the debugger.

Set a breakpoint at lines 23 and 28 (Fig. 9.18).

```
23    for(n=0;n<50;n++){
24
25      Result = __SMLAD(opl.Arry_word[n],op2.Arry_word[n],Result);
26
27    }
28  Result = 0;
```

Figure 9.18
Set breakpoints on either side of the SIMD code.

Run to the first breakpoint and make a note of the cycle count (Fig. 9.19).

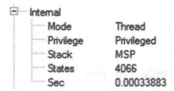

Internal	
Mode	Thread
Privilege	Privileged
Stack	MSP
States	4066
Sec	0.00033883

Figure 9.19
Note the start cycle count.

Run the code until it hits the second breakpoint. See how many cycles have been used to execute the SIMD instruction (Fig. 9.20).

Figure 9.20
Note the final cycle count.

Cycles used = 5172−4066 = 1106

Set a breakpoint at the final while loop (Fig. 9.21).

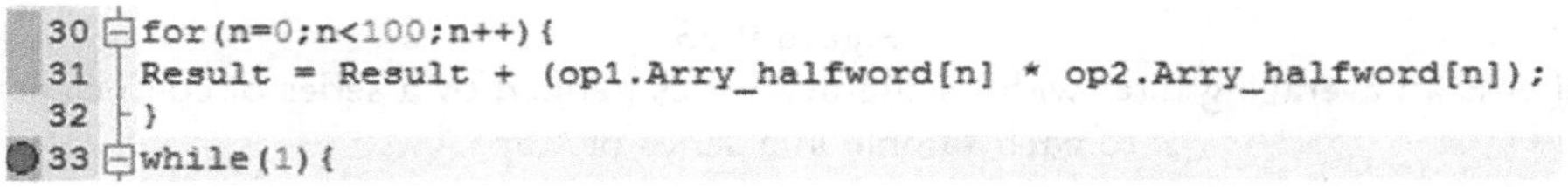

```
30 for(n=0;n<100;n++){
31     Result = Result + (op1.Arry_halfword[n] * op2.Arry_halfword[n]);
32 }
33 while(1){
```

Figure 9.21
Now set a breakpoint after the array copy routine.

Run the code and see how many cycles are used to perform the calculation without using the SIMD instruction (Fig. 9.22).

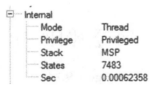

Figure 9.22
Note the final cycle count.

Cycles used = 7483−5172 = 2311

Compare the number of cycles used to perform the same calculation without using the SIMD instructions.

As expected, the SIMD instructions are much more efficient when performing calculations on large data sets.

The primary use for the SIMD instructions is to optimize the performance of DSP algorithms. In the next exercise, we will look at various techniques in addition to the SIMD instructions that can be used to increase the efficiency of a given algorithm.

Exercise 9.3: Optimizing DSP Algorithms

In this exercise, we will look at optimizing a Finite Impulse Response filter. This is a classic algorithm that is widely used in DSP applications (Fig. 9.23).

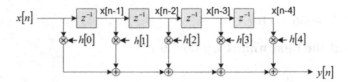

Figure 9.23
An FIR filter is an averaging filter with its characteristics defined by a series of coefficients applied to each sample in a series of "taps."

The FIR filter is an averaging filter that consists of a number of "taps." Each tap has a coefficient and as the filter runs each sample is multiplied against the coefficient in the first tap and then shifted to the next tap to be multiplied against its coefficient when the next sample arrives. The output of each tap is summed to give the filter output. Or to put it mathematically in Fig. 9.24.

$$y[n] = \sum_{k=0}^{N-1} h[k]x[n-k]$$

Figure 9.24
Mathematical expression for an FIR filter.

In this exercise, we will use the Cortex-M4 simulation model to look at several techniques that can be used to optimize a DSP algorithm running on a Cortex-M processor.

Open the Pack Installer.

Select the Boards::Designers Guide Tutorial.

Select the example tab and Copy "Ex 9.3 FIR Optimization."

Build the project and start the debugger.

The main function consists of four FIR functions that introduce different optimizations to the standard FIR algorithm.

```
int main (void){
fir                (data_in,data_out,  coeff,&index,  FILTERLEN, BLOCKSIZE);
fir_block          (data_in,data_out,  coeff,&index,  FILTERLEN, BLOCKSIZE);
fir_unrolling      (data_in,data_out,  coeff,&index,  FILTERLEN, BLOCKSIZE);
fir_SIMD           (data_in,data_out,  coeff,&index,  FILTERLEN, BLOCKSIZE);
fir_SuperUnrolling (data_in,data_out,  coeff,&index,  FILTERLEN, BLOCKSIZE);
while(1);
}
```

Step into the first function and examine the code.

The filter function is implemented in C as shown below. This is a standard implementation of an FIR filter written purely in C.

```
void fir(q31_t *in, q31_t *out, q31_t *coeffs, int *stateIndexPtr,
                    int filtLen, int blockSize)
{
  int sample;
  int k;
  q31_t sum;
  int stateIndex = *stateIndexPtr;
  for(sample = 0; sample < blockSize; sample++)
    {
      state[stateIndex++] = in[sample];
      sum = 0;
      for(k = 0;k<filtLen;k++)
          {
           sum += coeffs[k] * state[stateIndex];
           stateIndex--;
           if (stateIndex < 0)
              {
                stateIndex = filtLen-1;
              }
          }
      out[sample] = sum;
    }
    *stateIndexPtr = stateIndex;
}
```

While this compiles and runs fine, it does not take full advantage of the Cortex-M4 DSP enhancements. To get the best out of the Cortex-M4, we need to optimize this algorithm, particularly the inner loop. The inner loop performs the FIR multiply and accumulates for each tap.

```
for(k = 0;k<filtLen;k + + )
            {
         sum + = coeffs[k] * state[stateIndex];
         stateIndex--;
         if (stateIndex < 0)
  {
   stateIndex = filtLen-1;
   }
  }
```

The inner loop processes the samples by implementing a circular buffer in software. While this works ok we have to perform a test for each loop to wrap the pointer when it reaches the end of the buffer (Fig. 9.25).

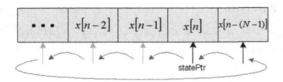

Figure 9.25
Processing data in a circular buffer requires the Cortex M4 to check for the end of the buffer on each iteration. This increases the execution time.

Run to the start of the inner loop and set a breakpoint (Fig. 9.26).

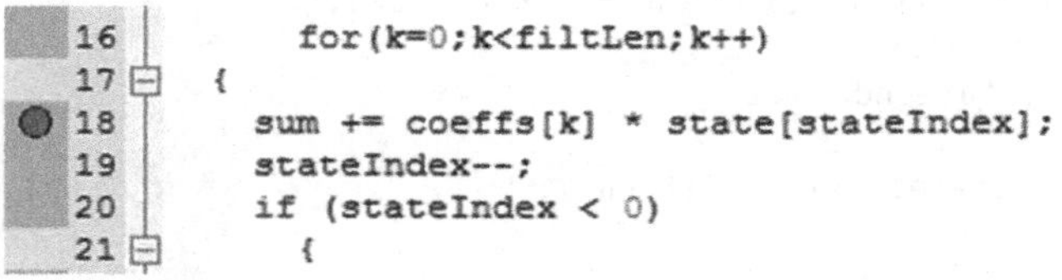

```
16          for (k=0;k<filtLen;k++)
17      {
18      sum += coeffs[k] * state[stateIndex];
19      stateIndex--;
20      if (stateIndex < 0)
21          {
```

Figure 9.26
Set a breakpoint at the start of the inner loop.

Run the code so it does one iteration of the inner loop and note the number of cycles used.

Circular addressing requires us to perform an "end of buffer" test on each iteration. A dedicated DSP device can support circular buffers in hardware without any such overhead, so this is one area we need to improve. By passing our FIR filter function a block of data rather than individual samples allows us to use block processing as an alternative to circular addressing. This improves the efficiency of the critical inner loop (Fig. 9.27).

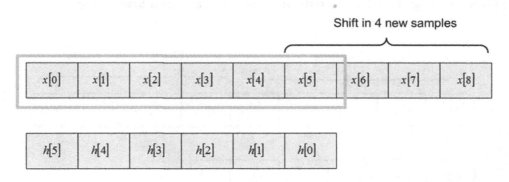

Figure 9.27
Block processing increases the size of the buffer but increases the efficiency of the inner processing loop.

By increasing the size of the state buffer to "Number of filter Taps + Processing block size" we can eliminate the need for circular addressing. In the outer loop, the block of samples is loaded into the top of the state buffer (Fig. 9.28).

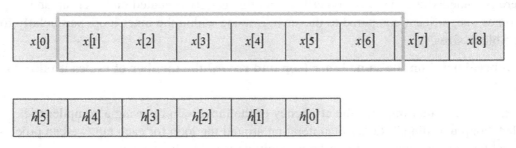

Figure 9.28
With block processing the fixed-size buffer is processed without the need to check for the end of the buffer.

The inner loop then performs the filter calculations for each sample of the block by sliding the filter window one element to the right for each pass through the loop. So the inner loop now becomes,

```
for(k = 0; k<filtLen; k + + )
{
 sum + = coeffs[k] * state[stateIndex];
 stateIndex + + ;
}
```

Once the inner loop has finished processing the current block of sample data held in the state buffer must be shifted to the right and a new block of data loaded (Fig. 9.29).

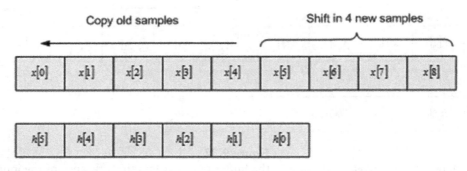

Figure 9.29
Once the block of data has been processed the outer loop shifts the samples one block to the left and adds a new block of data.

Now step into the second FIR function.

Examine how the code has been modified to process blocks of data.

There is some extra code in the outer loop but this is only executed once per tap and becomes insignificant compared to the savings made within the inner loop, particularly for large block sizes.

Set a breakpoint on the same inner loop and record the number of cycles it takes to run.

Next, we can further improve the efficiency of the inner loop by using a compiler trick called "loop unrolling." Rather than iterating around the loop for each tap, we can process several taps in each iteration by in-lining multiple tap calculations per loop.

```
I = filtLen >>2
for(k = 0;k<filtLen;k + +)
{
  sum + = coeffs[k] * state[stateIndex];
  stateIndex + +;
  sum + = coeffs[k] * state[stateIndex];
  stateIndex + +;
  sum + = coeffs[k] * state[stateIndex];
  stateIndex + +;
  sum + = coeffs[k] * state[stateIndex];
  stateIndex + +;
}
```

Now step into the third FIR function.

Set a breakpoint on the same inner loop and record the number of cycles it takes to run. Divide this by four and compare it to the previous implementations.

The next step is to make use of the SIMD instructions. By packing the coefficient and sample data into 32-bit words the single multiply accumulates can be replaced by dual signed multiply-accumulate which allows us to extend the loop unrolling from four summations to eight for the same number of cycles.

```
for(k = 0;k<filtLen;k + +)
{
  sum + = coeffs[k] * state[stateIndex];
  stateIndex + +;
  sum + = coeffs[k] * state[stateIndex];
  stateIndex + +;
  sum + = coeffs[k] * state[stateIndex];
  stateIndex + +;
  sum + = coeffs[k] * state[stateIndex];
  stateIndex + +;
}
```

Step into the fourth FIR function and again calculate the number of cycles used per tap for the inner loop.

Remember we are now calculating eight summations so divide the raw loop cycle count by eight.

To reduce the cycle count of the inner loop even further we can extend the loop unrolling to calculate several results simultaneously (Fig. 9.30).

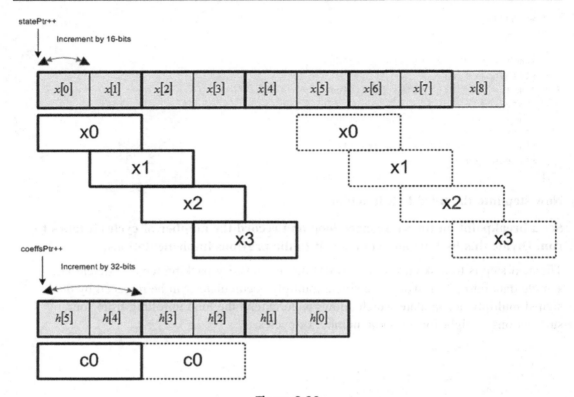

Figure 9.30
Super loop unrolling extends loop unrolling to process multiple output samples simultaneously.

This is kind of a "super loop unrolling" where we perform each of the inner loop calculations for a block of data in one pass.

```
sample = blockSize/4;
     do
     {
             sum0 = sum1 = sum2 = sum3 = 0;
             statePtr = stateBasePtr;
             coeffPtr = (q31_t *)(S->coeffs);
             x0 = *(q31_t *)(statePtr + +);
             x1 = *(q31_t *)(statePtr + +);
             i = numTaps>>2;
             do
             {
                     c0 = *(coeffPtr + +);
                     x2 = *(q31_t *)(statePtr + +);
                     x3 = *(q31_t *)(statePtr + +);
                     sum0 = __SMLALD(x0, c0, sum0);
                     sum1 = __SMLALD(x1, c0, sum1);
```

```
                              sum2 = __SMLALD(x2, c0, sum2);
                              sum3 = __SMLALD(x3, c0, sum3);
                              c0 = *(coeffPtr++);
                              x0 = *(q31_t *)(statePtr++);
                              x1 = *(q31_t *)(statePtr++);
                              sum0 = __SMLALD(x0, c0, sum0);
                              sum1 = __SMLALD(x1, c0, sum1);
                              sum2 = __SMLALD (x2, c0, sum2);
                              sum3 = __SMLALD (x3, c0, sum3);
                      } while(--i);
                      *pDst++ = (q15_t) (sum0>>15);
                      *pDst++ = (q15_t) (sum1>>15);
                      *pDst++ = (q15_t) (sum2>>15);
                      *pDst++ = (q15_t) (sum3>>15);

                      stateBasePtr = stateBasePtr + 4;
              } while(--sample);
```

Now step into the final FIR function and again calculate the number of cycles used by the inner loop per tap.

This time we are calculating eight summations for four taps simultaneously this brings us close to one cycle per tap which is comparable to a dedicated DSP device.

While you can code DSP algorithms in "C" and get reasonable performance these kinds of optimizations are needed to get performance levels comparable to a dedicated DSP device. This kind of code development needs experience of the Cortex-M4/M7 and the DSP algorithms you wish to implement. Fortunately Arm provides a free DSP library already optimized for the Cortex-M processors.

The CMSIS-DSP Library

While it is possible to code all of your own DSP functions, this can be time-consuming and requires a lot of domain-specific knowledge. To make it easier to add common DSP functions to your application, Arm has published a library of more than 100 common DSP functions which make up the CMSIS-DSP specification. Each of these functions is optimized for the Cortex-M4/M7 but can also be compiled to run on the Cortex-M3 and even on the Cortex-M0. The CMSIS-DSP library is a free download and is licensed for use in any commercial or noncommercial project. The CMSIS-DSP library is also included as part of the MDK-Arm installation and just needs to be added to your project by selecting the CMSIS::DSP option within the RTE manager. The installation includes a prebuilt library for each of the Cortex-M processors and all of the source code. Documentation for the library is included as part of the CMSIS help which may be opened from within the "Manage Run-Time Environment" window (Fig. 9.31).

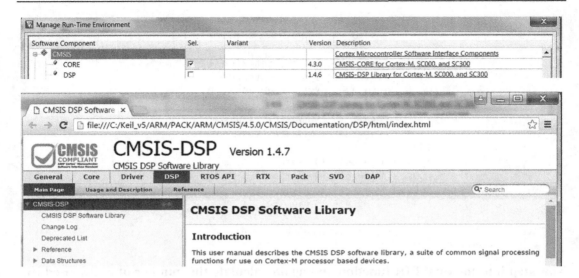

Figure 9.31
The CMSIS DSP documentation is accessed by clicking on the description link available in the Manage Run Time Environment window.

CMSIS-DSP Library Functions

The CMSIS-DSP library provides easy-to-use functions for the most commonly used signal processing algorithms. The key groups of functions included in the library are shown in Table 9.11.

Table 9.11: CMSIS DSP library functions

CMSIS-DSP Algorithm
Basic maths vector functions
Complex Maths Functions
Fast Maths functions
Interpolation functions
Matrix functions
Filtering functions
Statistical Functions
Support functions
Transform Functions
Controller functions
Quaternion
Classic Machine Learning functions

Exercise 9.4: Using the CMSIS-DSP Library

In this project, we will have a first look at using the CMSIS-DSP library by setting up a project to experiment with the PID Control algorithm (Fig. 9.32).

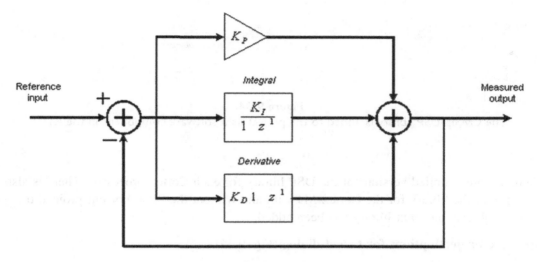

Figure 9.32
A PID control loop consists of Proportional, integral, and derivative control blocks.

Open the Pack Installer

Select the Boards::Designers Guide Tutorial.

Select the example tab and Copy "Ex 9.4 CMSIS-DSP PID."

The project is targeted at a Cortex-M4 with FPU (Fig. 9.33). It includes the CMSIS startup and system files for the Cortex-M4 and the CMSIS-DSP functions as a precompiled library.

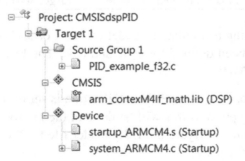

Figure 9.33
The PID example project is configured with the DSP added as a CMSIS software component.

The CMSIS-DSP library is located in c:\CMSIS\lib with subdirectories for the Arm compiler and GCC versions of the library and may be added from the CMSIS branch of the RTE (Fig. 9.34).

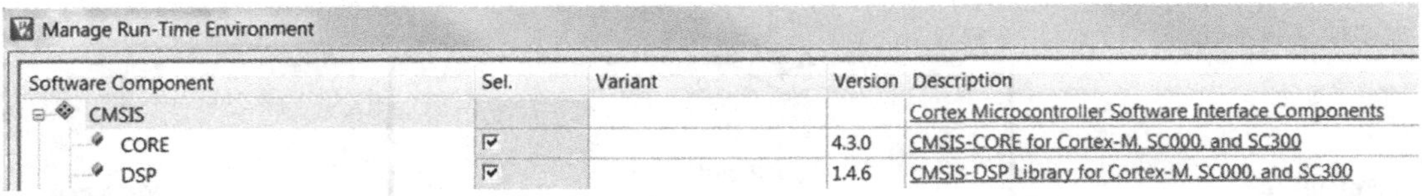

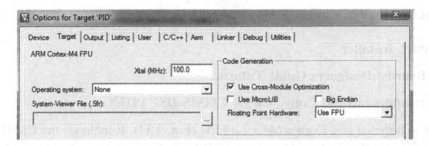

Figure 9.34
The CMSIS-DSP library is added to the project as a software component in the RTE.

There are precompiled versions of the DSP library for each Cortex processor. There is also a version of the library for the Cortex-M4 with and without the FPU. For our project, the Cortex-M4 floating-point library has been added.

Open the project\options for target dialog (Fig. 9.35).

Figure 9.35
The FPU is enabled in the Project Target settings.

In this example, we are using a simulation model for the Cortex-M4 processor only. A 512K memory region has been defined for code and data. The FPU is enabled and the CPU clock has been set to 100 MHz.

In the compiler options tab, the _FPU_PRESENT define is set to one. Normally you will not need to configure this option as this will be done in the microcontroller include file. We also need to add a #define to configure the DSP header file for the processor we are using (Fig. 9.36). The library defines are,

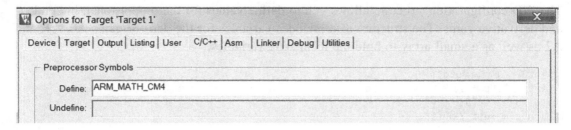

Figure 9.36
The FPU must be enabled by the startup code.

```
ARM_MATH_CM4
ARM_MATH_CM3
ARM_MATH_CM0
```

The source code for the library is located in C:\keil\arm\CMSIS\DSP_Lib\Source.

Open the file pid_example_f32.c which contains the application code.

To access the library we need to add its header file to our code.

```
#include "arm_math.h"
```

This header file is located in c:\keil\arm\cmsis\include.

In this project we are going to use the PID algorithm. All of the main functions in the DSP library have two function calls: an initializing function and a process function.

```
void arm_pid_init_f32 (arm_pid_instance_f32 *S,int32_t resetStateFlag)
__STATIC_INLINE void arm_pid_f32 (arm_pid_instance_f32 *s, float32_t in)
```

The initializing function is passed a configuration structure that is unique to the algorithm. The configuration structure holds constants for the algorithm, derived values, and arrays for state memory. This allows multiple instances of each function to be created.

```
typedef struct
  {
    float32_t A0;         /**< The derived gain, A0 = Kp + Ki + Kd. */
    float32_t A1;         /**< The derived gain, A1 = -Kp - 2Kd. */
    float32_t A2;         /**< The derived gain, A2 = Kd. */
    float32_t state[3];   /**< The state array of length 3. */
    float32_t Kp;         /**< The proportional gain. */
    float32_t Ki;         /**< The integral gain. */
    float32_t Kd;         /**< The derivative gain. */
  } arm_pid_instance_f32;
```

The PID configuration structure allows you to define values for the proportional, integral, and derivative gains. The structure also includes variables for the derived gains A0, A1, and A2 as well as a small array to hold the local state variables.

```
int32_t main(void){
int i; S.Kp = 1; S.Ki = 1; S.Kd = 1;
setPoint = 10;
arm_pid_init_f32 (&S,0);
while(1){
        error = setPoint-motorOut;
        motorIn = arm_pid_f32          (&S,error);
        motorOut = transferFunction(motorIn,time);
        time + = 1; for(i = 0;i<100000;i + + );
}}
```

The application code sets the PID gain values and initializes the PID function. The main loop calculates the error value before calling the PID process function. The PID output is fed into a simulated hardware transfer function. The output of the transfer function is fed back into the error calculation to close the feedback loop. The time variable provides a pseudo-time reference.

Build the project and start the debugger.

Add the key variables to the system analyzer and start the code running (Fig. 9.37).

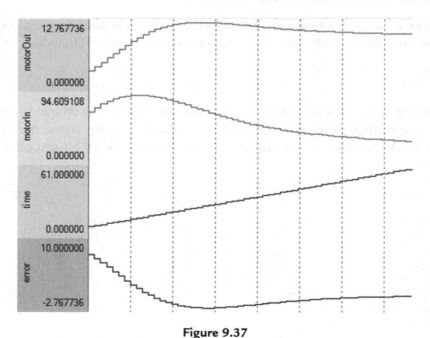

Figure 9.37
The Logic Analyzer is invaluable for visualizing real-time signals.

The system analyzer is invaluable for visualizing data in a real-time algorithm and can be used in the simulator or can capture data from the CoreSight Data Watch Trace Unit as we saw in Chapter 8, Debugging with CoreSight.

Experiment with the gain values to tune the PID function.

The performance of the PID control algorithm is tuned by adjusting the gain values. As a rough guide, each of the gain values has the following effects:

Kp Effects the rise time of the control signal
Ki Effects the steady-state error of the control signal
Kd Effects the overshoot of the control signal

DSP Data Processing Techniques

One of the major challenges of a DSP application is managing the flow of data from the sensors and ADC through the DSP algorithm and back out to the real world via the DAC (Fig. 9.38).

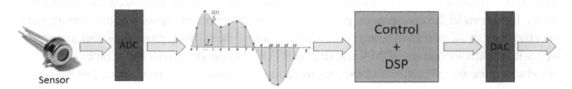

Figure 9.38
A typical DSP system consists of an analog sample stage, microcontroller with DSP algorithm, and an output DAC. In this chapter we microcontroller software in isolation from the hardware design.

In a typical system, each sampled value is a discrete value at a point in time. The sample rate must be at least twice the signal bandwidth or up to four times the bandwidth for a high-quality oversampled audio system. Clearly, the volume of data is going to ramp up very quickly and it becomes a major challenge to process the data in real time. In terms of processing the sampled data, there are two basic approaches, stream processing and block processing (Fig. 9.39).

Stream Processing

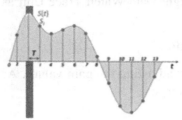

Block Processing

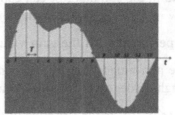

Figure 9.39
Analog data can be processed as single samples with minimum latency or as a block of samples for maximum processing efficiency.

In stream processing, each sampled value is processed individually. This gives the lowest signal latency and also minimum memory requirements. However, it has the disadvantage of making the DSP algorithm consume more processor cycles. The DSP algorithm has to be run every time an ADC conversion is made, which can cause problems with other high-priority interrupt routines. The alternative to stream processing is block processing. Here a number of ADC results are stored in a buffer, typically about 32 samples, and then this buffer is processed by the DSP algorithm as a block of data. This lowers the number of times that the DSP algorithm has to run. As we have seen in the optimization exercise, there are a number of techniques that can improve the efficiency of an algorithm when processing a block of data. Block processing also integrates well with the microcontroller DMA unit and an RTOS. On the downside, block processing introduces more signal latency and requires more memory than stream processing. For the majority of applications, block processing should be the preferred route.

Exercise 9.5: FIR Filter with Block Processing

In this exercise, we will implement an FIR filter, this time by using the CMSIS-DSP functions. This example uses the same project template as the PID program. The characteristics of the filter are defined by the filter coefficients. It is possible to calculate the coefficient values manually or by using a design tool. Calculating the coefficients is outside the scope of this book, but the appendices list some excellent design tools and DSP books for further reading.

Open the Pack Installer.

Select the Boards::Designers Guide Tutorial.

Select the example tab and Copy "Ex 9.5 CMSIS-DSP FIR."

This project can be built to run on different Cortex-M processors by selecting the built option from the drop down box on the toolbar (Fig. 9.40).

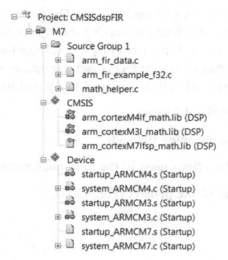

Figure 9.40
The FIR project has build options for the Cortex-M7, M4, and M3.

There is an additional data file that holds a sampled data set and an additional math_helper. c file which contains some ancillary functions.

```
int32_t main(void)
{
  uint32_t i;
  arm_fir_instance_f32 S;
  arm_status status;
  float32_t *inputF32, *outputF32;
  /* Initialize input and output buffer pointers */
  inputF32 = &testInput_f32_1kHz_15kHz[0];
  outputF32 = &testOutput[0];
  /* Call FIR init function to initialize the instance structure. */
  arm_fir_init_f32(&S, NUM_TAPS, (float32_t *)&firCoeffs32[0], &firStateF32[0],
blockSize);
  for(i = 0; i < numBlocks; i + +)
  {
   arm_fir_f32(&S, inputF32 + (i * blockSize), outputF32 + (i * blockSize), blockSize);
  }
  snr = arm_snr_f32(&refOutput[0], &testOutput[0], TEST_LENGTH_SAMPLES);
  if (snr < SNR_THRESHOLD_F32)
  {
   status = ARM_MATH_TEST_FAILURE;
  } else {
status = ARM_MATH_SUCCESS;
  }
```

The code first creates an instance of a 29-tap FIR filter. The processing block size is 32 bytes. An FIR state array is also created to hold the working state values for each tap. The size of this array is calculated as the block size + number of taps − 1. Once the filter has been initialized, we can pass it to the sample data in 32-byte blocks and store the resulting processed data in the output array. The final filtered result is then compared to a precalculated result.

Build the project and start the debugger.

Step through the project to examine the code.

Lookup the CMSIS-DSP functions used in the help documentation.

Set a breakpoint at the end of the project (Fig. 9.41).

```
209       if( status != ARM_MATH_SUCCESS)
210         {
211           while(1);
212         }
213
214       while(1);
215     }
```

Figure 9.41
Set a breakpoint on the final while() loop.

Reset the project and run the code until it hits the breakpoint.

Now look at the cycle count in the registers window (Fig. 9.42).

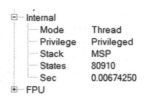

```
Internal
  Mode        Thread
  Privilege   Privileged
  Stack       MSP
  States      80910
  Sec         0.00674250
FPU
```

Figure 9.42
Use the register windows states counter to monitor the number of cycles executed.

Open the project in c:\exercises\CMSIS _FIR\CM3.

This is the same project built for the Cortex-M3.

Build the project and start the debugger.

Set a breakpoint in the same place as the previous Cortex-M4 example.

Reset the project and run the code until it hits the breakpoint.

Now compare the cycle count used by the Cortex-M3 to the Cortex-M4 version (Fig. 9.43).

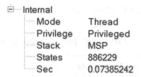

Figure 9.43
Compare the Cortex-M3 cycle count to the Cortex-M4.

When using floating-point numbers the Cortex-M4 is nearly and order faster than the Cortex-M3 using software floating-point libraries. When using fixed-point maths the Cortex-M4 still has a considerable advantage over the Cortex-M3 (Fig. 9.44).

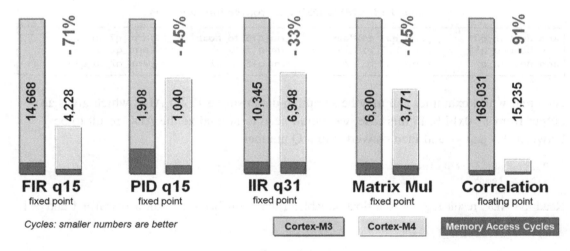

Figure 9.44
The bar graphs show a comparison between the Cortex M3 and Cortex M4 for common DSP algorithms.

Fixed Point DSP with Q Numbers

The functions in the DSP library support floating-point and fixed-point data. The fixed-point data is held in a Q number format. Q numbers are fixed-point fractional numbers held in integer variables. A Q number has a sign bit followed by a fixed number of bits to represent the integer value. The remaining bits represent the fractional part of the number. Signed = S IIIIIIIII. FFFFF. Signed Q numbers are stored as two's complement values. A Q number is typically referred to by the number of fractional bits it uses so Q10 has 10 fractional places. The CMSIS DSP library functions are designed to take input values between + 1 and − 1 so the fractional section takes all the bits of the data type minus one bit that is used as a sign bit. The supported integer values are shown in Table 9.12.

Table 9.12: Floating point unit performance

CMSIS DSP Type Def	Q number	C Type
Q31_t	Q31	Int 32
Q15_t	Q15	Int 16
Q7_t	Q7	Int 8

The library includes a group of conversion functions to change between floating-point numbers and the integer Q numbers. Support functions are also provided to convert between different Q number resolutions (Table 9.13). These functions are used to normalize the values between the different types.

Table 9.13: CMSIS DSP type conversion functions

arm_float_to_q31	arm_q31_to_float	arm_q15_to_float	arm_q7_to_float
arm_float_to_q15	arm_q31_to_q15	arm_q15_to_q31	arm_q7_to_q31
arm_float_to_q7	arm_q31_to_q7	arm_q15_to_q7	arm_q7_to_q15

As a real-world example, you may be sampling data using a 12-bit ADC which gives an output from 1−0xFFF. In this case, we would need to normalize the ADC result to be between +1 and -1 and then convert it to a Q number.

```
Q31_t   ADC_FixedPoint;
float   temp;
```

Read the ADC result register to a float variable. Then normalize the result between +1 and −1.

```
temp = ((float32_t)((ADC_DATA_REGISTER) & 0xFFF) / (0xFFF / 2)) - 1;
```

Convert the float value to a fixed point Q31 value.

```
arm_float_to_q31(&ADC_FixedPoint, &temp, 1);
```

Similarly, after the DSP function has run, it is necessary to convert back to a floating-point value before using the result. Here we are converting from a Q31 result to a 10-bit integer value prior to outputting the value to a DAC peripheral.

```
arm_q31_to_float(&temp, &DAC_Float, 1);
DAC_DATA_REGISTER = (((uint32_t)((DAC_Float))) & 0x03FF);
```

Exercise 9.6: Fixed Point FFT Transform

In this project, we will use the Fast Fourier transform to transform a signal into its component sine waves.

Open the Pack Installer.

Select the Boards::Designers Guide Tutorial.

Select the example tab and Copy "Ex 9.6 CMSIS-DSP FFT."

The FFT project uses the same template as the PID example with the addition of a data file that holds an array of sample data (Fig. 9.45). This data is a 10-kHz signal plus random "white noise."

Figure 9.45
The project is configured for the Cortex-M4. Signal data is included in the arm_fft_bin_data.c file.

```
int32_t main(void) {
        arm_status status;
        arm_cfft_radix4_instance_q31 S;
        q31_t maxValue;
        status = ARM_MATH_SUCCESS;
        /* Convert the floating point values the Q31 fixed point format */
        arm_float_to_q31 (testInput_f32_10khz,testInput_q31_10khz, 2048);
        /* Initialize the CFFT/CIFFT module */
        status = arm_cfft_radix4_init_q31(&S, fftSize, ifftFlag, doBitReverse);
        /* Process the data through the CFFT/CIFFT module */
        arm_cfft_radix4_q31(&S, testInput_q31_10khz);
        /* Process the data through the Complex Magnitude Module for
        calculating the magnitude at each bin */
        arm_cmplx_mag_q31(testInput_q31_10khz, testOutput, fftSize);
        /* Calculates maxValue and returns corresponding BIN value */
        arm_max_q31(testOutput, fftSize, &maxValue, &testIndex);
        if(testIndex! = refIndex)
        {
                status = ARM_MATH_TEST_FAILURE;
        }
```

The code initializes the Complex FFT transform with a block size of 1024 output bins. When the FFT function is called the sample data set is first converted to fixed point Q31 format and is then passed to the transform as one block. Next, the complex output is converted to a scalar magnitude and then scanned to find the maximum value. Finally, we compare this value to an expected result.

Work through the project code and look up the CMSIS DSP functions used.

Note: In an FFT transform the number of output bins will be half the size of the sample data size.

Build the project and start the debugger.

Execute the code function by function and record the run time required to perform the DSP functions.

Machine Learning

The CMSIS standards provide support for Machine Learning (ML) algorithms both within the CMSIS-DSP library and with a separate standard CMSIS-Neural Net. While a detailed discussion of these algorithms is outside the scope of this book, we can have a look at what support is provided by CMSIS-DSP and CMSIS-NN.

Most of the ML algorithms provided are standard functions that are customized for a particular application through a training process that is used to tune a set of constants. This training process is typically done on a PC or cloud computer. Algorithms such as a Neural Net will need a large quantity of labeled training data. For example, a set of images where the content of the image is known (cat, dog, etc.). This data has to be acquired and cleaned to make a suitable training data set. Once the neural net has been trained it can be used to recognize any of the objects it has been trained on. Other algorithms such as the distance functions are trained using unlabeled data. Here, we are detecting clusters of data and assigning a new data point to a cluster. These are very useful algorithms for spotting anomalies.

Classical Machine Learning

The Classical ML algorithms supported by CMSIS-DSP are outlined in Table 9.14.

Table 9.14: CMSIS DSP Classic machine learning algorithms

Algorithm	Description
Support Vector Machine	Supervised learning algorithm for classification of data groups
MFC Transform	Transform algorithm used to identify "features" in audio data streams
Distance Functions	Used to calculate distance from the centroid of a data class
Naive Gaussian Bayes Estimators	Estimate the likelihood of an outcome based on prior knowledge

Support Vector Machine

The Support Vector Machine (SVM) and distance functions are used for the classification of values within a dataset. The SVM is one of the most widely used ML algorithms and can provide linear and nonlinear classification of data within small and medium-sized datasets. The SVM in CMSIS-DSP is a binary classifier that can be used to assign a data point to one of two classes. The SVM classifies data by providing a decision boundary (Fig. 9.46) between different two clusters of data.

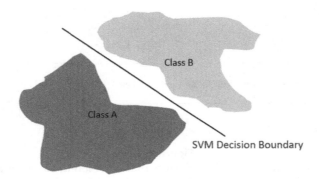

Figure 9.46
An SVM places a decision boundary between two classes within a data set.

When we want to classify a new data point, we simply need to see which side of the boundary our new point lies. However, many data sets are not linearly separable, in this case, the SVM can use nonlinear transforms to create a linearly separable dataset.

If we have a dataset that contains multiple classes, it is possible to take the cluster of data for each class and calculate its centroid. We can then use the distance functions to calculate the distance of a new data point to each class's centroid. The new data can then be assigned to the class with the nearest centroid.

Naive Gaussian Bayes Estimator

A Naïve Gaussian Bayes Estimator is used to determine the probability that an event such as a machine performance belongs to a class such as a breakdown based on prior knowledge of the distribution of elements that make up that event. For example, we can predict if a machine is faulty if we have a data set that contains probability distributions for features such as operating temperature, vibration, and cycle time. We can make measurements of its current operation and use the Naïve Gaussian Bayes estimator to calculate the likelihood that it is about to break down.

Mel Frequency Cepstral Transform

In sound processing, the Mel Frequency Cepstral Transform can be used to calculate the short-term power spectrum of a sound in the form of Cepstral coefficients. In speech processing, this technique can be used to identify the phonemes (units of sound) in a spoken word. The transform calculates the cepstral coefficients in a series of discrete steps. An outline of the algorithm is shown in Fig. 9.47.

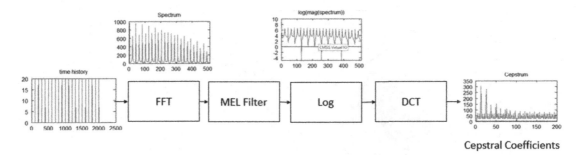

Cepstral Coefficients

Figure 9.47

The MFC transform generates a set of coefficients that correspond to the phonemes in an audio speech signal. *DCT*, Discrete Cosine Transform.

The MEL filter is used to scale the measured frequencies to the perceived frequency that a human ear can hear. The final output is a spectrum of a log of the spectrum of a time signal. The output peaks are the cepstral coefficients which represent a periodic element in the original time signal.

CMSIS-Neural Net

The CMSIS-NN specification provides a set of optimized kernel functions designed to efficiently implement neural network applications on Cortex-M processors. CMSIS-NN can be used to implement a range of common ML tasks such as keyword spotting and image recognition. An overview of the supported algorithms is shown in Table 9.15.

Table 9.15: CMSIS-NN functions

Algorithm	Description
Activation functions	Perceptron output activation functions
Concatenation functions	Supports stacking of channel layers in a CNN
Convolution functions	Feature extraction using convolution filters
Fully connected layer functions	A layer of perceptrons where every input is connected to the previous layer
Pooling functions	Generalizing features extracted by convolution filters
Reshape functions	Matrix transform
Softmax Functions	Output layer on a multiclass Neural Net.
SVDF Layer functions	Single Value Decomposition matrix functions

CMSIS-NN may be used with any ML framework such as Tensor flow, PyTouch, or Caffe. While these frameworks may have their own support for constrained devices, they mostly use floating-point numbers. Sine CMSIS-NN uses fixed-point Q numbers retargeting a model created in any of the above frameworks to use CMSIS-NN will result in a faster and more memory-efficient application.

While ML applications have been successfully developed to run on Armv7 processors and in particular the Cortex-M7 processor. The introduction of Armv8.1 and the Helium vector extension now provides a big boost in computing power that enables ML algorithms to execute on a new generation of Cortex-M processors.

Micro Neural processing Unit

Arm has also developed a coprocessor for Cortex-M processors which is designed to provide hardware support for ML algorithms. The Ethos-U55 is a first-generation Micro Neural Processing Unit. When used with a Cortex-M55 processor, ML and DSP algorithms can be accelerated by up to 32 times while still maintaining a low power consumption.

A NN model can be developed within an existing framework such as Tensor Flow. An Arm-provided NN Optimizer tool is then used to decide which layers of the NN are executed within the Ethos-U55 and which layers run on the Cortex-M processor using CMSIS-NN.

Conclusion

In this chapter, we have looked at the DSP extensions included in the Cortex-M4/M7 and how to make the best use of these features in a real application. The CMSIS-DSP library provides many common DSP functions that have been optimized for the Cortex-M4/M7. In Chapter 11, RTOS Techniques, we will look at how to integrate continuous real-time DSP processing and event-driven microcontroller code into the same project.

CHAPTER 10

Using a Real-Time Operating System

Introduction

In this chapter we will look at using a small footprint real-time operating system running on a Cortex-M microcontroller. If you are used to writing procedural-based "C" code on small 8- or 16-bit microcontrollers, you may be doubtful about the need for such an operating system. If you are not familiar with using an RTOS in real-time embedded systems, you should read this chapter before dismissing the idea. The use of an RTOS represents a more sophisticated design approach, inherently fostering structured code development which is enforced by the RTOS application programming interface (API).

The RTOS structure allows you to take a more object-orientated design approach while still programming in "C." The RTOS also provides you with Multithreaded support on a small microcontroller. These two features actually create quite a shift in design philosophy, moving us away from thinking about procedural "C" code and flow charts. Instead, we consider the fundamental program Threads and the flow of data between them. The use of an RTOS also has several additional benefits which may not be immediately apparent. Since an RTOS-based project is composed of well-defined Threads, it helps to improve project management, code reuse, and software testing.

The tradeoff for this is that an RTOS has additional memory requirements and increased interrupt latency. Typically, the Keil RTX5 RTOS will require 500 bytes of RAM and 5k bytes of code, but remember that some of the RTOS code would be replicated in your program anyway. We now have a generation of small, low-cost microcontrollers that have enough on-chip memory and processing power to support the use of an RTOS. Developing using this approach is therefore much more accessible.

In this chapter, we will first look at setting up an introductory RTOS project for a Cortex-M-based microcontroller. Next, we will go through each of the RTOS primitives and how they influence the design of our application code. Finally, when we have a clear understanding of the RTOS features, we will take a closer look at the RTOS configuration options. If you are used to programming a microcontroller without using an RTOS, that is, bare metal, there are two key things to understand as you work through this tutorial. In the first section, we will focus on creating and managing Threads. The key concept here is to consider them running as parallel concurrent objects. In the second section, we will look at how to communicate between Threads. In this section, the key concept is the synchronization of the concurrent Threads.

First Steps With CMSIS-RTOS2

The RTOS itself consists of a scheduler (Fig. 10.1) that supports round-robin, preemptive and cooperative multitasking of program Threads, as well as time and memory management services. Inter-Thread communication is supported by additional RTOS objects, including signal Thread and event flags, semaphores, mutex, message passing, and a memory pool system. As we will see, interrupt handling can also be accomplished by prioritized Threads that are scheduled by the RTOS kernel.

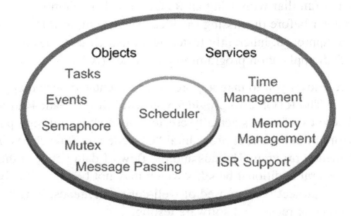

Figure 10.1
The RTOS kernel contains a scheduler that runs program code as Threads. Communication between Threads is accomplished by RTOS objects such as events, semaphores, mutexes, and mailboxes. Additional RTOS services include time and memory management and interrupt support.

Accessing the CMSIS-RTOS2 API

To access any of the CMSIS-RTOS2 features in our application code, it is necessary to include the following header file.

```
#include <cmsis_os2.h>
```

This header file is maintained by Arm as part of the CMSIS-RTOS2 standard. For the CMSIS-RTOS2 Keil RTX5, this is the default API. Other RTOS will have their own proprietary API but may provide a wrapper layer to implement the CMSIS-RTOS2 API so they can be used where compatibility with the CMSIS standard is required.

Threads

The building blocks of a typical "C" program are functions that we call to perform a
specific procedure and which then return to the calling function. In CMSIS-RTOS2, the
basic unit of execution is a thread.

```
unsigned int Procedure (void)   void Thread (void)      __NO_RETURN void Thread1(void*argument)
{                               {                       {
                                  while(1)                while(1)
    ......                        {                       {
                                    ......                  .......
    return(ch);                   }                       }
}                               }                       }
```

While we always return from our "C" function, once started, an RTOS Thread must contain
a loop so that it never terminates and thus runs forever. You can think of a Thread as a
mini self-contained program that runs within the RTOS. With the Arm Compiler, it is
possible to optimize a Thread by using a __NO_RETURN macro. This attribute reduces the
cost of calling a function that never returns.

Understanding the Scheduler

An RTOS program is made up of a number of Threads that are controlled by the RTOS
scheduler. This scheduler uses the SysTick timer to generate a periodic interrupt as a time base.
The scheduler will allot a certain amount of execution time to each thread. So Thread1 will run
for 5 ms and then be descheduled to allow Thread2 to run for a similar period; Thread2 will give
way to Thread3, and finally, control passes back to Thread1. By allocating these slices of runtime
to each thread in a round-robin fashion, we get the appearance of all three Threads running in
parallel to each other. It is also important to realize that the scheduler is also triggered by any
RTOS function call. This ensures that the scheduler responds to the application programs activity
to maintain its real-time performance. We will look at this more closely in the next chapter. So
for now conceptually we can think of each thread as performing a specific functional unit of our
program with all Threads running simultaneously and responding to RTOS system calls instantly.
This leads us to a more object-orientated design, where each functional block can be coded and
tested in isolation and then integrated into a fully running program. This not only imposes a
structure on the design of our final application but also aids debugging, as a particular bug can be
easily isolated to a specific Thread. It also aids code reuse in later projects. When a Thread is
created, it is also allocated its own Thread ID. This is a variable that acts as a handle for each
thread and is used when we want to manage the activity of the thread.

```
osThreadId_t id1,id2,id3;
```

In order to make the Thread-switching process happen, we have the code overhead of the
RTOS, and we have to dedicate a CPU hardware timer to provide the RTOS time reference.

In addition, each thread is allocated two blocks of memory (Fig. 10.2). A Thread stack space is used to hold all of the variables declared within the Thread code. Each time we switch and resume execution of a Thread, we also switch to its associated stack space. Second, all the runtime information about a Thread is stored in a Thread control block, which is managed by the RTOS kernel. Thus the "context switch time," that is, the time to save the current Thread state and load up and start the next thread, is a crucial figure and will depend on both the RTOS kernel and the design of the underlying hardware.

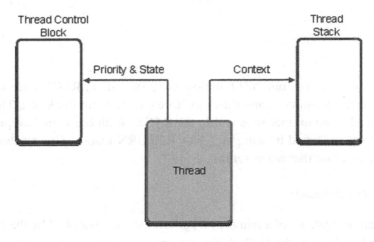

Figure 10.2
Each Thread has its own stack for saving its data during a context switch. The Thread control block is used by the kernel to manage the active Thread.

The Thread Control Block contains information about the status of a Thread. Part of this information is its run state. In a given system, only one thread can be running, and all the others will be suspended but ready to run. The RTOS has various methods of inter-Thread communication (signals, semaphores, messages). Here, a Thread may be suspended to wait to be signaled by another Thread or interrupt before it resumes its ready state, at which point it can be placed into running state by the RTOS scheduler. Each thread has a lifecycle that switches between these different states as the application runs (Table 10.1).

Table 10.1: At any given moment a single thread may be running. The remaining threads will be ready to run and will be scheduled by the kernel. Threads may also be waiting pending an OS event. When this occurs, they will return to the ready state and be scheduled by the kernel

State	Description
Running	The Currently Running Thread
Ready	Threads ready to Run
Wait	Blocked Threads waiting for an OS Event

Starting the RTOS

To build a simple RTOS program, we declare each thread as a standard "C" function and also declare a Thread ID variable for each function.

```
void Thread1 (void);
void Thread2 (void);

osThreadId thrdID1, thrdID2;
```

Once the processor leaves the reset vector, we will enter the main() function as normal. Once in main(), we must call osKernelInitialize() to setup the RTOS. It is not possible to call an RTOS function before the osKernelInitialize() function has successfully completed. Once osKernelInitialize() has completed we can create further Threads and other RTOS objects. This can be done by creating a launcher Thread. In the example below, this is called app_main(). Inside the app_main() thread, we create all the RTOS Threads and objects we need to start our application running. As we will see later, it is also possible to dynamically create and destroy RTOS objects as the application is running. Next, we can call osKernelstart() to start the RTOS and the scheduler task switching. You can run any initializing code you want before starting the RTOS to setup peripherals and initialize the hardware.

```
void app_main (void *argument)
{
        T_led_ID1 = osThreadNew(led_Thread1, NULL, &ThreadAttr_LED1);
        T_led_ID2 = osThreadNew(led_Thread2, NULL,&ThreadAttr_LED2);
        osDelay(osWaitForever);
        while(1);
}
void main (void)
{
  IODIR1 = 0x00FF0000;                // Do any C code you want
  osKernelInitialize ();
  osThreadNew(app_main, NULL, NULL);  //Create the app_main() launcher Thread
  osKernelStart();                    //Start the RTOS
}
```

When Threads are created, they are also assigned a priority. If there are a number of Threads ready to run and they all have the same priority, they will be allotted run time in a round-robin fashion (Fig. 10.3). However, if a Thread with a higher priority becomes ready to run, the RTOS scheduler will de-schedule the currently running thread and start the high-priority Thread running. This is called preemptive priority-based scheduling. When assigning priorities, you have to be careful because the high-priority Thread will continue to run until it enters a waiting state or until a Thread of equal or higher priority is ready to run.

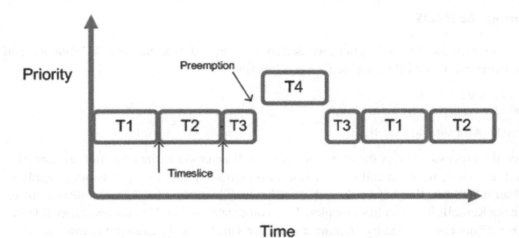

Figure 10.3
Threads of equal priority will be scheduled in a round-robin fashion. High priority Thread will preempt low priority Threads and enter the running state "on demand."

Exercise 10.1: A First CMSIS-RTOS2 Project

This project will take you through the steps necessary to create and debug a CMSIS-RTOS2-based project.

In the pack installer, select "Ex 10.1" and press the copy button.

This first project is a multiproject workspace. The shell project is set as the active project. A prebuilt working project is included as a reference. If you want to build this project, highlight the project, right-click and select "Set as active project" (Fig. 10.4). Any compile and debug actions will work on the active project.

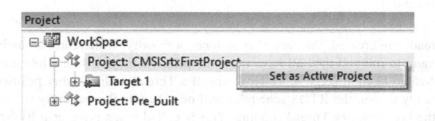

Figure 10.4
Select the first project workspace and make it the active project.

Open the Run Time Environment (RTE) by selecting the green diamond on the toolbar (Fig. 10.5).

Figure 10.5
Select and click the Run Time Environment icon.

We can now add the RTOS code to our project.

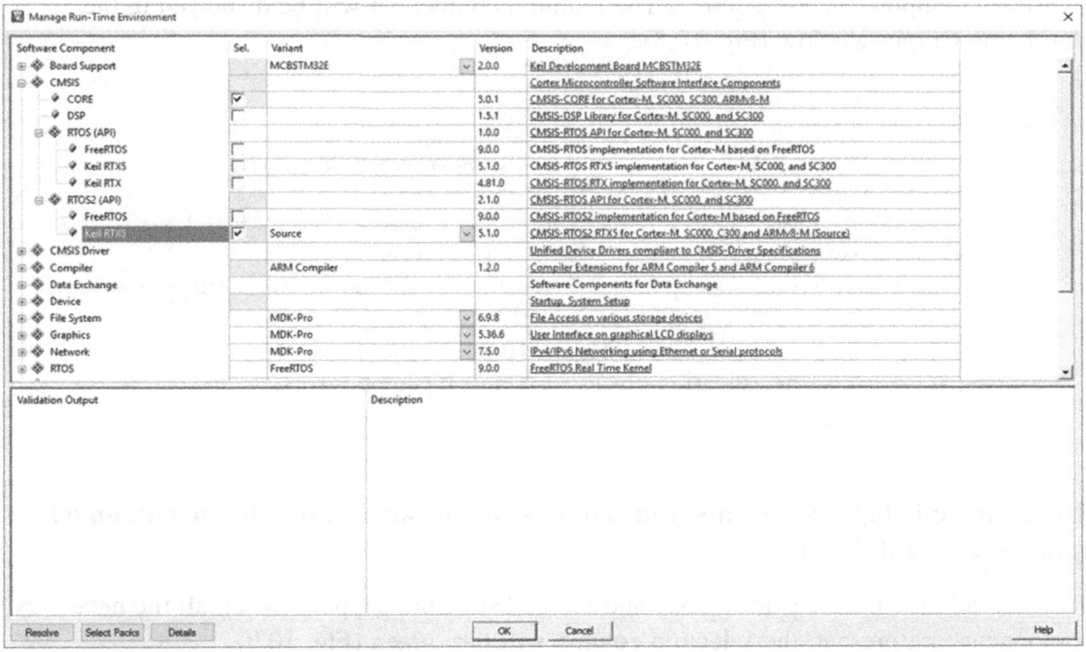

Software Component	Sel.	Variant	Version	Description
⊞ ◆ Board Support		MCBSTM32E	∨ 2.0.0	Keil Development Board MCBSTM32E
⊟ ◆ CMSIS				Cortex Microcontroller Software Interface Components
◆ CORE	✓		5.0.1	CMSIS-CORE for Cortex-M, SC000, SC300, ARMv8-M
◆ DSP	☐		1.5.1	CMSIS-DSP Library for Cortex-M, SC000, and SC300
⊟ ◆ RTOS (API)			1.0.0	CMSIS-RTOS API for Cortex-M, SC000, and SC300
◆ FreeRTOS	☐		9.0.0	CMSIS-RTOS implementation for Cortex-M based on FreeRTOS
◆ Keil RTX5	☐		5.1.0	CMSIS-RTOS RTX5 implementation for Cortex-M, SC000, and SC300
◆ Keil RTX	☐		4.81.0	CMSIS-RTOS RTX implementation for Cortex-M, SC000, and SC300
⊟ ◆ RTOS2 (API)			2.1.0	CMSIS-RTOS API for Cortex-M, SC000, and SC300
◆ FreeRTOS	☐		9.0.0	CMSIS-RTOS2 implementation for Cortex-M based on FreeRTOS
◆ Keil RTX5	✓	Source	∨ 5.1.0	CMSIS-RTOS2 RTX5 for Cortex-M, SC000, C300 and ARMv8-M (Source)
⊞ ◆ CMSIS Driver				Unified Device Drivers compliant to CMSIS-Driver Specifications
⊞ ◆ Compiler		ARM Compiler	1.2.0	Compiler Extensions for ARM Compiler 5 and ARM Compiler 6
⊞ ◆ Data Exchange				Software Components for Data Exchange
⊞ ◆ Device				Startup, System Setup
⊞ ◆ File System		MDK-Pro	∨ 6.9.8	File Access on various storage devices
⊞ ◆ Graphics		MDK-Pro	∨ 5.36.6	User Interface on graphical LCD displays
⊞ ◆ Network		MDK-Pro	∨ 7.5.0	IPv4/IPv6 Networking using Ethernet or Serial protocols
⊞ ◆ RTOS		FreeRTOS	9.0.0	FreeRTOS Real Time Kernel

Validation Output		Description

Resolve | Select Packs | Details   OK | Cancel   Help

Figure 10.6
Add the RTOS.

To configure the project for use with the CMSIS-RTOS2 Keil RTX, simply tick the CMSIS::RTOS2 (API):Keil RTX5 box (Fig. 10.6).

Switch the Keil RTX5 dropdown variant box from "Source" to "Library."

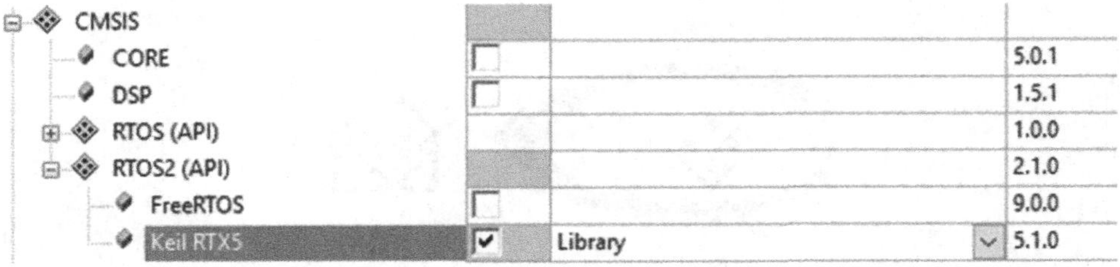

Figure 10.7
If the Sel column elements turn Orange then the RTOS requires other components to be added.

Addition of the RTOS will cause the selection box to turn orange (Fig. 10.7), meaning that additional components are required. The required component will be displayed in the Validation Output window (Fig. 10.8).

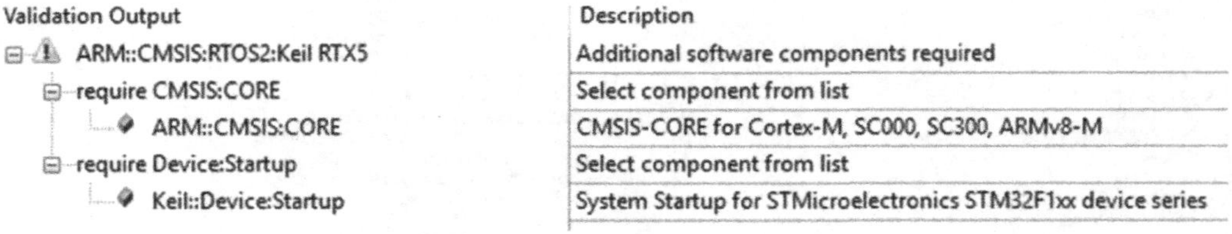

Figure 10.8
The validation box lists the missing components.

To add the missing components, you can press the Resolve button in the bottom left-hand corner of the RTE.

This will add the device startup code and the CMSIS Core support. When all the necessary components are present, the selection column will turn green (Fig. 10.9).

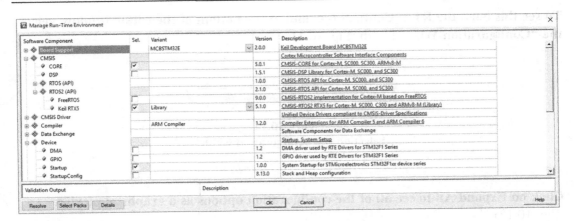

Figure 10.9
Pressing the resolve button adds the missing components and the Sel. column turns green.

Now press the OK button, and all the selected components will be added to the new project (Fig. 10.10).

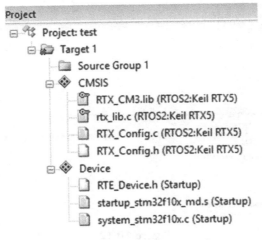

Figure 10.10
The configured project platform.

The CMSIS components are added to folders displayed as a green diamond. There are two types of file here. The first type is a library file that is held within the tool chain and is not editable. This file is shown with a yellow key to show that it is "locked" (read-only). The second type of file is a configuration file. These files are copied to your project directory and can be edited as necessary. Each of these files may be displayed as text files, but it is also possible to view the configuration options as a set of pick lists and drop-down menus.

To see this open the RTX_Config.h file, and at the bottom of the editor window, select the "Configuration Wizard" tab (Fig. 10.11).

Figure 10.11
Selecting the configuration wizard.

Click on Expand All to see all of the configuration options as a graphical picklist (Fig. 10.12).

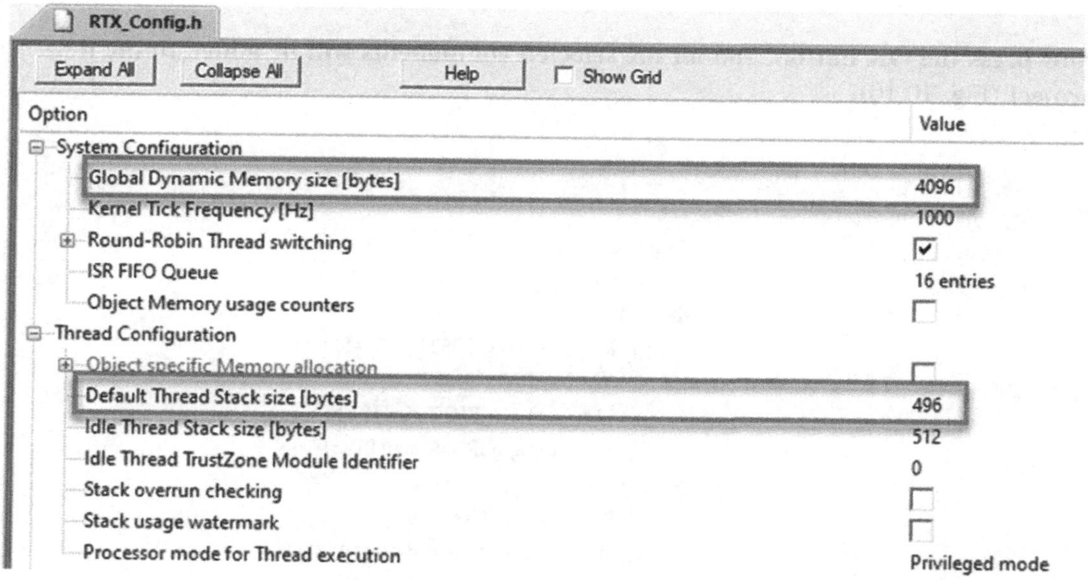

Option	Value
□ System Configuration	
Global Dynamic Memory size [bytes]	4096
Kernel Tick Frequency [Hz]	1000
⊞ Round-Robin Thread switching	☑
ISR FIFO Queue	16 entries
Object Memory usage counters	□
□ Thread Configuration	
⊞ Object specific Memory allocation	□
Default Thread Stack size [bytes]	496
Idle Thread Stack size [bytes]	512
Idle Thread TrustZone Module Identifier	0
Stack overrun checking	□
Stack usage watermark	□
Processor mode for Thread execution	Privileged mode

Figure 10.12
The RTX configuration options.

We need to make two changes to the default RTX configuration file. The default Global Dynamic Memory size is 32 K. This is too big for the simulated microcontroller that we are using.

Set the SystemConfiguration\Global Dynamic Memory pool to 4096 bytes.

We also need to reduce the amount of memory allocated to each Thread.

Set the Thread Configuration\Default Thread Stack size to 496 bytes.

Our project contains four configuration files, three of which are standard CMSIS files (Table 10.2).

Table 10.2: Project configuration files

File Name	Description
Startup_STM32F10x_md.s	Assembler vector table
System_STM32F10x.c	C code to initialize key system peripherals, such as clock tree, PLL external memory interface.
RTE_Device.h	Configures the pin multiplex for CMSIS Drivers
RTX_Config.h	Configures Keil RTX

Now that we have the basic platform for our project in place, we can add some user source code that will start the RTOS and create a running Thread.

Right-click the "Source Group 1" folder and select "Add new item to Source Group 1" (Fig. 10.13).

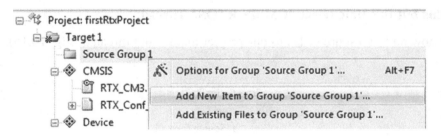

Figure 10.13
Adding a source module.

In the Add new Item dialog select the "User code template" Icon and in the CMSIS section, select the "CMSIS-RTOS 'main' function" then click Add (Fig. 10.14).

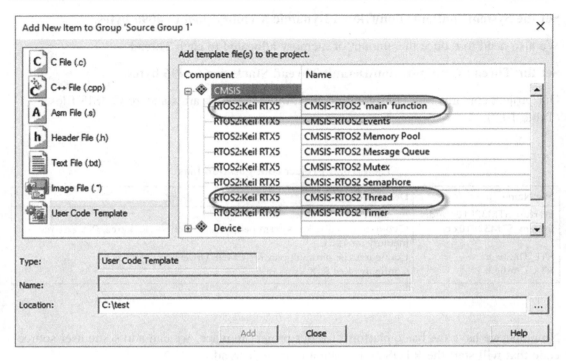

Figure 10.14
Selecting a CMSIS RTOS template.

Repeat this but this time select "CMSIS-RTOS2 Thread."

This will now add two source files to our project main.c and thread.c (Fig. 10.15).

Figure 10.15
The project with main and Thread code.

Open thread.c in the editor.

We will look at the RTOS definitions in this project in the next section. For now, this file contains two functions Init_Thread() which are used to start the thread running and the actual Thread function.

Copy the Init_Thread function prototype and then open main.c.

Main contains the functions to initialize and start the RTOS kernel. Then unlike a bare-metal project main is allowed to terminate rather than enter an endless loop. However, this is not really recommended, and we will look at a more elegant way of terminating a Thread later.

In main.c add the Init_Thread prototype as an external declaration and then call it after the osKernelInitialize() function as shown below.

```
extern int Init_Thread (void);
void app_main (void *argument) {
        Init_Thread ();
        for (;;) {}
}
```

Build the project (F7).

Exit the debugger.

While this project does not actually do anything, it demonstrates the few steps necessary to start using CMSIS-RTOS2.

Advanced Debug Features

The Microvision debugger provides extended debug support for the RTX RTOS in the form of a component view and also event recorder messages. These features can be used to gain a detailed view of how the RTOS code is running on the microcontroller.

Exercise 10.1: Continued—RTOS Debugger Support

Now that we have built a first RTOS project, we can load it into the debugger and see the additional debug support provided for the RTOS.

Start the debugger (Ctrl + F5).

This will run the code up to main().

Open the Debug \View \Watch Windows\RTX RTOS (Fig. 10.16).

Start the code running (F5).

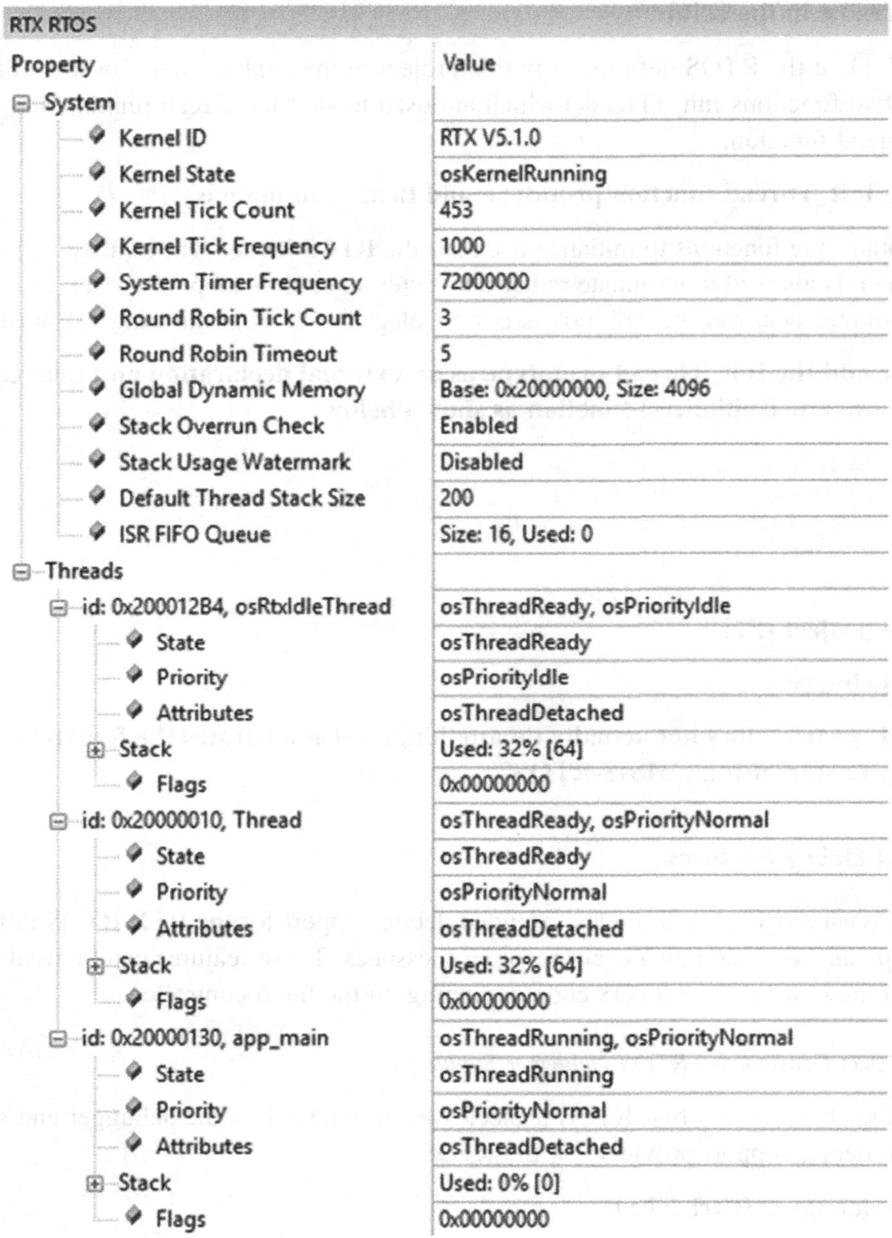

RTX RTOS	
Property	**Value**
⊟ System	
🔷 Kernel ID	RTX V5.1.0
🔷 Kernel State	osKernelRunning
🔷 Kernel Tick Count	453
🔷 Kernel Tick Frequency	1000
🔷 System Timer Frequency	72000000
🔷 Round Robin Tick Count	3
🔷 Round Robin Timeout	5
🔷 Global Dynamic Memory	Base: 0x20000000, Size: 4096
🔷 Stack Overrun Check	Enabled
🔷 Stack Usage Watermark	Disabled
🔷 Default Thread Stack Size	200
🔷 ISR FIFO Queue	Size: 16, Used: 0
⊟ Threads	
⊟ id: 0x200012B4, osRtxIdleThread	osThreadReady, osPriorityIdle
🔷 State	osThreadReady
🔷 Priority	osPriorityIdle
🔷 Attributes	osThreadDetached
⊞ Stack	Used: 32% [64]
🔷 Flags	0x00000000
⊟ id: 0x20000010, Thread	osThreadReady, osPriorityNormal
🔷 State	osThreadReady
🔷 Priority	osPriorityNormal
🔷 Attributes	osThreadDetached
⊞ Stack	Used: 32% [64]
🔷 Flags	0x00000000
⊟ id: 0x20000130, app_main	osThreadRunning, osPriorityNormal
🔷 State	osThreadRunning
🔷 Priority	osPriorityNormal
🔷 Attributes	osThreadDetached
⊞ Stack	Used: 0% [0]
🔷 Flags	0x00000000

Figure 10.16
The RTX5 component viewer.

This window is a component view that we saw in Chapter 8, Debugging With CoreSight. It provides a detailed diagnostic view of the current RTOS state.

The event recorder has also been configured in the original project. The RTOS source code is instrumented with a set of event recorder messages which provide a detailed log of its activity as the code runs.

Open view\Analysis\Event Recorder (Fig. 10.17).

This will display a detailed trace of RTOS events.

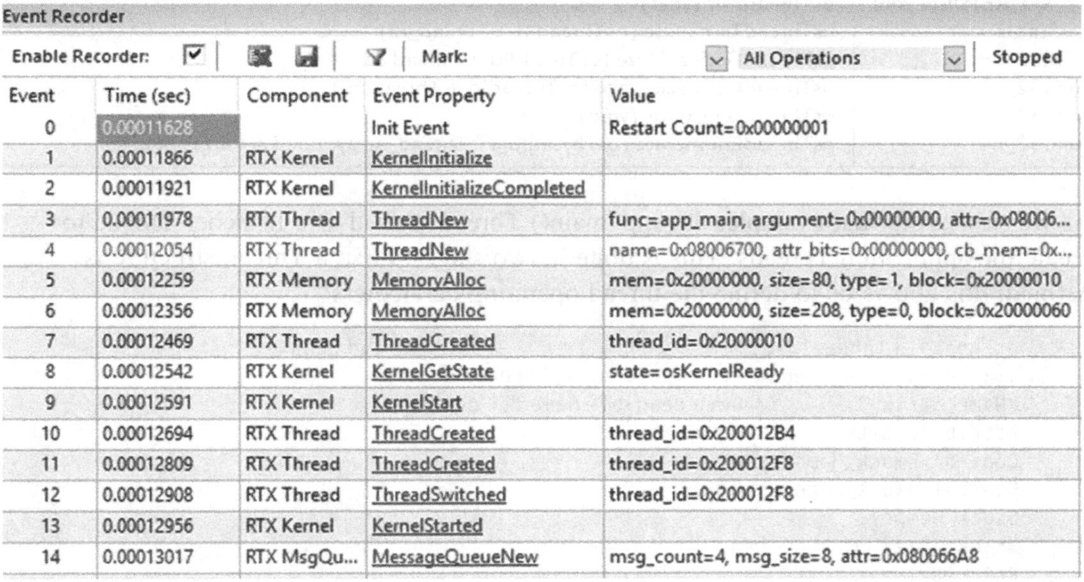

Event	Time (sec)	Component	Event Property	Value
0	0.00011628		Init Event	Restart Count=0x00000001
1	0.00011866	RTX Kernel	KernelInitialize	
2	0.00011921	RTX Kernel	KernelInitializeCompleted	
3	0.00011978	RTX Thread	ThreadNew	func=app_main, argument=0x00000000, attr=0x08006...
4	0.00012054	RTX Thread	ThreadNew	name=0x08006700, attr_bits=0x00000000, cb_mem=0x...
5	0.00012259	RTX Memory	MemoryAlloc	mem=0x20000000, size=80, type=1, block=0x20000010
6	0.00012356	RTX Memory	MemoryAlloc	mem=0x20000000, size=208, type=0, block=0x20000060
7	0.00012469	RTX Thread	ThreadCreated	thread_id=0x20000010
8	0.00012542	RTX Kernel	KernelGetState	state=osKernelReady
9	0.00012591	RTX Kernel	KernelStart	
10	0.00012694	RTX Thread	ThreadCreated	thread_id=0x200012B4
11	0.00012809	RTX Thread	ThreadCreated	thread_id=0x200012F8
12	0.00012908	RTX Thread	ThreadSwitched	thread_id=0x200012F8
13	0.00012956	RTX Kernel	KernelStarted	
14	0.00013017	RTX MsgQu...	MessageQueueNew	msg_count=4, msg_size=8, attr=0x080066A8

Figure 10.17
The event viewer shows a trace of the RTOS activity.

Creating Threads

Once the RTOS is running, there are a number of system calls that are used to manage and control the active Threads as shown in Table 10.3.

Table 10.3: RTOS thread management functions

Thread Management Functions	
osThreadId_t	osThreadNew (os_Thread_func_t func, void *argument, const osThreadAttr_t *attr)
osThreadId_t	osThreadGetId (void)
osThreadState_t	osThreadGetState (osThreadId_t Thread_id)
osStatus_t	osThreadSetPriority (osThreadId_t Thread_id, osPriority_t priority)
osPriority_t	osThreadGetPriority (osThreadId_t Thread_id)
osStatus_t	osThreadYield (void)
osStatus_t	osThreadSuspend (osThreadId_t Thread_id)
osStatus_t	osThreadResume (osThreadId_t Thread_id)
osStatus_t	osThreadDetach (osThreadId_t Thread_id)
osStatus_t	osThreadJoin (osThreadId_t Thread_id)
__NO_RETURN void	osThreadExit (void)
osStatus_t	osThreadTerminate (osThreadId_t Thread_id)
uint32_t	osThreadGetStackSize (osThreadId_t Thread_id)
uint32_t	osThreadGetStackSpace (osThreadId_t Thread_id)
uint32_t	osThreadGetCount (void)
uint32_t	osThreadEnumerate (osThreadId_t *Thread_array, uint32_t array_items)

As we saw in the first example, the app_main() Thread is used as a launcher Thread to create the application Threads. This is done in two stages. First, a Thread structure is defined; this allows us to define the thread operating parameters.

```
osThreadId Thread1_id;                        //Thread handle
static const osThreadAttr_t ThreadAttr_Thread1 = {
"Name_String",       //Human readable Name for debugger
Attribute_bits
Control_Block_Memory,
Control_Block_Size,
Stack_Memory,
Stack_Size,
Priority,
TrustZone_ID,
reserved
};
```

The Thread structure requires us to define the name of the Thread function, its Thread priority, any special attribute bits, its TrustZone_ID, and its memory allocation. This is quite a lot of detail to go through, but we will cover everything by the end of this chapter. Once the Thread structure has been defined, the thread can be created using the osThreadNew() API call. Then the thread is created from within the application code, this is often within the app_main() thread, but a Thread can be created at any point within any Thread.

```
Thread1_id = osThreadNew(name_Of_C_Function, argument,&ThreadAttr_Thread1);
```

This creates the thread and starts it running. It is also possible to pass a parameter to the thread when it starts.

```
uint32_t startupParameter = 0x23;
Thread1_id = osThreadNew(name_Of_C_Function, (uint32_t)startupParameter,
  &ThreadAttr_Thread1);
```

Exercise 10.2: Creating and Managing Threads

In this project, we will create and manage some additional Threads. Each of the Threads created will toggle a GPIO pin on GPIO port B to simulate flashing an LED. We can then view this activity in the simulator.

In the pack installer, select the example tab and Copy "Ex 10.2 CMSIS-RTOS2 Threads."

This will install the project to a directory of your choice and open the project in Microvision.

Open the Run-Time Environment Manager.

In the board support section ensure that the MCBSTM32E:LED box is ticked (Fig. 10.18) **and resolve its subcomponents so the sel column is green.**

This adds support functions to control the state of a bank of LEDs on the Microcontroller's GPIO port B.

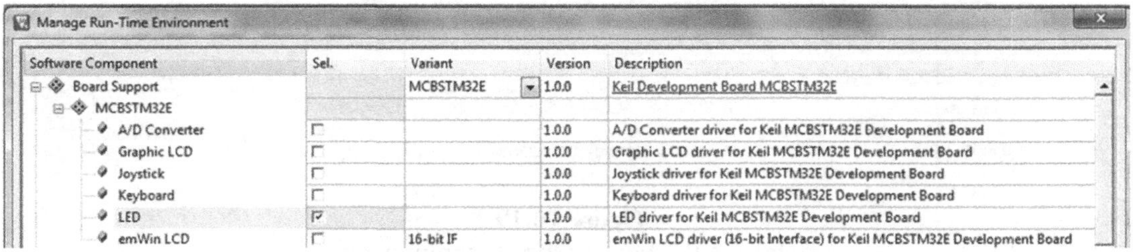

Figure 10.18
Selecting the board support components.

As in the first example main() creates app_main() and starts the RTOS. Inside app_main(), we create two additional Threads. First, we create handles for each of the Threads and then define the structures for each thread. The structures are defined in two different ways, for app_main we define the full structure and use NULL to inherit the default values.

```
static const osThreadAttr_t ThreadAttr_app_main = {
    "app_main",
    NULL,
    NULL,
    NULL,
```

```
        NULL,
        NULL,
        osPriorityNormal,
        NULL,
        NULL
    };
```

For the Thread LED1 a truncated syntax is used as shown below:

```
static const osThreadAttr_t ThreadAttr_LED1 = {
    .name = "LED_Thread_1",
};
```

In order to use this syntax, the compiler options must be changed to allow C99 declarations.

Select the Project\Options for Target\ C/C++ tab and make sure the C99 language option is selected (Fig. 10.19).

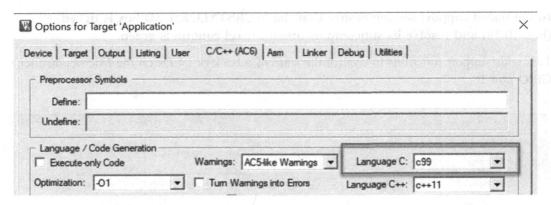

Figure 10.19
Selecting the board support components.

Now app_main() is used to first initialize the bank of LEDs and then create the two Threads. Finally app_main() is terminated with the osThreadExit() api call.

```
void app_main (void *argument) {
    LED_Initialize ();
    led_ID1 = osThreadNew(led_Thread1, NULL, &ThreadAttr_LED1);
    led_ID1 = osThreadNew(led_Thread2, NULL, &ThreadAttr_LED2);
    osThreadExit();
}
```

Build the project and start the debugger.

Start the code running and open the View\watch windows\RTX_RTOS (Fig. 10.20).

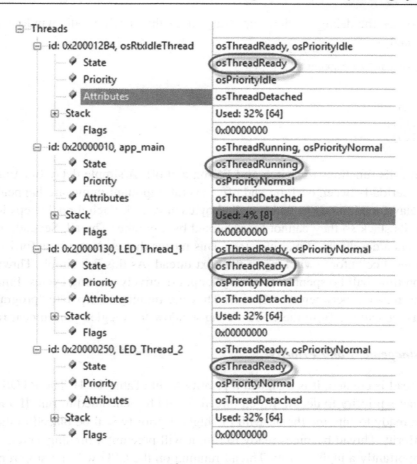

Figure 10.20
The running Threads.

Now we have four active Threads with one running and the others ready to run.

Now open the Peripherals\General Purpose IO\GPIOB window (Fig. 10.21).

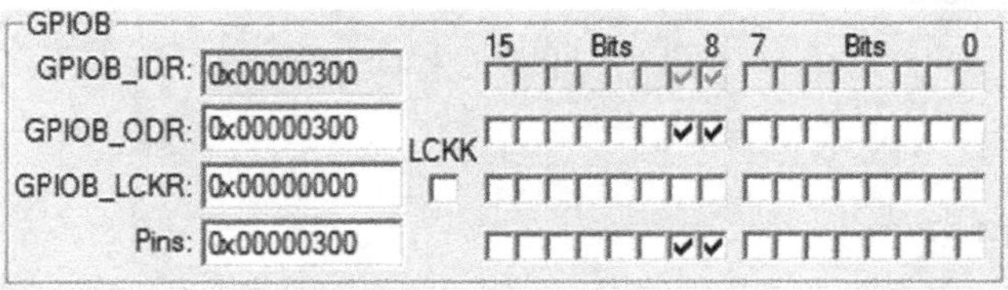

Figure 10.21
The peripheral window shows the LED pin activity.

Our two led Threads are each toggling a GPIO port pin. Leave the code running and watch the pins toggle for a few seconds.

If you do not see the debug windows updating check the view\periodic window update option is ticked.

```
void led_Thread2 (void const *argument) {
for (;;) {
  LED_On(1);
  delay(500);
  LED_Off(1);
  delay(500);
}}
```

Each thread calls functions to switch an LED on and off. A simple delay function is used to provide the period between each on and off. Several important things are happening here. First, the delay function can be safely called by each thread. Each thread keeps local variables in its stack so they cannot be corrupted by any other Thread. Second, none of the Threads enters a descheduled waiting state. This means that each one runs for its full allocated time slice before switching to the next thread. As this is a simple Thread, most of its execution time will be spent in the delay loop, effectively wasting cycles. Finally, there is no synchronization between the Threads. They are running as separate "programs" on the CPU, and as we can see from the GPIO debug window the toggled pins appear random.

Thread Management and Priority

When a Thread is created, it is assigned a priority level (Table 10.4). The RTOS scheduler uses a Thread's priority to decide which thread should be scheduled to run. If a number of Threads are ready to run, the thread with the highest priority will be placed in the run state. If a high-priority Thread becomes ready to run, it will preempt a running Thread of lower priority. Importantly a high-priority Thread running on the CPU will not stop running unless it blocks on an RTOS API call or is preempted by a higher priority Thread. A Thread's priority is defined in the Thread structure, and the following priority definitions are available. The default priority is osPriorityNormal.

Table 10.4: RTOS priority levels

CMSIS RTOSv2 priority levels
osPriorityIdle
osPriorityLow
osPriorityLow 1—7
osPriorityBelowNormal
osPriorityBelowNormal 1—7
osPriorityNormal
osPriorityNormal 1—7
osPriorityHigh
osPriorityHigh 1—7
osPriorityRealTime
osPriorityRealTime 1—7
osPriorityError

It is also then possible to elevate or lower a Thread's priority either from another function or from within its own code.

```
osStatus  osThreadSetPriority(ThreadID, priority);
osPriority osThreadGetPriority(ThreadID);
```

Once the Threads are running, there are a small number of RTOS system calls that are used to manage the running Threads. As well as creating Threads, it is also possible for a Thread to delete another active Thread from the RTOS. Again, we use the Thread ID rather than the function name of the thread.

```
osStatus = osThreadTerminate (ThreadID1);
```

If a Thread wants to terminate itself, then there is a dedicated exit function.

```
osThreadExit (void)
```

Finally, there is a special case of Thread switching where the running thread passes control to the next ready thread of the same priority. This is used to implement a third form of scheduling called cooperative Thread switching.

```
osStatus osThreadYield();        //switch to next ready to run Thread at the same priority
```

Exercise 10.2: Continued—Creating and Managing Threads

In this exercise, we will look at assigning different priorities to Threads and also how to create and terminate Threads dynamically.

Go back to the project "Ex 10.2 Threads."

Change the priority of LED Thread 1 to Above Normal.

```
static const osThreadAttr_t ThreadAttr_LED1 = {
    "LED_Thread_1",
     NULL,
     NULL,
     NULL,
     NULL,
     NULL,
     osPriorityAboveNormal,
     NULL,
     NULL
};
```

Build the project and start the debugger.

Start the code running.

Open main.c in the editor and locate Led_thread1().

The coverage monitor will mark code that has been executed with a green block in the margin (Fig. 10.22).

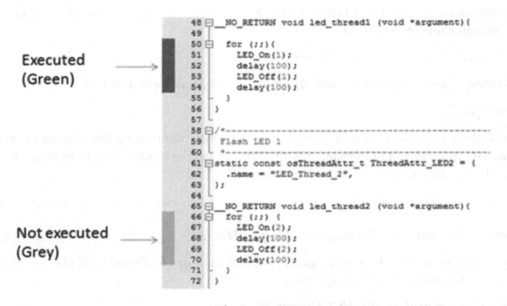

```
48 □ __NO_RETURN void led_thread1 (void *argument){
49
50 □   for (;;){
51         LED_On(1);
52         delay(100);
53         LED_Off(1);
54         delay(100);
55     }
56 }
57
58 □ /*---------------------------------------------------
59    Flash LED 1
60    *---------------------------------------------------
61 □ static const osThreadAttr_t ThreadAttr_LED2 = {
62       .name = "LED_Thread_2",
63 };
64
65 □ __NO_RETURN void led_thread2 (void *argument){
66 □   for (;;) {
67         LED_On(2);
68         delay(100);
69         LED_Off(2);
70         delay(100);
71     }
72 }
```

Executed (Green)

Not executed (Grey)

Figure 10.22
The coverage monitor shows what code has been executed by coloring the margin green.

Led_Thread2 is running at normal priority, and led_Thread1 is running at a higher priority, so has preempted led_Thread2 (Fig. 10.23). To make it even worse, led_Thread1 never blocks, so it will run forever, preventing the lower priority Thread from ever running.

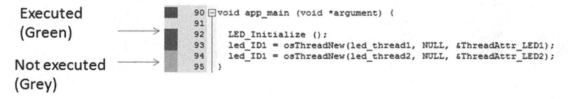

Executed (Green)

Not executed (Grey)

```
90 □ void app_main (void *argument) {
91
92      LED_Initialize ();
93      led_ID1 = osThreadNew(led_thread1, NULL, &ThreadAttr_LED1);
94      led_ID1 = osThreadNew(led_thread2, NULL, &ThreadAttr_LED2);
95 }
```

Figure 10.23
Led Thread 1 is running at a higher priority than appmain. This means it preempts app main as soon as it is created.

If we look at the app_main() Thread, we can see that as soon as led_Thread2() was created at osNormalPriority, it too was preempted, so led_Thread2() was never even created.

Although such an error may seem obvious in the above example, this kind of mistake is very common when designers first start to use an RTOS.

Memory Management

When each thread is created, it is assigned its own stack for storing data during the context switch. This should not be confused with the native Cortex-M processor stack; it is really a block of memory that is allocated to the thread. A default stack size is defined in the RTOS configuration file (we will see this later), and this amount of memory will be allocated to each thread unless we override it to allocate a custom size. The default stack size will be assigned to a Thread if the stack size value in the Thread definition structure is set to zero. If necessary, a Thread can be given additional memory resources by defining a bigger stack size in the Thread structure. RTX-RTOS2 supports several memory models to assign this Thread memory. The default model is a global memory pool. In this model, each RTOS object that is created (Threads, message queues, semaphores, etc.) is allocated memory from a single block of memory (Fig. 10.24).

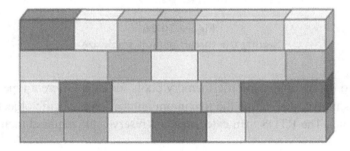

Figure 10.24
The global object pool provides a single region of memory that can be allocated to different RTOS Objects.

If an object is destroyed, the memory it has been assigned is returned to the memory pool. This has the advantage of memory reuse but also introduces the possible problem of memory fragmentation.

The size of the global memory pool is defined in the configuration file RTX_Config.h (Fig. 10.25). This defaults to 32 K, so may cause a linker error if your target device has a limited amount of RAM. If this happens, you simply need to shrink the memory pool size down to something more suitable.

```
⊟ System Configuration
    ┌─────────────────────────────────────────────────┐
    │ Global Dynamic Memory size [bytes]      32768    │
    └─────────────────────────────────────────────────┘
    ┄ Kernel Tick Frequency [Hz]              1000
    ⊞ Round-Robin Thread switching            ☑
    ┄ ISR FIFO Queue                          16 entries
    ┄ Object Memory usage counters            ☐
```

Figure 10.25
The size of the global memory pool is defined in RTX_Config.h.

And the default stack size for each thread is defined in the threads section (Fig. 10.26). Again like the global memory pool the default stack size can be adjusted to suit your needs.

```
⊟ Thread Configuration
    ⊞ Object specific Memory allocation       ☐
    ┌─────────────────────────────────────────────────┐
    │ Default Thread Stack size [bytes]        3072    │
    └─────────────────────────────────────────────────┘
      Idle Thread Stack size [bytes]          512
    ┄ Idle Thread TrustZone Module Identifier 0
    ┄ Stack overrun checking                  ☐
    ┄ Stack usage watermark                   ☐
    ┄ Processor mode for Thread execution     Privileged mode
```

Figure 10.26
Setting the thread default stack size.

It is also possible to define object-specific memory pools for each different type of RTOS object (Fig. 10.27). In this model, you define the maximum number of a specific object type and its memory requirements. The RTOS then calculates and reserves the required memory usage.

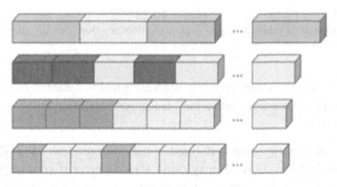

Figure 10.27
The object memory pool creates individual memory pools for each different RTOS object.

The object-specific model is again defined in the RTOS configuration file by enabling the "object specific memory" option provided in each section of the configuration file (Fig. 10.28).

Figure 10.28
Enabling the Object specific memory pool for an RTOS object.

In the case of a simple object which requires a fixed memory allocation, we just need to define the maximum number of a given object type. In the case of more complex objects such as Threads, we will need to define the required memory usage (Fig. 10.29).

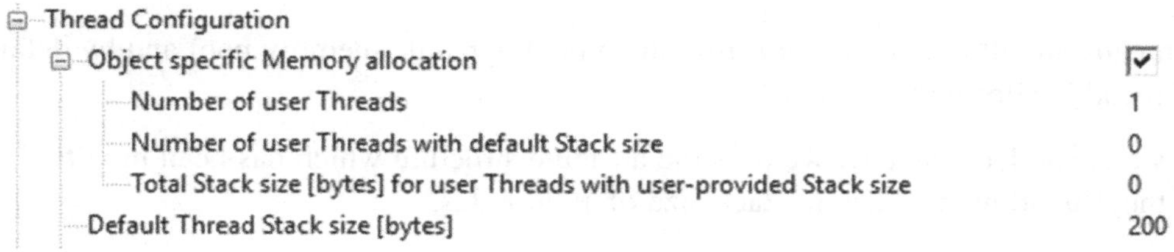

```
⊟ Thread Configuration
   ⊟ Object specific Memory allocation                                        ✔
      Number of user Threads                                                   1
      Number of user Threads with default Stack size                          0
      Total Stack size [bytes] for user Threads with user-provided Stack size 0
   Default Thread Stack size [bytes]                                         200
```

Figure 10.29
The Thread Object specific memory pool needs to know the overall Thread configuration to calculate the required memory allocation.

To use the object-specific memory allocation model with Threads, we must provide details of the overall Thread memory usage.

Finally, it is possible to statically allocate the Thread stack memory. This is important for safety-related systems where memory usage has to be rigorously defined. You can also use statically allocated thread memory if you are going to reuse a thread in a future project. This way when a thread is reused, it will allocate its own resources and require minimal configuration in the new project.

Exercise 10.3: Memory Model

In this exercise, we will create a Thread with a custom memory allocation and also create a Thread with a static memory allocation.

In the Pack Installer select "Ex 10.3 memory model" and press the copy button.

This exercise uses the same two LED flasher Threads as the previous exercise.

Open cmsis::rtx_config.c (Fig. 10.30).

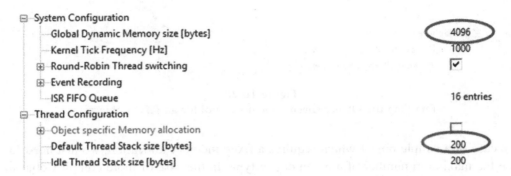

Figure 10.30
Thread memory configuration for the LED flasher threads.

The Threads are allocated memory from the global dynamic memory pool and by default each Thread is allocated 200 bytes.

When we create led-Thread1, we pass the attribute structure which has been modified to create the Thread with a custom stack size of 1024 bytes.

```
static const osThreadAttr_t ThreadAttr_LED1 = {
    "LED_Thread_1",
    NULL,//attributes
    NULL,//cb memory
    NULL,//cb size
    NULL,//stack memory
    1024,//stack size          This memory is allocated from the global memory pool
    osPriorityNormal,
    NULL,//trust zone id
    NULL//reserved
};
```

The second thread is created with a statically defined Thread control block and a statically defined stack space. First, we need to define an array of memory for the stack space:

```
static uint64_t LED2_Thread_stk[64];
```

Followed by a custom RTX Thread control block;

```
static osRtxThread_t LED2_Thread_tcb;
```

The custom type osRtxThread is defined in rtx_os.h.

Now, we can create a Thread attribute that statically allocates both the stack and the task control block.

```
static const osThreadAttr_t ThreadAttr_LED2 = {
    "LED_Thread_2",
    NULL,                           //attributes
```

Here, the control block and user stack space are statically allocated.

```
    &LED2_Thread_tcb,               //cb memory
    sizeof(LED2_Thread_tcb),        //cb size
    &LED2_Thread_stk[0],            //stack memory
    sizeof(LED2_Thread_stk),        //stack size
    osPriorityNormal,
    NULL,           //trust zone id
    NULL            //reserved
};
```

Build the code.

Start the debugger and check it runs.

Exit the debugger.

In the CMSIS:RTX_Conf.c file, we can change the memory model to use "Object Specific" memory allocation.

Set the Global Dynamic memory size to zero.

In Thread configuration enable the Object-specific memory model.

Set the number of Threads to two.

Number of user Threads with default stack size to 1 and total stack size for Threads with use provided stack to 1024 (Fig. 10.31).

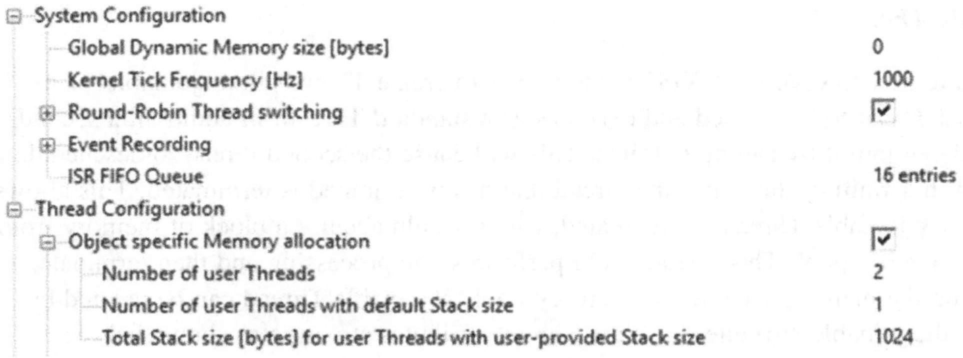

System Configuration	
Global Dynamic Memory size [bytes]	0
Kernel Tick Frequency [Hz]	1000
Round-Robin Thread switching	☑
Event Recording	
ISR FIFO Queue	16 entries
Thread Configuration	
Object specific Memory allocation	☑
Number of user Threads	2
Number of user Threads with default Stack size	1
Total Stack size [bytes] for user Threads with user-provided Stack size	1024

Figure 10.31
Defining the memory usage for a custom Thread stack size.

In total, we have three user Threads but one has statically allocated memory so our Thread object pool only needs to accommodate two. One of those Threads (Led_Thread1) has a custom stack size of 1024 bytes. We need to provide this information to the RTOS so it can work out the total amount of memory to allocate for Thread use.

Enable the MUTEX object.

Set the number of mutex objects to 5 (Fig. 10.32).

Figure 10.32
Allocating an object specific memory pool for the Mutex objects.

We will use mutexes later but they are concerned with protecting access to resources. The RTOS creates a number to protect access to the run time "C" library from different Threads. As we are not using the global memory pool, we must provide memory for these RTOS objects.

Build the code.

Start the debugger.

Run the code.

Now, we have one thread using statically located memory and object using object specific memory.

Joinable Threads

A feature new to CMSIS RTOS2 is the ability to create Threads in a "joinable" state. This allows a Thead to be created and executes as a standard Thread. In addition, a second Thread can join it by calling osJoin(). This will cause the second thread to deschedule and remain in a waiting state until the thread that has been joined is terminated. This allows a temporary joinable Thread to be created, which would acquire a block of memory from the global memory pool. This thread could perform some processing and then terminate, releasing the memory back to the memory pool. A joinable Thread can be created by setting the joinable attribute bit in the Thread attributes structure as shown below:

```
static const osThreadAttr_t ThreadAttr_worker = {
        .attr_bits = osThreadJoinable,
        };
```

Once the thread has been created, it will execute following the same rules as "normal" Threads. Any other Thread can then join it by using the os call.

```
osThreadJoin(<joinable_Thread_handle>);
```

Once osThread Join has been called, the thread will deschedule and enter a waiting state until the joinable thread has terminated.

Exercise 10.4: Joinable Threads

In this exercise, we will create a Thread that in turn spawns two joinable Threads. The initial thread will then call osThreadJoin() to wait until each of the joinable Threads has terminated.

In the Pack Installer select "Ex 10.4 Join" and copy it to your tutorial directory.

Open main.c.

In main.c we create a Thread called worker_Thread and define it as joinable in the Thread attribute structure.

When the RTOS starts we create the led_Thread() as normal.

```
__NO_RETURN void led_Thread1 (void *argument) {
for (;;) {
        worker_ID1 = osThreadNew(worker_Thread,(void *) LED1_ON, &ThreadAttr_worker);
        LED_On(2);
        osThreadJoin(worker_ID1);
        ........................
```

In this thread, we create an instance of the worker Thread and then call osJoin() to join it. At this point the led_Thread enters a waiting state and the worker Thread runs.

```
void worker_Thread (void *argument) {
if((uint32_t)argument == LED1_ON) {
LED_On(1);
}
else if ((uint32_t)argument == LED1_OFF){
LED_Off(1);
}
delay(500);
osThreadExit();
}
```

When the worker Thread runs it flashes the LED but instead of having an infinite loop, it calls osExit(); to terminate its runtime, which will cause led_Thread1 to leave the waiting state and enter the ready state and in this example, then enter the run state.

Build the code.

Start the debugger.

Open the View\watch\RTOS window.

Run the code and watch the behavior of the Threads.

Multiple Instances

One of the interesting possibilities of an RTOS is that you can create multiple running instances of the same base Thread code. So, for example, you could write a Thread to control a UART and then create two running instances of the same Thread code. Here each instance of the UART code could manage a different UART.

Then we can create two instances of the thread assigned to different Thread handles. A parameter is also passed to allow each instance to identify which UART it is responsible for.

```
#define UART1 (void *) 1UL
#define UART2 (void *) 2UL

ThreadID_1_0 = osThreadNew (Thread1, UART1,&ThreadAttr_Task1);
ThreadID_1_1 = osThreadNew (thred1, UART0,&ThreadAttr_Task1);
```

Exercise 10.5: Multiple Thread Instances

In this project, we will look at creating one thread and then create multiple runtime instances of the same thread.

In the Pack Installer select "Ex 10.5 Multiple Instances" and copy it to your tutorial directory.

This project performs the same function as the previous LED flasher program. However, we now have one led switcher function that uses an argument passed as a parameter to decide which LED to flash.

```
void ledSwitcher (void const *argument) {
    for (;;) {
            LED_On((uint32_t)argument);
            delay(500);
            LED_Off((uint32_t)argument);
            delay(500);
    }
}
```

Then in the main thread, we create two Threads that are different instances of the same base code. We pass a different parameter that corresponds to the led that will be toggled by the instance of the thread.

First, we can create two different Thread attribute definitions with different debug names.

```
static const osThreadAttr_t ThreadAttr_LedSwitcher1 = {
    .name = "LedSwitcher1",
  };
static const osThreadAttr_t ThreadAttr_LedSwitcher2 = {
  .name = "LedSwitcher2",
  };
```

Next, we can create two instances of the same Thread code.

```
led_ID1 = osThreadNew(ledSwitcher,(void *) 1UL, &ThreadAttr_LedSwitcher1);
led_ID2 = osThreadNew(ledSwitcher,(void *) 2UL, &ThreadAttr_LedSwitcher2);
```

Build the code and start the debugger.

Start the code running and open the View Watch RTX component window (Fig. 10.33).

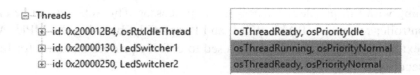

Figure 10.33
Multiple instances of Thread running.

Here we can see both instances of the ledSwitcher task, each with a different ID.

Examine the Call stack + locals window (Fig. 10.34).

⊟ ● ledSwitcher : 0x20000250	0x080001BD	Task
⊟ ● delay	0x080001B8	void f(unsigned int)
●● count	0x000001F4	param - unsigned int
● index	0x0000011F	auto - unsigned int
⊟ ● ledSwitcher	0x080001CE	void f(void *)
⊞ ●● argument	0x00000002	param - void *
⊟ ● ledSwitcher : 0x20000130	0x080001BD	Task
⊟ ● delay	0x080001B4	void f(unsigned int)
●● count	0x000001F4	param - unsigned int
● index	0x0000003D	auto - unsigned int
⊟ ● ledSwitcher	0x080001DC	void f(void *)
⊞ ●● argument	0x00000001	param - void *

Figure 10.34
The watch window is Thread aware.

By expanding each component branch, we can see both instances of the ledSwitcher
Threads and the state of their variables.

Understanding RTOS API Calls

Now that we are familiar with RTOS threads we can begin to look at the RTOS API calls
that can be used within our application code. It is important to understand that the os
function calls differ from standard function calls in that they use a supervisor instruction to
call functions in the RTOS code rather than a branch instruction. This allows the RTOS
code to be called as an exception and forces the processor to switch to Handler mode and
use the main stack. Once the RTOS code has been executed, it will return to the application
thread running in Thread mode and using the process stack. This creates a division between
the application code and the "system level" RTOS code. We can further enforce this
division by configuring the application threads to run in unprivileged mode while the RTOS
code is running with full privileged access to the processor. This scheme can be extended to
the microcontroller resources (RAM, ROM, and Peripherals) by using the MPU. As we will
see in the next chapter, this scheme can be used to enforce process isolation for functional
safety and security.

Time Management

As well as running your application code as Threads, the RTOS also provides some timing
services which can be accessed through RTOS system calls.

Time Delay

The most basic of these timing services is a simple timer delay function. This is an easy
way of providing timing delays within your application. Although the RTOS kernel size is
quoted as 5k bytes, features such as delay loops and simple scheduling loops are often part
of a non-RTOS application and would consume code bytes anyway, so the overhead of the
RTOS can be less than it immediately appears.

```
void osDelay (uint32_t ticks )
```

This call will place the calling thread into the WAIT_DELAY state for the specified
number of scheduler ticks (the default value for each tick is 1 Ms). The scheduler will pass
execution to the next thread in the READY state during the delay period (Fig. 10.35).

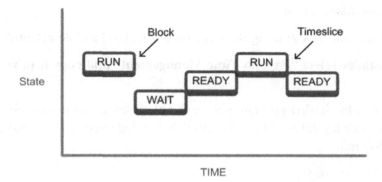

Figure 10.35
During their lifetime Threads move through many states. Here a running Thread is blocked by an osDelay call so it enters a wait state. When the delay expires, it moves to ready. The scheduler will place it in the run state. If its timeslice expires, it will move back to ready.

When the timer expires, the thread will leave the WAIT_DELAY state and move to the READY state. The thread will resume running when the scheduler moves it to the RUNNING state. If the thread then continues executing without any further blocking OS calls, it will be descheduled at the end of its time slice and be placed in the ready state, assuming another Thread of the same priority is ready to run.

Absolute Time Delay

In addition to the osDelay() function, which gives a relative time delay starting from the instant it is called there is also a delay function that halts a Thread until a specific point in time.

```
osStatus osDelayUntil (uint32_t ticks)
```

The osDelayUntil() function will halt a Thread until a specific value of kernel timer ticks is reached. There are a number of kernel functions that allow you to read both the current SysTick count and the kernel ticks count as shown in Table 10.5.

Table 10.5: Kernel time functions

Kernel time functions	
uint64_t	osKernelGetTickCount(void)
uint32_t	osKernelGetTickFreq(void)
uint32_t	osKernelGetSysTimerCount(void)
uint32_t	osKernelGetSysTimerFreq(void)

Exercise 10.6: Time Management

In this exercise we will look at using the basic time osDelay() and delayUntil() functions.

In the Pack Installer select "Ex 10.6 Time Management" and copy it to your tutorial directory.

This is our original led flasher program but the simple delay function has been replaced by the osDelay and osDelayUntil() API calls. LED2 is toggled every 100 ms and LED1 is toggled every 500 mS.

```
void ledOn (void *argument) {
for (;;) {
        LED_On(1);
        osDelay(50);
        LED_Off(1);
        osDelay(50);
}}
```

In the Led2 Thread, we use the osDelayUntil() function to create a 1000 tick delay.

```
__NO_RETURN void led2 (void *argument) {
    for (;;) {
            ticks = osKernelGetTickCount();
            LED_On(2);
            osDelayUntil((ticks + 1000));          //Toggle LED 2 with an absolute
                                                     delay
            LED_Off(2);
            osDelayUntil((ticks + 2000));
    }
}
```

Build the project and start the debugger.

Now we can see that the activity of the code is very different. When each of the LED tasks reaches the osDelay() API call it "blocks" and moves to a waiting state. The appMain thread will be in a ready state, so the scheduler will start it running. When the delay period has timed out, the led tasks will move to the ready state and will be placed into the running state by the scheduler. This gives us a multi-Threaded program where CPU runtime is efficiently shared between Threads.

Virtual Timers

The CMSIS-RTOS API can be used to define any number of virtual timers which act as count down timers. When they expire, they will run a user callback function to perform a specific action. Each timer can be configured as a one-shot or repeat timer. A virtual timer is created by first defining a timer structure.

```
static const osThreadAttr_t ThreadAttr_app_main = {
const char * name        //symbolic name of the timer
uin32_tattr_bits         //None
void*cb_mem              //pointer to memory for control block
uint32_tcb_size          //size of memory control block
```

This defines a name for the timer and the name of the call back function. The timer must then be instantiated by an RTOS Thread.

```
osTimerId_t timer0_handle;
timer0_handle = osTimerNew(&callback, osTimerPeriodic,(void *)<parameter>,
  &timerAttr_timer0);
```

This creates the timer and defines it as a periodic timer or a single shot timer (osTimerOnce). The next parameter passes an argument to the call back function when the timer expires.

```
osTimerStart (timer0_handle,0x100);
```

The timer can then be started at any point in a Thread. The timer start function invokes the timer by its handle and defines a count period in kernel ticks.

Exercise 10.7: Virtual Timer

In this exercise, we will configure a number of virtual timers to trigger a callback function at various frequencies.

In the Pack Installer select "Ex 10.7 Virtual Timers" and copy it to your tutorial directory.

This is our modified LED flasher program using osDelay(). Code has been added to create four virtual timers to trigger a callback function. Depending on which timer has expired, this function will toggle an additional LED.

The timers are defined at the start of the code.

```
osTimerId_t timer0,timer1,timer2,timer3;
static const osTimerAttr_t timerAttr_timer0 = {
      .name = "timer_0",
};
static const osTimerAttr_t timerAttr_timer1 = {
      .name = "timer_1",
};
static const osTimerAttr_t timerAttr_timer2 = {
      .name = "timer_2",
};
static const osTimerAttr_t timerAttr_timer3 = {
      .name = "timer_3",
};
```

They are then initialized in the main function:

```
timer0 = osTimerNew(&callback, osTimerPeriodic,(void *)0, &timerAttr_timer0);
timer1 = osTimerNew(&callback, osTimerPeriodic,(void *)1, &timerAttr_timer1);
timer2 = osTimerNew(&callback2, osTimerPeriodic,(void *)2, &timerAttr_timer2);
timer3 = osTimerNew(&callback2, osTimerPeriodic,(void *)3, &timerAttr_timer3);
```

Each timer has a different handle and ID and passed a different parameter to the common callback function:

```
void callback(void const *param){
switch( (uint32_t) param){
        case 0:
                GPIOB->ODR ^= 0x8;
        break;
        case 1:
                GPIOB->ODR ^= 0x4;
        break;
        case 2:
                GPIOB->ODR ^= 0x2;
        break;
}}
```

When triggered, the callback function uses the passed parameter as an index to toggle the desired LED.

In addition to configuring the virtual timers in the source code, the timer Thread must be enabled in the RTX5 configuration file.

Open the RTX_Config.h file and press the configuration wizard tab (Fig. 10.36).

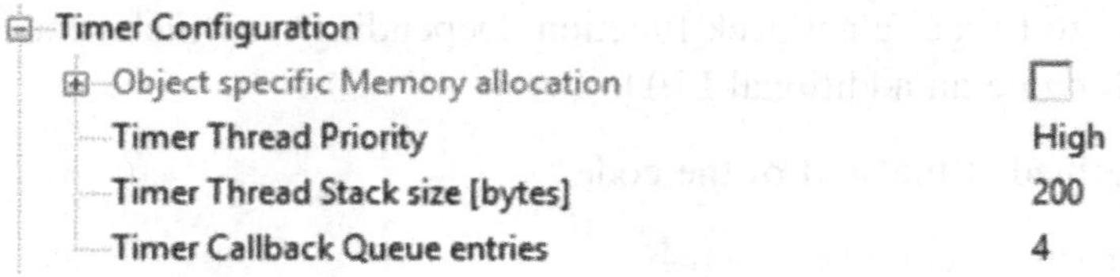

⊟ Timer Configuration	
⊞ Object specific Memory allocation	☐
Timer Thread Priority	High
Timer Thread Stack size [bytes]	200
Timer Callback Queue entries	4

Figure 10.36
Configuring the virtual timers.

In the system configuration section, make sure the User Timers box is ticked. If this thread is not created the RTOS timers will not work.

Build the project and start the debugger.

Run the code and observe the activity of the GPIOB pins in the peripheral window (Fig. 10.37).

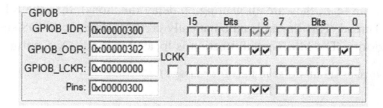

Figure 10.37
The user timers toggle additional LED pins.

There will also be an additional Thread running in the System and Thread Viewer window (Fig. 10.38).

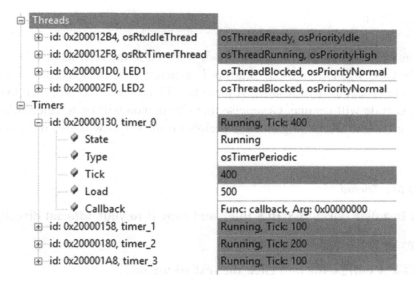

Figure 10.38
The user timers create an additional osTimerThread.

The osDelay() function provides a relative delay from the point at which the delay is started. The virtual timers provide an absolute delay that allows you to schedule code to run at fixed intervals.

Idle Thread

The final timer service provided by the RTOS isn't really a timer, but this is probably the best place to discuss it. If during our RTOS program we have no Thread running and no

Thread ready to run (e.g., they are all waiting on delay functions), then the RTOS will start to run the Idle Thread. This thread is automatically created when the RTOS starts and runs at the lowest priority. The Idle Thread function is located in the RTX_Config.c file.

```
__NO_RETURN void osRtxIdleThread (void *argument)
{
    for (;;) {
    /* HERE: include here optional user code to be executed when no Thread runs.          */
    }
}
```

You can add any code to this thread, but it has to obey the same rules as user Threads. The simplest use of the idle demon is to place the microcontroller into a low-power mode when it is not doing anything.

```
__NO_RETURN void osRtxIdleThread (void *argument){
for (;;) {
        __wfe();
}}
```

What happens next depends on the power mode selected in the microcontroller. At a minimum, the CPU will halt until an interrupt is generated by the SysTick timer and execution of the scheduler will resume. If there is a Thread ready to run, then execution of the application code will resume. Otherwise, the idle demon will be re-entered, and the system will go back to sleep. Any enabled peripheral interrupts will also wake up the processor.

Exercise 10.8 Idle Thread

In the Pack Installer select "Ex 10.8 Idle" and copy it to your tutorial directory.

This is a copy of the virtual timer project.

Open the RTX_Config.c file and click the text editor tab.

Locate the idle Thread.

```
__NO_RETURN void osRtxIdleThread (void *argument){
for (;;) {
        //wfe();
}}
```

Build the code and start the debugger.

Run the code and observe the activity of the Threads in the event Viewer.

This is a simple program that spends most of its time in the idle demon, so this code will be run almost continuously.

Open the View/Analysis Windows/Performance Analyzer (Fig. 10.39).

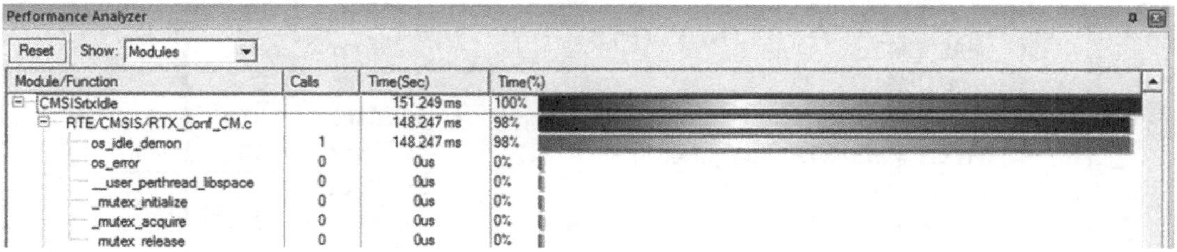

Performance Analyzer				
Reset	Show: Modules			
Module/Function	Calls	Time(Sec)	Time(%)	
⊟ CMSISrtxIdle		151.249 ms	100%	
⊟ RTE/CMSIS/RTX_Conf_CM.c		148.247 ms	98%	
os_idle_demon	1	148.247 ms	98%	
os_error	0	0us	0%	
__user_perthread_libspace	0	0us	0%	
_mutex_initialize	0	0us	0%	
_mutex_acquire	0	0us	0%	
mutex_release	0	0us	0%	

Figure 10.39
The performance analyzer shows that most of the run time is being spent in the idle loop.

This window shows the cumulative run time for each function in the project. In this simple project, the idle thread is using most of the runtime because there is very little application code.

Exit the debugger.

Uncomment the __wfe() instruction in the for loop, so the code now looks like this.

```
__NO_RETURN void osRtxIdleThread (void *argument){
for (;;) {
    __wfe();
}}
```

Rebuild the code, restart the debugger.

Now when we enter the idle thread, the __wfe() (wait for event) instruction will halt the CPU until there is a peripheral or SysTick interrupt. This greatly reduces wasted cycles (Fig. 10.40).

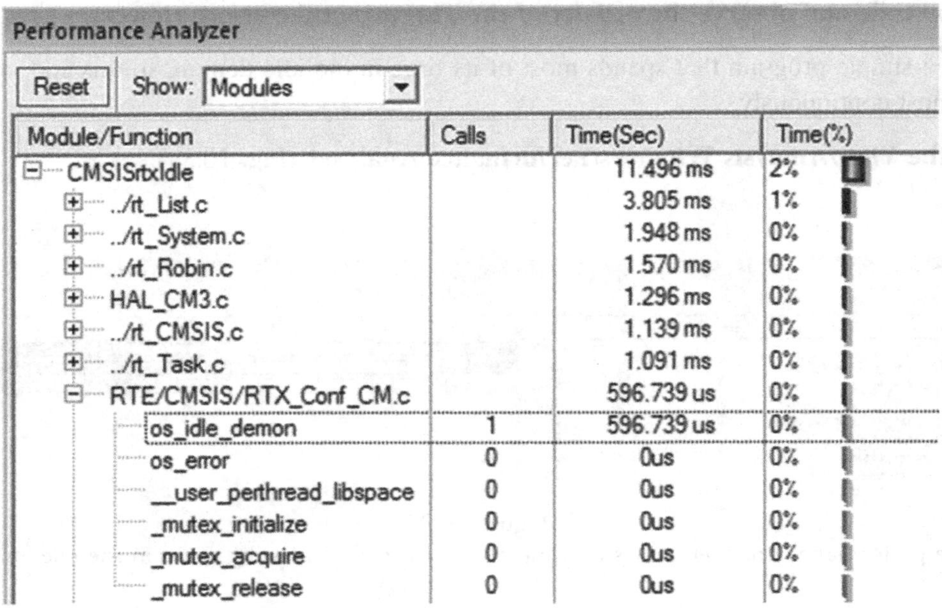

Figure 10.40

The __wfe() intrinsic halts the CPU when it enters the idle loop. Saving cycles and runtime energy.

Performance Analysis During Hardware Debugging

The code coverage and performance analysis tools are available when you are debugging on real hardware rather than simulation. However, to use these features, you need two things: First, you need a microcontroller that has been fitted with the optional Embedded Trace Macrocell (ETM). Second, you need to use Keil ULINK*pro* debug adapter, which supports instruction trace via the ETM.

Inter-Thread Communication

So far, we have seen how application code may be defined as independent Threads and how we can access the timing services provided by the RTOS. In a real application, we need to be able to communicate between Threads in order to make an application useful. To this end, a typical RTOS supports several different communication objects which can be used to link the Threads together to form a meaningful program. The CMSIS-RTOS2 API supports inter-Thread communication with Thread and event flags, semaphores, mutexes, mailboxes, and message queues. In the first section of this chapter the key concept was concurrency. In this section, the key concept is synchronizing the activity of multiple Threads.

Thread Flags

CMSIS-RTOS2 Keil RTX5 supports up to 32 Thread flags for each thread (Fig. 10.41). These Thread flags are stored in the Thread control block. It is possible to halt the execution of a Thread until a particular Thread flag or group of Thread flags are set by another Thread in the system.

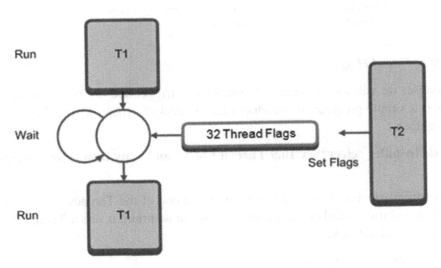

Figure 10.41
Each Thread has 32 signal flags. A Thread may be placed into a waiting state until a pattern of flags is set by another Thread. When this happens, it will return to the ready state and wait to be scheduled by the kernel.

The osThreadFlagsWait() system calls will suspend execution of the thread and place it into the wait_evnt state. Execution of the Thread will not start until at least one of the flags set in the osThreadFlagsWait() API call has been set. It is also possible to define a periodic timeout after which the waiting Thread will move back to the ready state, so that it can resume execution when selected by the scheduler. A value of osWaitForever (0xFFFFFFFF) defines an infinite timeout period.

```
osEvent osThreadFlagsWait (int32_t flags,int32_t options, uint32_t timeout);
```

The Thread flag options are shown in Table 10.6.

Table 10.6: Thread flag options

Options	Description
osFlagsWaitAny	Wait for any flag to be set(default)
osFlagsWailAll	Wait for all flags to be set
osFlagsNoClear	Do not clear flags that have been specified to wait for

If a pattern of flags is specified, the thread will resume execution when any one of the specified flags is set (Logic OR). If the osFlagsWaitAll option is used, then all the flags in the pattern must be set (Logic AND).

Any Thread can set a flag on any other Thread and a Thread may clear its own flags.

```
int32_t osThredFlagsSet (osThreadId_t Thread_id, int32_t flags);
int32_t osThreadFlagsClear (int32_t signals);
```

Exercise 10.9: Thread Flags

In this exercise, we will look at using Thread flags to trigger activity between two Threads. While this is a simple program, it introduces the concept of synchronizing the activity of Threads together.

In the Pack Installer select "Ex 10.9 Thread Flags" and copy it to your tutorial directory.

This is a modified version of the led flasher program one of the Threads calls the same LED function and uses osDelay() to pause the task. In addition, it sets a Thread flag to wake up the second led task.

```
void led_Thread2 (void *argument) {
for (;;) {
        LED_On(2);
        oThreadFlagSet (T_led_ID1,0x01);
        osDelay(500);
        LED_Off(2);
        osThreadFlagSet (T_led_ID1,0x01);
        osDelay(500);}}
```

The second led function waits for the signal flags to be set before calling the led functions.

```
void led_Thread1 (void *argument) {
for (;;) {
        osThreadFlagsWait (0x01,osWaitForever);
        LED_On(1);
        osSignalWait (0x01,osWaitForever);
        LED_Off(1);
}}
```

Build the project and start the debugger.

Open the GPIOB peripheral window and start the code running.

Now the port pins will appear to be switching on and off together. Synchronizing the Threads gives the illusion that both Threads are running in parallel.

This is a simple exercise, but it illustrates the key concept of synchronizing activity between Threads in an RTOS-based application.

Event Flags

Event flags operate in a similar fashion to Thread flags but must be created and then act as a global RTO object that can be used by all the running Threads (Fig. 10.42).

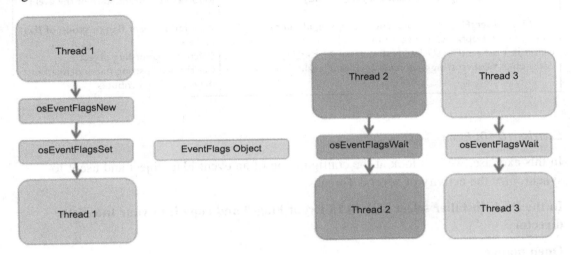

Figure 10.42
Event flags are similar to thread flags but are created as a global object.

Creating a set of event flags, this is a similar process to creating a Thread. First, we define an event flag attribute structure. The attribute structure defines an ASCII name string, attribute bits, and memory detention if we are using the static memory model.

```
osEventFlagsAttr_t {
  const char       *name;   ///< name of the event flags
  uint32_t         attr_bits;   ///< attribute bits (none)
  void             *cb_mem;   ///< memory for control block
  uint32_t         cb_size;   ///< size of provided memory for control block
};
```

Next, we need a handle to control access the event flags.

```
osEventFlagsId_t EventFlag_LED;
```

Then, we can create the event flag object.

```
EventFlag_LED = osEventFlagsNew(&EventFlagAttr_LED);
```

Now any thread can use the event flags as a global object using the functions in Table 10.7.

Table 10.7: Event flag functions

Function	Description
osEventFlagsId_t osEventFlagsNew (const osEventFlagsAttr_t *attr)	Create a new event flag object
uint32_t osEventFlagsSet (osEventFlagsId_t ef_id, uint32_t flags)	Set an event flag or group of flags
uint32_t osEventFlagsClear (osEventFlagsId_t ef_id, uint32_t flags)	Clear an event flag or group of flags
uint32_t osEventFlagsGet (osEventFlagsId_t ef_id)	Returns the current state of the event flags
uint32_t osEventFlagsWait (osEventFlagsId_t ef_id, uint32_t flags, uint32_t options, uint32_t timeout)	Wait for an event flag or group of flags to be set
osStatus_t osEventFlagsDelete (osEventFlagsId_t ef_id)	Delete the event flag object
const char * osEventFlagsGetName (osEventFlagsId_t ef_id)	Get the ASCII string name defined in the event flag attributes

Exercise 10.10: Event Flags

In this exercise, we will look at the configuration of an event Flag object and use it to synchronize the activity of several Threads.

In the Pack Installer select "Ex 10.10 Event Flags" and copy it to your tutorial directory.

Open main.c.

The code in main.c creates and event flag object and instantiates it in appMain().

```
static const osEventFlagsAttr_t EventFlagAttr_LED = {
        .name = "LED_Events",
};
void app_main (void *argument)
{
        LED_Initialize();
        EventFlag_LED = osEventFlagsNew(&EventFlagAttr_LED);
```

The code then creates three Threads. Two of the Threads wait for an event flag to be set.

```
_NO_RETURN void led_Thread1 (void *argument) {
for (;;) {
osEventFlagsWait (EventFlag_LED,0x01,osFlagsWaitAny,osWaitForever);
LED_On(1);

_NO_RETURN void led_Thread2 (void *argument) {
for (;;) {
        osEventFlagsWait (EventFlag_LED,0x01,osFlagsWaitAny,osWaitForever);
        LED_On(2);
```

The remaining Thread is used to set the flag.

```
__NO_RETURN void led_Thread3 (void *argument) {
for (;;) {
        osEventFlagsSet      (EventFlag_LED,0x01);
        LED_On(3);
```

Build the code.

Start the debugger and run the code.

Observe the activity of the LEDs.

Why does the code not run as expected?

When the event flag is set, one of the waiting Threads will wake up and clear the flag. The second waiting Thread is not triggered. Each thread should be waiting on a separate Event flag within the event flag object.

Change the code so that the waiting Threads are waiting on separate flags and the remaining Thread sets both flags.

```
_NO_RETURN void led_Thread1 (void *argument) {
for (;;) {
osEventFlagsWait (EventFlag_LED,0x01,osFlagsWaitAny,osWaitForever);
LED_On(1);

_NO_RETURN void led_Thread2 (void *argument) {
for (;;) {
        osEventFlagsWait (EventFlag_LED,0x02,osFlagsWaitAny,osWaitForever);
        LED_On(2);
```

The remaining Thread is used to set both flags.

```
__NO_RETURN void led_Thread3 (void *argument) {
for (;;) {
        osEventFlagsSet      (EventFlag_LED,0x03);
        LED_On(3);
```

Semaphores

Like Thread flags, semaphores are a method of synchronizing activity between two or more Threads. Put simply, a semaphore is a container that holds a number of tokens (Fig. 10.43). As a Thread executes, it will reach an RTOS call to acquire a semaphore token. If the semaphore contains one or more tokens, the thread will continue executing, and the number of tokens in the semaphore will be decremented by one. If there are currently no tokens in the semaphore, the thread will be placed in a waiting state until a token becomes available. At any point in its execution, a Thread may add a token to the semaphore causing its token count to increment by one.

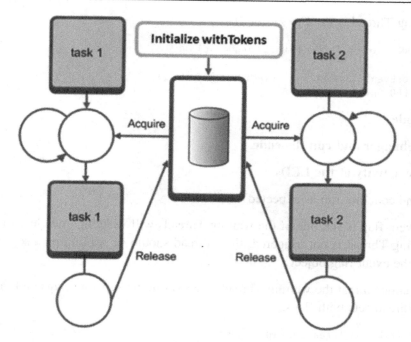

Figure 10.43

Semaphores help to control access to program resources. Before a Thread can access a resource, it must acquire a token. If none is available, it waits. When it is finished with the resource, it must return the token.

The diagram above illustrates the use of a semaphore to synchronize two Threads. First, the semaphore must be created and initialized with an initial token count. In this case, the semaphore is initialized with a single token. Both Threads will run and reach a point in their code where they will attempt to acquire a token from the semaphore. The first thread to reach this point will acquire the token from the semaphore and continue execution. The second thread will also attempt to acquire a token, but as the semaphore is empty, it will halt execution and be placed into a waiting state until a semaphore token is available.

Meanwhile, the executing thread can release a token back to the semaphore. When this happens, the waiting Thread will acquire the token and leave the waiting state for the ready state. Once in the ready state, the scheduler will place the thread into the run state so that Thread execution can continue. While semaphores have a simple set of OS calls, they can be one of the more difficult OS objects to fully understand. In this section, we will first look at how to add semaphores to an RTOS program and then go on to look at the most useful semaphore patterns.

To use a semaphore in the CMSIS-RTOS you must first declare a semaphore attributes structure:

```
osSemaphoreAttr_t {
        const char          *name;    ///< name of the semaphore
        uint32_t        attr_bits;    ///< attribute bits (none)
        void            *cb_mem;      ///< memory for control block
        uint32_t         cb_size;     ///< size of provided memory for control block
};
```

Next, declare the semaphore handle.

```
osSemaphoreId_t sem1;
```

Then within a Thread, the semaphore container can be initialized with a number of tokens.

```
sem1 = osSemaphoreNew(maxTokenCount,initalTokencount,&osSemaphoreAttr_t);
```

It is important to understand that semaphore tokens may also be created and destroyed as Threads run. So, for example, you can initialize a semaphore with zero tokens and then use one thread to create tokens into the semaphore while another Thread removes them. This allows you to design a system that has producer and consumer Threads.

Once the semaphore is initialized, tokens may be acquired and sent to the semaphore in a similar fashion to event flags. The os_sem_acquire call is used to block a Thread until a semaphore token is available, like the os_evnt_acquire call. A timeout period may also be specified with osWaitForever being an infinite wait.

```
osStatus osSemaphoreAcquire(osSemaphoreId_t semaphore_id, uint32_t ticks);
```

Once the thread has finished using the semaphore resource, it can send a token to the semaphore container.

```
osStatus osSemaphoreRelease(osSemaphoreId_t semaphore_id);
```

The semaphore functions are shown in Table 10.8.

Table 10.8: Semaphore functions

Function	Description
osSemaphoreNew (uint32_t max_count, uint32_t initial_count, const osSemaphoreAttr_t *attr)	Create a new semaphore
const char *osSemaphoreGetName(osSemaphoreId_t semaphore_id)	Get the semaphore name defined in the attribute structure
osStatus_t osSemaphoreAcquire (osSemaphoreId_t semaphore_id, uint32_t timeout)	Acquire a semaphore token
osStatus_t osSemaphoreRelease (osSemaphoreId_t semaphore_id)	Release a semaphore token
uint32_t osSemaphoreGetCount (osSemaphoreId_t semaphore_id)	Get the count of available semaphore tokens
osStatus_t osSemaphoreDelete (osSemaphoreId_t semaphore_id)	Delete the semaphore

Exercise 10.11 Semaphore Signaling

In this exercise, we will look at the configuration of a semaphore and use it to signal between two Threads.

In the Pack Installer select "Ex 10.11 Interrupt Signals" and copy it to your tutorial directory.

First, the code creates a semaphore called sem1 and initializes it with zero tokens and a maximum count of five tokens.

```
osSemaphoreId_t sem1;
static const osSemaphoreAttr_t semAttr_SEM1 = {
.name = "SEM1",
};

void app_main (void *argument) {
    sem1 = osSemaphoreNew(5, 0, &semAttr_SEM1 );
```

The first task waits for a token to be sent to the semaphore.

```
__NO_RETURN void led_Thread1 (void *argument) {
for (;;) {
    osSemaphoreAcquire(sem1, osWaitForever);
    LED_On(1);
    osSemaphoreAcquire(sem1, osWaitForever);
    LED_Off(1);
}
```

While the second task periodically sends a token to the semaphore.

```
__NO_RETURN void led_Thread2 (void *argument) {
for (;;) {
    osSemaphoreRelease(sem1);
    LED_On(2);
    osDelay(500);
    osSemaphoreRelease(sem1);
    LED_Off(2);
    osDelay(500);
}}
```

Build the project and start the debugger.

Set a breakpoint in the led_Thread2 task (Fig. 10.44).

Figure 10.44
Breakpoint on the semaphore release call in led_Thread2.

Run the code and observe the state of the Threads when the breakpoint is reached (Fig. 10.45).

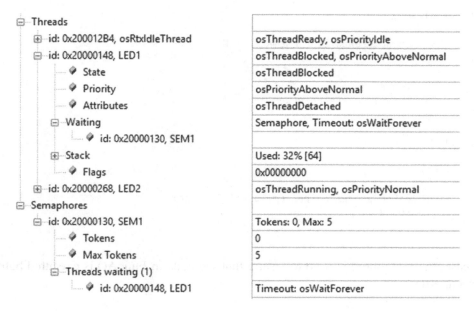

Figure 10.45
Led_Thread1 is waiting to acquire a semaphore.

Now, led_Thread1 is blocked, waiting to acquire a token from the semaphore. Led_Thread1 has been created with a higher priority than led_Thread2, so as soon as a token is placed in the semaphore, it will move to the ready state and preempt the lower priority task and start running. When it reaches the osSemaphoreAcquire() call it will again block.

Now block step the code (F10) and observe the action of the Threads and the semaphore.

Using Semaphores

Although semaphores have a simple set of OS calls, they have a wide range of synchronizing applications. This makes them perhaps the most challenging RTOS object to understand. In this section we will look at the most common uses of semaphores. These are taken from *The Little Book of Semaphores* by Allen B. Downey. This book may be freely downloaded from the URL given in the bibliography at the end of this book.

Signaling

Synchronizing the execution of two Threads is the simplest use of a semaphore:

```
     osSemaphoreId_t sem1;
     static const osSemaphoreAttr_t semAttr_SEM1 = {
     .name = "SEM1",
     };

void Thread1 (void)
{
     sem1 = osSemaphoreNew(5, 0,& semAttr_SEM1);
     while(1)
     {
             FuncA();
             osSemaphoreRelease(sem1)
     }
}
void task2 (void)
{
     while(1)
     {
             osSemaphoreAcquire(sem1,osWaitForever)
             FuncB();
     }
}
```

In this case, the semaphore is used to ensure that the code in FuncA() is executed before the code in FuncB().

Multiplex

A multiplex is used to limit the number of Threads that can access a critical section of code. For example, this could be a routine that accesses memory resources and can only support a limited number of processes.

```
     osSemaphoreId_t multiplex;
     static const osSemaphoreAttr_t semAttr_Multiplex = {
     .name = "SEM1",
     };

void Thread1 (void)
{
     multiplex = osSemaphoreCreate(5,5,& semAttr_Multiplex);
     while(1){
             osSemaphoreAcquire(multiplex,osWaitForever)
             processBuffer();
             osSemaphoreRelease(multiplex);
}}
```

In this example, we initialize the multiplex semaphore with five tokens. Before a Thread can call the processBuffer() function, it must acquire a semaphore token. Once the function has been completed, the token is sent back to the semaphore. If more than five Threads are attempting to call processBuffer(), the sixth must wait until a Thread has finished with processBuffer() and returns its token. Thus the multiplex semaphore ensures that a maximum of five Threads can call the processBuffer() function "simultaneously."

Exercise 10.12: Multiplex

In this exercise, we will look at using a semaphore to control access to a function by creating a multiplex.

In the Pack Installer select "Ex 10.12 Multiplex" and copy it to your tutorial directory.

The project creates a semaphore called semMultiplex, which contains one token. Next, six instances of a Thread containing a semaphore multiplex are created.

Build the code and start the debugger.

Open the Peripherals\General Purpose IO\GPIOB window.

Run the code and observe how the tasks set the port pins.

As the code runs, only one thread at a time can access the LED functions, so only one port pin is set.

Exit the debugger and increase the number of tokens allocated to the semaphore when it is created.

```
semMultiplex = osSemaphoreNew(5, 3,&semAttr_Multiplex);
```

Build the code and start the debugger.

Run the code and observe the GPIOB pins.

Now three Threads can access the led functions "concurrently."

Rendezvous

A more generalized form of semaphore signaling is a rendezvous. A rendezvous ensures that two Threads reach a certain point of execution. Neither may continue until both have reached the rendezvous point.

```
        osSemaphoreId_t arrived1,arrived2;
static const osSemaphoreAttr_t semAttr_Arrived1 = {
        .name = "Arr1",
        };
static const osSemaphoreAttr_t semAttr_Arrived2 = {
        .name = "Arr2",
        };
    void Thread1 (void){
    arrived1 = osSemaphoreNew(2,0);
    arrived1 = osSemaphoreNew(2,0);
```

```
        while(1){
                FuncA1();
                osSemaphoreRelease(arrived1);
                osSemaphoreAcquire(arrived2,osWaitForever);
                FuncA2();
        }}
        void Thread2 (void) {
          while(1){
                FuncB1();
                os_sem_Release(arrived2);
                os_sem_Acquire(arrived1,osWaitForever);
                FuncB2();
        }}
```

In the above case, the two semaphores will ensure that both Threads will rendezvous and then proceed to execute FuncA2() and FuncB2().

Exercise 10.13: Rendezvous

In this project, we will create two tasks and make sure that they have reached a semaphore rendezvous before running the LED functions.

In the Pack Installer select "Ex 10.13 Rendezvous" and copy it to your tutorial directory.

Build the project and start the debugger.

Open the Peripherals General Purpose IO\GPIOB window.

Run the code.

Initially, the semaphore code in each of the LED tasks is commented out. Since the Threads are not synchronized, the GPIO pins will toggle randomly.

Exit the debugger.

Un-comment the semaphore code in the LED tasks.

Built the project and start the debugger.

Run the code and observe the activity of the pins in the GPIOB window.

Now the tasks are synchronized by the semaphore and run the LED functions "concurrently."

Barrier Turnstile

Although a rendezvous is very useful for synchronizing the execution of code, it only works for two functions. A barrier is a more generalized form of rendezvous which works to synchronize multiple Threads.

```
        osSemaphoreId_t count, barrier;
        static const osSemaphoreAttr_t semAttr_Counter = {
        .name = "Counter",
        };
        static const osSemaphoreAttr_t semAttr_Barier = {
        .name = "Barrier",
        };
        unsigned int count;
void Thread1 (void)
{
        Turnstile_In = osSemaphoreNew(5, 0,&semAttr_SEM_In);
        Turnstile_Out = osSemaphoreNew(5, 1,&semAttr_SEM_Out);
        Mutex = osSemaphoreNew(1, 1,&semAttr_Mutex);

        while(1){
          osSemaphoreAcquire(Mutex,osWaitForever);      //Allow one task at a time to access
                                                                the first turnstile
          count = count + 1;                            // Increment count
          if( count == 5)
          {
            osSemaphoreAcquire (Turnstile_Out,osWaitForever);    //Lock the second turnstile
            osSemaphoreRelease(Turnstile_In);                    //Unlock the first turnstile
          }
            osSemaphoreRelease(Mutex);                           //Allow other tasks to
                                                                 access the turnstile

            osSemaphoreAcquire(Turnstile_In,osWaitForever);      //Turnstile Gate
            osSemaphoreRelease(Turnstile_In);
            critical_Function();
}}
```

In this code, we use a global variable to count the number of Threads that have arrived at the barrier. As each function arrives at the barrier, it will wait until it can acquire a token from the counter semaphore. Once acquired, the count variable will be incremented by one. Once we have incremented the count variable, a token is sent to the counter semaphore so that other waiting Threads can proceed. Next, the barrier code reads the count variable. If this is equal to the number of Threads that are waiting to arrive at the barrier, we send a token to the barrier semaphore.

In the example above, we are synchronizing five Threads. The first four Threads will increment the count variable and then wait at the barrier semaphore. The fifth and last thread to arrive will increment the count variable and send a token to the barrier semaphore. This will allow it to immediately acquire a barrier semaphore token and continue execution. After passing through the barrier, it immediately sends another token to the barrier semaphore. This allows one of the other waiting Threads to resume execution. This thread places another token in the barrier semaphore, which triggers another waiting Thread and so on. This final section of the barrier code is called a turnstile because it allows one thread at a time to pass the barrier. In our model of concurrent execution, this means that each thread waits at the barrier until the last arrives then they all resume simultaneously. In the following exercise, we create five instances of one thread containing barrier code. However, the barrier could be used to synchronize five unique Threads.

Exercise 10.14: Semaphore Barrier

In this exercise, we will use semaphores to create a barrier to synchronize multiple tasks.

In the Pack Installer select "Ex 10.14 Barrier" and copy it to your tutorial directory.

Build the project and start the debugger.

Open the Peripherals \General Purpose IO\GPIOB window.

Run the code.

Initially, the semaphore code in each of the Threads is commented out. Since the Threads are not synchronized the GPIO pins will toggle randomly like in the rendezvous example.

Exit the debugger.

Remove the comments on lines 62, 75, 80, and 93 to enable the barrier code.

Built the project and start the debugger.

Run the code and observe the activity of the pins in the GPIOB window.

Now the tasks are synchronized by the semaphore and run the LED functions "concurrently."

Semaphore Caveats

Semaphores are an extremely useful feature of any RTOS. However, semaphores can be misused. You must always remember that the number of tokens in a semaphore is not fixed. During the runtime of a program, semaphore tokens may be created and destroyed. Sometimes this is useful, but if your code depends on having a fixed number of tokens available to a semaphore, you must be very careful to always return tokens back to it. You should also rule out the possibility of accidentally creating additional new tokens.

Mutex

Mutex stands for "Mutual Exclusion." In reality, a mutex is a specialized version of semaphore. Like a semaphore, a mutex is a container for tokens. The difference is that a mutex can only contain one token, which cannot be created or destroyed. The principal use of a mutex is to control access to a chip resource such as a peripheral. For this reason, a mutex token is binary and bounded. Apart from this, it really works in the same way as a semaphore. First of all, we must declare the mutex container and initialize the mutex:

```
osMutexId_t uart_mutex;
osMutexAttr_t {
const char        *name;    ///< name of the mutex
uint32_t          attr_bits;  ///< attribute bits
void              *cb_mem;   ///< memory for control block
uint32_           t cb_size;  ///< size of provided memory for control block
};
```

When a mutex is created, its functionality can be modified by setting the attribute bits shown in Table 10.9.

Table 10.9: Mutex attribute

Bit Mask	Description
osMutexRecursive	The same thread can consume a mutex multiple times without locking itself
osMutexPrioInherit	While a Thread owns the mutex it cannot be preempted by a higher priority Thread
osMutexRobust	Notify Threads that acquire a mutex that the previous owner was terminated

Once declared the mutex must be created in a Thread.

```
uart_mutex = osMutexNew(&MutexAttr);
```

Then any Thread needing to access the peripheral must first acquire the mutex token:

```
osMutexAcquire(osMutexId_t mutex_id,uint32_t ticks);
```

Finally, when we are finished with the peripheral, the mutex must be released:

```
osMutexRelease(osMutexId_t mutex_id);
```

Mutex use is much more rigid than semaphore use but is a much safer mechanism when controlling absolute access to underlying chip registers.

Exercise 10.15: Mutex

In this exercise, our program writes streams of characters to the microcontroller UART from different Threads. We will declare and use a mutex to guarantee that each thread has exclusive access to the UART until it has finished writing its block of characters.

In the Pack Installer select "Ex 10.15 Mutex" and copy it to your tutorial directory.

This project declares two Threads which both write blocks of characters to the UART. Initially, the mutex is commented out.

```
void uart_Thread1 (void *argument) {
uint32_t i;
for (;;) {
        //osMutexAcquire(uart_mutex, osWaitForever);
        for( i = 0;i<10;i++) SendChar('1');
        SendChar('\n');
        SendChar('\r');
        //osMutexRelease(uart_mutex);
}}
```

In each thread, the code prints out the Thread number. At the end of each block of characters, it then prints the carriage return and newline characters.

Build the code and start the debugger.

Open the UART1 console window with View\Serial Windows\UART #1 (Fig. 10.46).

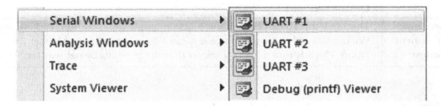

Figure 10.46
Open the UART console window.

Start the code running and observe the output in the console window (Fig. 10.47).

Figure 10.47
The mis-ordered serial output.

Here, we can see that the output data stream is corrupted by each thread writing to the UART without any accessing control.

Exit the debugger.

Uncomment the mutex calls in each thread.

Build the code and start the debugger.

Observe the output of each task in the console window (Fig. 10.48).

Figure 10.48
Order restored by using a mutex.

Now the mutex guarantees each task exclusive access to the UART while it writes each block of characters.

Mutex Caveats

Clearly, you must take care to return the mutex token when you are finished with the chip resource, or you will have effectively prevented any other Thread from accessing it. You must also be extremely careful about using the osThreadTerminate() call on functions that control a mutex token. Keil RTX5 is designed to be a small footprint RTOS so that it can run on even very constrained Cortex-M microcontrollers. Consequently, there is no Thread deletion safety. This means that if you delete a Thread that is controlling a mutex token, you will destroy the mutex token and prevent any further access to the guarded peripheral.

Data Exchange

So far, all of the inter-Thread communication methods have only been used to trigger the execution of Threads; they do not support the exchange of program data between Threads. In a real program, we will need to move data between Threads. This could be done by reading and writing to globally declared variables. In anything but a very simple program, guaranteeing data integrity would be extremely difficult and prone to unforeseen errors. The exchange of data between Threads needs a more formal asynchronous method of communication.

CMSIS-RTOS provides two methods of data transfer between Threads. The first method is a message queue which creates a buffered data "pipe" between two Threads (Fig. 10.49). The message queue is designed to transfer integer values.

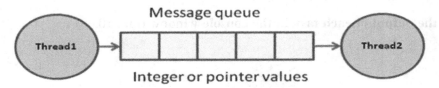

Figure 10.49
Message queue acts as a FIFO buffer between Threads.

The second form of data transfer is a mail queue (Fig. 10.50). This is very similar to a message queue except that it transfers blocks of data rather than a single integer.

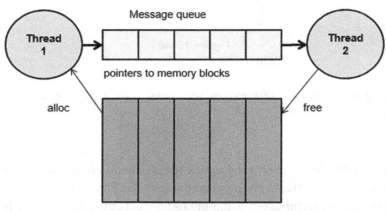

Figure 10.50
A mail queue can transfer blocks of structured data between Threads.

Message and mail queues both provide a method for transferring data between Threads. This allows you to view your design as a collection of objects (Threads) interconnected by data flows. The data flow is implemented by message and mail queues. Message and Mail Queues provide both a buffered transfer of data and a well-defined communication interface between Threads (Fig. 10.51). Starting with a system-level design based on Threads connected by mail and message queues allows you to code different subsystems of your project, especially useful if you are working in a team. Also, as each thread has well-defined inputs and outputs, it is easy to isolate for testing and code reuse.

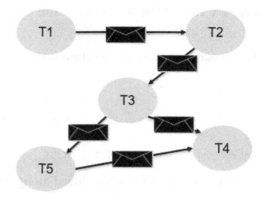

Figure 10.51
The system-level view of an RTOS-based project consists of Thread objects connected by data flows in the form of message and mail queues.

Message Queue

To setup a message queue, we first need to allocate the memory resources.

```
osMessageQId_t Q_LED;
osMessageQueueAttr_t {
        const char    *name;        ///< name of the message queue
        uint32_t      attr_bits;  ///< attribute bits
        void          *cb_mem;    ///< memory for control block
        uint32_t      cb_size;    ///< size of provided memory for control block
        void          *mq_mem;    ///< memory for data storage
        uint32_t      mq_size;    ///< size of provided memory for data storage
};
```

Once the message queue handle and attributes have been declared, we can create the message queue in a Thread.

```
Q_LED = osMessageNew(DepthOfMesageQueue,WidthOfMessageQueue,&osMessageQueueAttr);
```

Once the message queue has been created, we can put data into the queue from one thread

```
osMessageQueuePut(Q_LED,&dataIn,messagePrioriy,osWaitForever);
```

and then read if from the queue in another.

```
result = osMessageQueueGet(Q_LED,&dataOut,messagePriority,osWaitForever);
```

Exercise 10.16: Message Queue

In this exercise, we will look at defining a message queue between two Threads and then use it to send process data.

In the Pack Installer select "Ex 10.16 Message Queue" and copy it to your tutorial directory.

Open Main.c and view the message queue initialization code.

```
osMessageQId Q_LED;
osMessageQDef (Q_LED,0x16,unsigned char);
osEvent result;
int main (void) {
     LED_Init ();
     Q_LED = osMessageCreate(osMessageQ(Q_LED),NULL);
```

We define and create the message queue in the main thread along with the event structure.

```
osMessagePut(Q_LED,0x1,osWaitForever);
osDelay(100);
```

Then in one of the Threads, we can post data and receive it in the second.

```
result = osMessageGet(Q_LED,osWaitForever);
LED_On(result.value.v);
```

Build the project and start the debugger.

Set a breakpoint in led_Thread1 (Fig. 10.52).

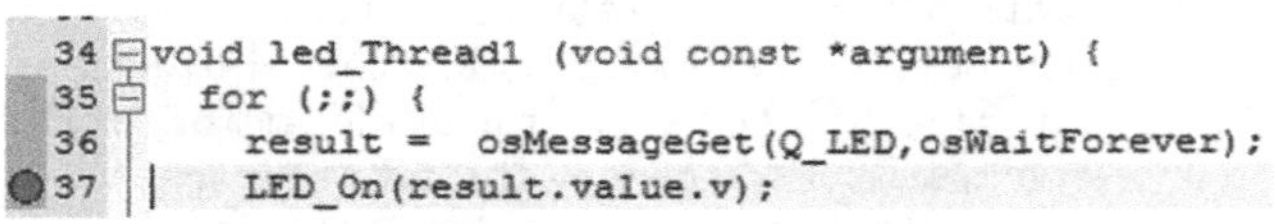

```
34 ⊟void led_Thread1 (void const *argument) {
35 ⊟  for (;;) {
36        result =  osMessageGet(Q_LED,osWaitForever);
37 ||     LED_On(result.value.v);
```

Figure 10.52
Set a breakpoint on the receiving Thread.

Now run the code and observe the data as it arrives.

Extended Message Queue

In the last example, we defined a word-wide message queue. If you need to send a larger amount of data, it is also possible to define a message queue where each slot can hold more complex data. First, we can define a structure to hold our message data.

```
typedef struct {
 uint32_t duration;
 uint32_t ledNumber;
       uint8_t priority;
} message_t;
```

Then, we can define a message queue that is formatted to receive this type of message.

```
Q_LED = osMessageQueueNew(16,sizeof(message_t),&queueAttr_Q_LED);
```

Exercise 10.17: Message Queue

In the Pack Installer select "Ex 10.17 Extended Message Queue" and copy it to your tutorial directory.

Open Main.c and view the message queue initialization code.

Led_Thread2 updates the message structure and posts a new message into the queue.

The second message is sent with a high priority which will cause it to arrive first.

Led_Thread1 reads the queue and writes the transferred data to the LED.

Run the code and observe the order of the message data arriving in led_Thread1.

Message Queue API

In addition to the put and get functions, each message queue has a number of support functions that allow you to manage a message queue (Table 10.10).

Table 10.10: Message queue functions

Function	Description
osMessageQueueNew	Create a new message queue
osMessageQueueGet	Receive a message from a queue
osMessageQueuePut	Send a message to a queue
osMessageQueueGetName	Get the message queue name string defined in the queue attribute structure
osMessageQueueGetCapacity	Get the number of message queue slots in a message queue
osMessageQueueGetMsgSize	Get the maximum message size in bytes for a message queue
osMessageQueueGetcount	Get the number of currently queued messages
osMessageQueueGetSpace	Get number of currently available slots in the message queue
osMessageQueueReset	Reset a given message queue
osMessageQueueDelete	Delete the message queue

Memory Pool

We can design a message queue to support the transfer of large amounts of data. However, this method has an overhead in that we are "moving" the data in the queue. In this section we will look at designing a more efficient "zero copy" mailbox where the data remains static. CMSIS-RTOS2 supports the dynamic allocation of memory in the form of a fixed block memory pool. First, we can declare the memory pool attributes:

```
osMemoryPoolAttr_t {
        const char      *name;      ///< name of the memory pool
        uint32_t        attr_bits;  ///< attribute bits
        void            *cb_mem;    ///< memory for control block
        uint32_t        cb_size;    ///< size of provided memory for control block
        void            *mp_mem;    ///< memory for data storage
        uint32_t        mp_size;    ///< size of provided memory for data storage
} osMemoryPoolAttr_t;
```

and a handle for the memory pool;

```
osMemoryPoolId_t mpool;
```

For the memory pool itself, we need to declare a structure that contains the memory elements we require in each memory pool lot.

```
typedef struct {
                uint8_t LED0;
                uint8_t LED1;
                uint8_t LED2;
                uint8_tLED3;
        } memory_block_t;
```

Then, we can create a memory pool in our application code.

```
mpool = osMemoryPoolNew(16, sizeof(message_t),&memorypoolAttr_mpool);
```

Now we can allocate a memory pool slot within a Thread

```
memory_block_t *led_data;
*led_data = (memory_block_t *) osMemoryPoolAlloc(mPool,osWaitForever);
```

and then populate it with data:

```
led_data->LED0 = 0;
led_data->LED1 = 1;
led_data->LED2 = 2;
led_data->LED3 = 3;
```

It is then possible to place the pointer to the memory block in a message queue.

```
osMessagePut(Q_LED,(uint32_t)led_data,osWaitForever);
```

so the data can be accessed by another task.

```
osEvent event; memory_block_t * received;
event = osMessageGet(Q_LED,osWatiForever);
*received = (memory_block *)event.value.p;
led_on(received->LED0);
```

Once the data in the memory block has been used, the block must be released back to the memory pool for reuse.

```
osPoolFree(led_pool,received);
```

To create a zero-copy mailbox system, we can combine a memory pool to store the data with a message queue which is used to transfer a pointer to the allocated memory pool slot. This way, the message data stays static, and we pass a pointer between Threads.

Exercise 10.18: Zero Copy Mailbox

This exercise demonstrates the configuration of a memory pool and message queue to transfer complex data between Threads.

In the Pack Installer select "Ex 10.18 Memory Pool" and copy it to your tutorial directory.

This exercise creates a memory pool and a message queue. A producer Thread acquires a buffer from the memory pool and fills it with data. A pointer to the memory pool buffer is then placed in the message queue. A second Thread reads the pointer from the message queue and then accesses the data stored in the memory pool buffer before freeing the buffer back to the memory pool. This allows large amounts of data to be moved from one thread to another in a safe synchronized way. This is called a "zero copy" memory queue as only the pointer is moved through the message queue. The actual data does not move memory locations.

At the beginning of main.c the memory pool and message queue are defined.

```
static const osMemoryPoolAttr_t memorypoolAttr_mpool = {
    .name = "memory_pool",
};
void app_main (void *argument) {
  mpool = osMemoryPoolNew(16, sizeof(message_t),&memorypoolAttr_mpool );
  queue = osMessageQueueNew(16,4, NULL);
  osThreadNew(producer_Thread, NULL, &ThreadAttr_producer);
  osThreadNew(consumer_Thread, NULL, &ThreadAttr_consumer);
}
```

In the producer Thread acquire a message buffer, fill it with data and post a testData++.

```
while (1){
        if(testData == 0xAA){
                testData = 0x55;
        }
        else{
                testData = 0xAA;
        }
        message = (message_t*)osMemoryPoolAlloc(mpool,osWaitForever);//Allocate a
                                                                    memory pool
                                                                    buffer

        for(index =0;index<8;index++){
                message->canData[index] = testData;
        }
        osMessageQueuePut(queue, &message,NULL, osWaitForever); osDelay(1000);
}
```

Then in the consumer Thread, we can read the message queue to get the next pointer and then access the memory pool buffer. Once we have used the data in the buffer, it can be released back to the memory pool.

```
while (1) {
        osMessageQueueGet(queue,&message,NULL,osWaitForever);

        LED_SetOut((uint32_t)message->canData[0]);
        osMemoryPoolFree(mpool, message);
}
```

Build the code and start the debugger.

Place breakpoints on the osMessagePut and osmessageGet functions (Fig. 10.53).

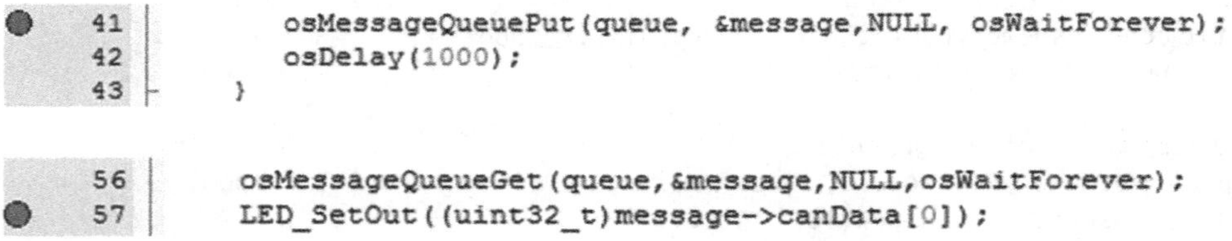

```
41        osMessageQueuePut(queue, &message,NULL, osWaitForever);
42        osDelay(1000);
43    }

56        osMessageQueueGet(queue,&message,NULL,osWaitForever);
57        LED_SetOut((uint32_t)message->canData[0]);
```

Figure 10.53
Set breakpoints on the sending and receiving Threads.

Run the code and observe the data being transferred between the Threads.

Configuration

So far, we have looked at the CMSIS-RTOS2 API. This includes Thread management functions, time management, and inter-Thread communication. Now that we have a clear idea of exactly what the RTOS kernel is capable of, we can take a more detailed look at the configuration file (Fig. 10.54). There is one configuration file for all of the Cortex-M-based microcontrollers.

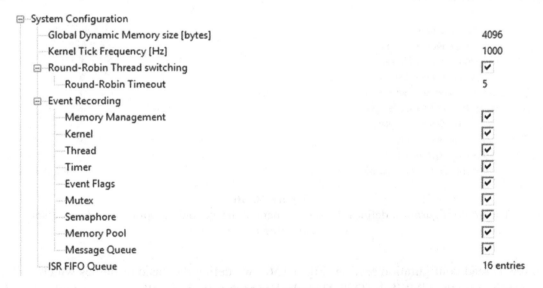

⊞ System Configuration
⊞ Thread Configuration
⊞ Timer Configuration
⊞ Event Flags Configuration
⊞ Mutex Configuration
⊞ Semaphore Configuration
⊞ Memory Pool Configuration
⊞ Message Queue Configuration

Figure 10.54
RTX is configured using one central configuration file.

Like the other configuration files, the RTX_Config.h file is a template file that presents all the necessary configurations as a set of menu options.

System Configuration

⊟ System Configuration
　　Global Dynamic Memory size [bytes]　　　　　　　　　　4096
　　Kernel Tick Frequency [Hz]　　　　　　　　　　　　　1000
　⊟ Round-Robin Thread switching　　　　　　　　　　　　☑
　　　Round-Robin Timeout　　　　　　　　　　　　　　　5
　⊟ Event Recording
　　　Memory Management　　　　　　　　　　　　　　　☑
　　　Kernel　　　　　　　　　　　　　　　　　　　　　☑
　　　Thread　　　　　　　　　　　　　　　　　　　　　☑
　　　Timer　　　　　　　　　　　　　　　　　　　　　☑
　　　Event Flags　　　　　　　　　　　　　　　　　　☑
　　　Mutex　　　　　　　　　　　　　　　　　　　　　☑
　　　Semaphore　　　　　　　　　　　　　　　　　　☑
　　　Memory Pool　　　　　　　　　　　　　　　　　☑
　　　Message Queue　　　　　　　　　　　　　　　　☑
　　ISR FIFO Queue　　　　　　　　　　　　　　16 entries

Figure 10.55
RTX Configuration options.

Before we discuss the settings in the system configuration section (Fig. 10.55), it is worth mentioning what is missing. In earlier versions of CMSIS-RTOS, it was necessary to define the CPU frequency as part of the RTOS configuration. In CMSIS-RTOS2, the CPU frequency is now taken from the "SystemCoreClock" variable which is set as part of the CMSIS-Core system startup code. If you are working with a new microcontroller, you will need to check that this value is being set correctly.

Within the system configuration, we can define the key RTOS kernel features. As we have seen earlier, we can set the amount of memory allocated to the "Global Dynamic Memory Pool." By default, this is set to 32 K. You will need to adjust this if you are using a very constrained device with a small amount of RAM. Next, we can define the tick frequency in Hertz. This defines the SysTick interrupt rate and is set to 1 ms by default. Generally, I would leave the tick frequency at its default setting. However, processor clock speeds are getting ever faster. If you are using a high-performance device, you may consider using a faster tick rate. "Round Robin Thread" switching is also enabled by default in this section. Again I would recommend leaving these settings in their default state unless you have a strong requirement to change them. The system configuration settings also allow us to control the range of messages sent to the event recorder as the RTOS runs. Finally, if we are setting Thread flags from an interrupt, they are held in a queue until they are processed. Depending on your application, you may need to increase the size of this queue.

Thread Configuration

Thread Configuration	
Object specific Memory allocation	☑
Number of user Threads	1
Number of user Threads with default Stack size	0
Total Stack size [bytes] for user Threads with user-provided Stack size	0
Default Thread Stack size [bytes]	200
Idle Thread Stack size [bytes]	200
Stack overrun checking	☑
Stack usage watermark	☐
Processor mode for Thread execution	Privileged mode

Figure 10.56
Thread configuration defines the thread memory usage, debug options, and processor operating mode.

In the Thread configuration section (Fig. 10.56), we define the basic resources which will be required by the CMSIS-RTOS2 Threads. For each thread, we allocate a "default Thread stack space" (in the above example, this is 200 bytes.) As you create Threads, this memory will be allocated from the Global Dynamic memory pool. However, if we enable Object-specific memory allocation, the RTO will define a memory region that is dedicated to Thread usage only. If you switch to object-specific memory allocation, it is necessary to provide details about the number and size of Threads memory so the RTOS can calculate the maximum memory requirement. For object-specific memory allocation, we must define the maximum number of user Threads (don't count the idle or timer Threads) that will be running. We must also define the number of Threads that have a default stack size and also the total amount of memory required by Threads with custom stack sizes. Once we have defined the memory used by user Threads, we can allocate memory to the idle thread. During development, CMSIS-RTOS can trap stack overflows. When this option is enabled, an overflow of a Thread stack space will cause the RTOS kernel to call the os_error function which is located in the RTX_Conf_CM.c file. This function gets an error code and then sits in an infinite loop. The stack checking option is intended for use during debugging and should be disabled on the final application to minimize the kernel overhead. However, it is possible to modify the os_error() function if enhanced error protection is required in the final release.

```
        uint32_t osRtxErrorNotify (uint32_t code, void *object_id) {
(void)object_id;
switch (code) {
  case osRtxErrorStackUnderflow:
  // Stack underflow detected for Thread (Thread_id = object_id)
  break;
case osRtxErrorISRQueueOverflow:
  // ISR Queue overflow detected when inserting object (object_id)
  break;
case osRtxErrorTimerQueueOverflow:
  // User Timer Callback Queue overflow detected for timer (timer_id = object_id)
  break;
case osRtxErrorClibSpace:
  // Standard C/C++ library libspace not available: increase OS_THREAD_LIBSPACE_NUM
  break;
case osRtxErrorClibMutex:
  // Standard C/C++ library mutex initialization failed
  break;
default:
  break;
}
 for (;;) {}
//return 0U;
}
```

It is also possible to monitor the maximum stack memory usage during run time. If you check the "Stack Usage Watermark" option, a pattern (0xCC) is written into each stack space. During runtime, this watermark is used to calculate the maximum memory usage. This figure is reported in the Threads section of the view\watch window\RTX RTOS (Fig. 10.57).

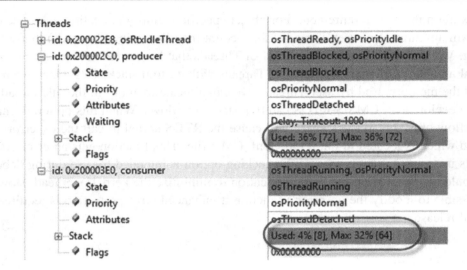

Figure 10.57
Thread stack watermarking allows the debugger to calculate maximum memory usage for each Thread.

The Thread definition section also allows us to select whether the Threads are running in privileged or unprivileged mode. The last option allows us to define the processor operating mode for the user Threads. If you want an easy life, leave this set to "privileged mode," and you will have full access to all the processor features. However, if you are writing a safety-critical or secure application, then "unprivileged mode" can be used to prevent Thread access to critical processor registers limiting run time errors or attempts at intrusion. We will have a closer look at this in the next chapter.

System Timer Configuration

The default timer for use with CMSIS-RTOS is the Cortex-M SysTick timer which is present on nearly all Cortex-M processors. The input to the SysTick timer will generally be the CPU clock. It is possible to use a different timer by overloading the kernel timer functions as outlined in Table 10.11.

Table 10.11: SysTick timer functions

Function	Description
uint32_t osRtxSysTimerSetup(void)	Sets up an alternate timer
void osRtxSysTimerEnable(void)	Enable the alternate timer
void osRtxSysTimerDisable(void)	Disable the alternate timer
void osRtxSysTimerAckIRQ(void)	Acknowledge the alternative timer IRQ
uint32_t osRtxSysTimerGetCount(void)	Returns the alternative timer count
uint32_t osRtxSysTimerGetFreq(void)	Returns the alternative timer frequency

RTX5 License

CMSIS-RTOS Keil RTX5 is provided under a three-clause BSD license and may be used freely without cost for commercial and noncommercial projects. RTX5 will also compile using the IAR and GCC tools. For more information, use the URL below.

https://www.keil.com/demo/eval/rtx.htm

Conclusion

In this chapter, we have worked our way through the CMSIS-RTOS2 API and introduced some of the key concepts associated with using an RTOS. The only real way to learn how to develop with an RTOS is to actually use one in a real project. In the next chapter, we will look at some proven techniques that can be used when developing using an RTOS running on a Cortex-M microcontroller.

CHAPTER 11

RTOS Techniques

Introduction

This chapter will look through some techniques that can be used when developing with an RTOS. These techniques have been used in a real project, so they are tried and tested. First, we will look at how to design a system that integrates RTOS threads and peripheral interrupt routines while maintaining the real-time response of the RTOS. We will also look at how to add power management and watchdog support to a multithreaded RTOS project. Next, we will see how to design a system that maintains real-time processing of continuous data but is also able to respond to event-driven tasks such as a user interface. Finally, we will have a look at the Functional Safety version of RTX and examine the additional features that can be used to develop high-integrity safety-critical systems.

RTOS and Interrupts

In the last chapter, we saw that our application code could execute in RTOS threads. However, in a real system, we will also have a number of Interrupt Service Routines (ISRs) which will be triggered by events in the microcontroller peripherals. The RTOS does not affect the raw interrupt latency and allows you to service an interrupt in precisely the same way you would on a Non-RTOS-based system. However, as we have seen, the scheduler and any RTOS API calls are also generating Cortex-M processor exceptions (Fig. 11.1).

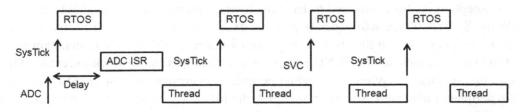

Figure 11.1

The RTOS will generate SysTick and SVC exceptions. These have to be integrated with the user peripheral interrupts to make a successful system.

The RTOS scheduler uses the SysTick timer exception, and the RTOS API calls use the Supervisor Call (SVC) instruction. These exceptions also contend with the peripheral interrupts through the NVIC priority scheme. If the SysTick and SVC exceptions are configured with a high priority, they will preempt the user peripherals and cause a delay in the servicing of the peripheral ISR. If we increase the priority of the user peripherals interrupts, then the peripheral ISR will be served, but the RTOS exceptions will be delayed, and this will destroy the real-time features of the RTOS (Fig. 11.2).

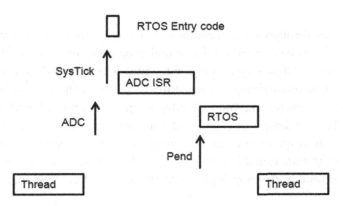

Figure 11.2
The SysTick exception runs the minimal amount of RTOS code to maintain its real-time features. It then sets the PEND exception and quits. This allows any peripheral interrupts to execute followed by the PEND exception. The PEND exception will execute the remainder of the RTOS code.

Fortunately, the Cortex-M processors have an additional exception called the PEND exception that is used to mitigate his problem. The PEND interrupt is simply an interrupt channel that is connected to a software register in the Cortex-M system control block (SCB) (USERSETMPEND bit in the CFG_CTRL register) rather than a device peripheral. When the software register is written to a PEND exception is raised and the NVIC will trigger execution on the PEND exception in the same fashion as any other interrupt source. The PEND exception is also configured to have the lowest interrupt priority available. So in an RTOS the SysTick Timer will generate periodic exceptions at the highest priority. In the SysTick exception, we run the absolute minimum amount of RTOS code to maintain its real-time features, then set the PEND exception, and then quit the SysTick exception. Once we quit the SysTick exception any pending peripheral interrupts are now able to run. When all the peripheral interrupts have been serviced, the PEND exception will be serviced. In the PEND ISR, we can now run the bulk of the RTOS code. By using this mechanism, we can guarantee to maintain the real-time features of the RTOS without adversely affecting the peripheral interrupt response. So it is possible to design a system that uses RTOS threads combined with ISR running at the native processor interrupt latency. In the next section, we will see how to more fully integrate peripheral interrupt handling with the RTOS threads.

RTOS Interrupt Handling

When working with an RTOS, it is desirable to keep ISR functions as short as possible. Ideally, an ISR will just be a few lines of code, taking a minimal amount of run time. One way to achieve this is to move the ISR code and place it in a dedicated high-priority thread, a servicing thread. The servicing Thread will be designed to sit in a blocked state as soon as it starts allowing the application threads to run. When the peripheral interrupt occurs, we will enter the ISR as usual, but rather than process the interrupt, the ISR will use an RTOS object (Thread flag, message queue semaphore, etc.) to wake up the blocked servicing thread. Any of the RTOS interprocessor communication objects can be used to link an ISR to a thread which can, in turn, service the peripheral interrupt. The servicing thread code will then process the interrupt and when finished go back to the blocked state. This method keeps the ISR code to a minimum and allows the RTOS scheduler to decide which Thread should be running. The downside is the increased context switch time required to schedule the servicing Thread. However, for many applications, this is not a real problem. If you do need the absolute minimum interrupt latency for key peripherals, it is still possible to service this interrupt with a dedicated ISR.

In the example below, we will look at integrating an ISR and servicing thread execution using the RTOS Thread flag functions. The first line of code in the servicing Thread should make it wait for a Thread flag. When an interrupt occurs, the ISR simply sets the Thread flag and terminates. This schedules the servicing Thread, which in turn runs the necessary code to manage the peripheral (Fig. 11.3). Once finished, the Thread will again hit the osThreadFlagsWait() call, forcing it to block and wait for the next peripheral interrupt.

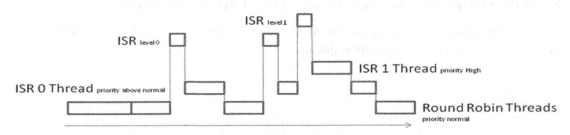

Figure 11.3
Within the RTOS, interrupt code is run as threads. The interrupt handlers signal the tasks when an interrupt occurs. The task priority level defines which task gets scheduled by the kernel.

A typical interrupt servicing thread will have the following structure:

```
__NO_RETURN void servicing_thread (void *argument){

    for (;;) {
        osThreadFlagsWait(0x01,NULL,osWaitForever);     //Block here until signaled by
                                                        an interrupt
        ....................                            //Run the interrupt service
                                                        code

    }
}
```

The actual interrupt handler will contain a minimal amount of code, and this is used to signal the servicing Thread.

```
void IRQ_Handler (void)
{
        osThreadFlagsSet(servicingThread_ID,0x01);      // Signal the servicing
                                                        thread with an event
        Peripheral->SR &= ~(1 << 1);                    //Clear the peripheral
                                                        status flag

}
```

Exercise 11.1: RTOS Interrupt Handling

CMSIS RTOS does not introduce any latency in serving interrupts generated by user peripherals. However, the operation of the RTOS may be disturbed if you lock out the SysTick interrupt for a long period of time. This exercise demonstrates a technique of signaling a thread from an interrupt and servicing the peripheral interrupt with a thread rather than using a standard ISR.

Open the Pack Installer.

Select the Boards::Designers Guide Tutorial.

Select the example tab and Copy "Ex 11.1 RTOS Interrupt Handling."

In the main() function, we initialize the ADC and create an ADC thread that has a higher priority than all the other application threads.

```
void app_main (void *argument) {
LED_Initialize();
init_ADC();
T_led_ID1 = osThreadNew(led_Thread1, NULL,&ThreadAttr_LED1);    //osPriorityNormal
T_led_ID2 = osThreadNew(led_Thread2, NULL,&ThreadAttr_LED2);    //osPriorityNormal
T_adc_ID = osThreadNew(adc_Thread, NULL,&ThreadAttr_ADC);       //osPriorityAboveNormal
for (;;);
}
```

In this project each thread priority is set to osPriorityNormal except for the ADC thread which is set to a higher priority.

```
static const osThreadAttr_t ThreadAttr_ADC = {
        .name = "ADC",
        .priority = osPriorityAboveNormal,
};
```

Build the code and start the debugger.

Set breakpoints in led_Thread2, ADC_Thread, and ADC1_2_IRQHandler (Figs. 11.4 and 11.5).

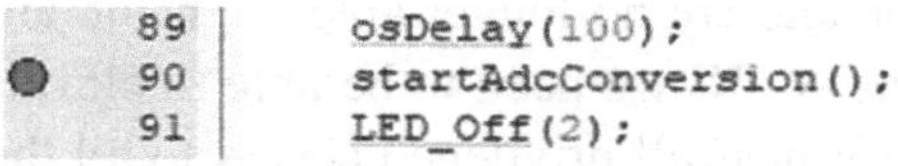

```
        89 │    osDelay(100);
   ●    90 │    startAdcConversion();
        91 │    LED_Off(2);
```

Figure 11.4
Breakpoint on startADC conversion.

Also in adc_Thread().

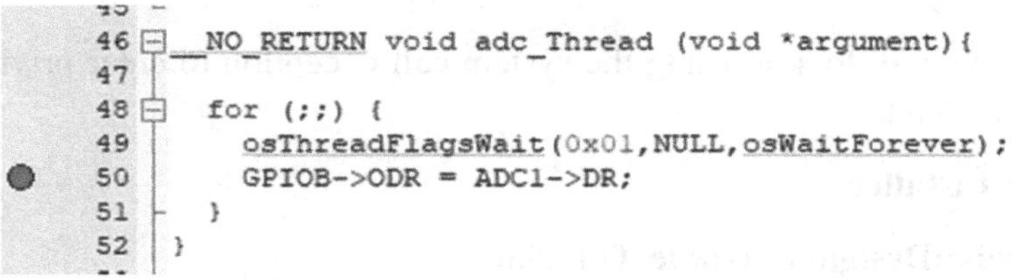

```
      46 ⊟__NO_RETURN void adc_Thread (void *argument){
      47 │
      48 ⊟   for (;;) {
      49 │      osThreadFlagsWait(0x01,NULL,osWaitForever);
   ● 50 │      GPIOB->ODR = ADC1->DR;
      51 ├   }
      52 │ }
```

Figure 11.5
Breakpoint in the adc_Thread task.

And in ADC1_2_Handler (Fig. 11.6).

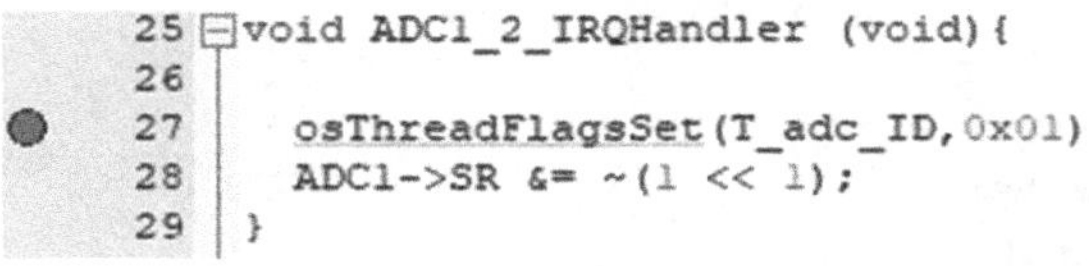

```
      25 ⊟void ADC1_2_IRQHandler (void){
      26 │
   ● 27 │    osThreadFlagsSet(T_adc_ID,0x01)
      28 │    ADC1->SR &= ~(1 << 1);
      29 │ }
```

Figure 11.6
Breakpoint in the adc interrupt.

Run the code.

You will hit the first breakpoint, which starts the ADC conversion, then run the code again, and you will enter the ADC interrupt handler. The handler sets the adc_Thread flag and quits. Setting the Thread flag will cause the ADC thread to preempt any other running Thread, run the ADC service code and then block waiting for the next interrupt signal.

User Supervisor Functions

In the last example, we were able to configure threads and the interrupt structure without any additional consideration because the RTOS was configured to execute thread code as privileged code. In some projects, it may be necessary to configure the RTOS so that threads are running in unprivileged mode. This means that Thread code will no longer be able to write to the NVIC since the RTOS is running when we enter main(). We are stuck in unprivileged Thread mode and are no longer able to enable any interrupt source. In order for Thread code to access the NVIC we need to be able to elevate our execution mode to run in Handler mode so that it has full privileged access to all the Cortex-M processor registers. As we saw in Chapter 5, Advanced Architecture Features this can be done by executing an SVC instruction.

Exercise 11.2: RTOS and User SVC Exceptions

In this exercise, we will look at using the system call exception to enter privileged mode to run "system level" code.

Open the Pack Installer.

Select the Boards::Designers Guide Tutorial.

Select the example tab and copy "Ex 11.2 RTOS USER SVC" to your tutorial directory.

Examine the RTX_Conf_CM.c file (Fig. 11.7).

⊟ Thread Configuration	
⊞ Object specific Memory allocation	☐
Default Thread Stack size [bytes]	1024
Idle Thread Stack size [bytes]	512
Idle Thread TrustZone Module Identifier	0
Stack overrun checking	☑
Stack usage watermark	☐
Processor mode for Thread execution	Unprivileged mode

Figure 11.7
RTX Configuration options.

In the Thread Configuration section, the operating mode for Thread execution is set to "Unprivileged mode."

In the project, we have added a new file called svc_user.c (Fig. 11.8). This file is available as a "User Code Template" (CMSIS-RTOS User SVC) from the "Add New Item" dialog.

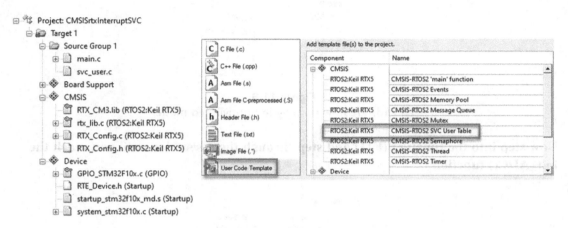

Figure 11.8
Adding RTX SVC user templates.

Open the SVC_Tables.c file.

This is the look up table for the SVC interrupts.

```
#define USER_SVC_COUNT 1    // Number of user SVC functions
extern void __SVC_1 (void);
extern void * const osRtxUserSVC[1 + USER_SVC_COUNT];
    void * const osRtxUserSVC[1 + USER_SVC_COUNT] = {
 (void *)USER_SVC_COUNT,
  (void *) __SVC_1,
};
```

In this file, we need to add the import name and table entry for each __SVC function we are going to use. In our example, we only need __SVC_1.

Now, we can convert the init_ADC() function to a service call exception:

```
__svc(1) void init_ADC (void);    //function prototype
        void __SVC_1 (void) {     //function
              .............
    }
```

Build the project and start the debugger.

Step the code (F11) to the call to the init_ADC() function and examine the operating mode in the register window.

Here, we are in Thread mode, unprivileged, and using the process stack (Fig. 11.9).

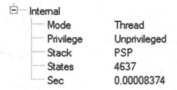

⊟ Internal	
Mode	Thread
Privilege	Unprivileged
Stack	PSP
States	4637
Sec	0.00008374

Figure 11.9
Threads are now running in unprivileged mode.

Now step into the function (F11) and step through the assembler until you reach the init_ADC() function.

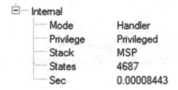

⊟ Internal	
Mode	Handler
Privilege	Privileged
Stack	MSP
States	4687
Sec	0.00008443

Figure 11.10
The adc interrupt is running in Handler mode with privileged access.

Now, we are running in Handler mode with privileged access and are using the main stack pointer (Fig. 11.10).

This allows us the set up the ADC and also access the NVIC. When the function terminates, the processor will return to the unprivileged state and will continue to use the process stack.

Power Management

One of the advantages of developing with an RTOS is that we can consider each of the application threads to be running in parallel. This allows us to develop independent threads of code which each perform a dedicated task and communicate together to create the desired application. This level of abstraction provides lots of benefits as projects get more complex. However, adding power management code to this multithreaded environment can seem daunting at first because you have to ensure that each Thread is ready to enter a low-power state.

A typical Cortex-M-based microcontroller will have several low-power modes. While these will vary between different silicon manufacturers, Table 11.1 summarizes the types of low-power mode you are likely to encounter.

Table 11.1: Typical microcontroller low-power modes

Low-power State	Power State	Wakeup Source
CPU Sleep	CPU in low-power state	Peripheral interrupt, SysTick timer
CPU & Peripheral	CPU and peripherals in low-power state	Reset, Wake up pin, or external interrupt pin
Deep Sleep	Full power down of CPU, Peripherals, SRAM, and Flash	Reset or Wake up pin

The low-power states are entered by executing the __wfi() instruction. By default, this will only affect the Cortex-M processor by placing the CPU into its sleep mode. The Cortex-M processor also has a SLEEPDEEP signal which is connected to the microcontroller power management hardware. If this is enabled in the Cortex-M "Processor System Control" register, then the SLEEPDEEP signal will be asserted when the __wfi() instruction is executed. To enter a given low-power state, the application code must configure the microcontroller power management registers to select the desired low-power state. Generally, the deeper the low-power mode, the more restrictive the exit conditions.

Power Management First Steps

We need to place the microcontroller into the lowest power mode possible without disturbing the performance of the application code. Fortunately, the first step in power management is very simple. When all of the application threads are in a blocked state, for example, waiting for an interrupt to occur or a delay to expire the RTOS scheduler will place the idle Thread into a running state. By default, the osRtxIdleThread()contains an empty for(;;) loop, which is executed until an application thread becomes ready to run. In other words, we simply sit in the osRtxIdleThread() burning energy.

```
__NO_RETURN void osRtxIdleThread (void *argument) {
(void)argument;
for (;;) {}
}
```

When we are in the osRtxIdleThread(), there is nothing for the CPU to do. This is the exact condition we need to enter the CPU low power using __wfi() or __wfe(). When the low-power instruction has executed, the clock to the Cortex-M processor will be stopped, but the microcontroller peripherals will still be active. Typically, the microcontroller low-power

operation will default to the basic Sleep mode, so all we need to do is execute the __wfi()
instruction when the code is in the osRtxIdleThread().

```
__NO_RETURN void osRtxIdleThread (void *argument) {
 for (;;) {
__wfi();
 }}
```

Now, the CPU clock is halted until a peripheral generates an interrupt or the next SysTick
interrupt occurs to run the RTOS scheduler.

This one line of code will stop the CPU from wasting energy without affecting the performance
of the application code and, depending on your application, can have a significant impact on
overall power consumption. This is an effective form of runtime power management.

The RTOS also provides an additional pair of power management functions that can be
used to decide when to place the device into a deep, low, power state. These functions are
designed to only be used in the osRtxIdleThread(). On entry, we can call the first function.

```
maximumSleepTicks = osKernelSuspend();
```

This will halt the scheduler and return the number of scheduler timer ticks that it expects to remain
in the idle loop before the RTOS needs to resume execution. The second function is used to restart
the scheduler; however, to keep the real-time performance of the RTOS, we must confirm how
long it has actually been asleep. This means that we need a low-power timer such as a real-time
clock to keep track of time while the rest of the microcontroller is asleep. When the device wakes
up, we need to resync the RTOS by telling it how much time has passed while it was asleep.

```
osKernelResume(actualSleepTicks);
RTC_Start(maximumSleepTicks);//
Power_SetDeepSleepMode();
__wfe();                                  /* Enter Power-down mode    */
actualSleepTicks = RTC_GetSleepTicks();
osKernelResume(actualSleepTicks);         /* Resume thread scheduler */
```

Making use of the more advanced power modes does require a bit more thought. When we
enter a deeper low-power mode, the wake-up methods are more restricted. This means that
we have to ensure that the application code is ready to enter the low-power state. Consider
the code shown below:

```
Application_thread(void *argument){
while(1)
{
      osMessageQueueGet ();
      function_1();
      osDelay(10);
      function_2();
}}
```

The Thread starts by entering a blocked state, waiting for a message to arrive. When the message data arrives, the Thread will wake up and call function_1(), and then it will again enter a blocked state when the osDelay() function is called. Once the delay has expired, the Thread will resume and call function_2() before once more entering a blocked state until the next message arrives. This Thread has two points at which it will enter a blocked state. If all the other threads are also blocked will cause the osRtxIdleThread() to run. While for the RTOS scheduler, there is no real difference between the two OS calls but for power management, there is an important difference. When the Thread is blocked by the os_message_wait() call, it is truly idle and cannot proceed until another part of the system gives it some data to process. When the Thread is blocked by the osDelay() function for power management purposes, it is still awake in the sense that it will resume processing without input from any other part of the system. We need to construct our application threads so that they wait on an "event" triggered by some other part of the system, perform their function and then wait for the next event to occur. This means that each Thread now has a key blocking point (Fig. 11.11).

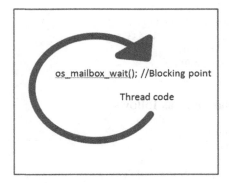

Figure 11.11
In each thread you must identify a blocking point. At the blocking point, a thread has finished its current task and is waiting for the next event to resume processing.

If we construct all our system threads like this, we can identify when all the threads are waiting on their key blocking points. Then the application is truly idle, and we can place it into a deeper low-power state.

Power Management Strategy

To build our power management strategy, we first need to declare a global variable that will hold two flags for each Thread running on the system (Fig. 11.12).

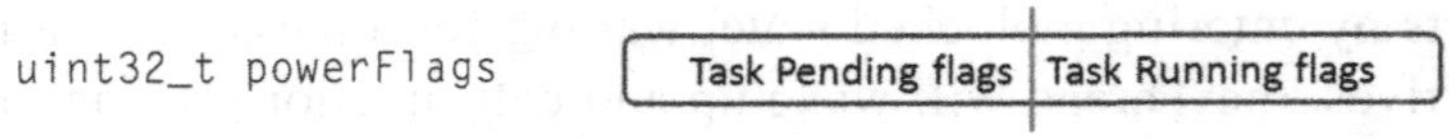

uint32_t powerFlags | Task Pending flags | Task Running flags |

Figure 11.12
Power management flags are held in a global variable and monitored in the idle thread.

Each Thread has one "Thread running" bit and one "thread pending" bit. Next, we need to write some simple functions that will control these bits. When the Thread wakes up to start the execution of its code, we need to set the thread-running bit. As multiple threads will be accessing the power flags variable, it should be protected by a mutex.

```
void ThreadIsRunning(uint16_t Thread) {
        osMutexWait (powerMutex,osWaitForever);
        powerFlags | = Thread;
        osMutexRelease (powerMutex);
}}
```

When it has completed its task and is about to go back into the waiting state, we need to clear the thread running bit.

```
void ThreadIsSuspended(uint16_t Thread) {
        osMutexWait (powerMutex,osWaitForever);
        powerFlags & = ~Thread;
        osMutexRelease (powerMutex);
}}
```

So now the application thread looks as follows:

```
#define APPLICATION_THREAD 1
Application_thread(void){
while(1){
        ThreadIsSuspended(APPLICATION_THREAD)
        osThreadFlagsWait(0x0001,osFlagsWaitAny, osWaitForever);
        ThreadIsRunning(APPLICATION_THREAD);
        function_1();
        osDelay(10);
        function_2();
}}
```

When another part of the system is about to trigger the Thread, in this case, by sending a message, we can set the threads pending bit. This ensures that we avoid any problems with the scheduler.

```
void ThreadIsPending(uint16_t Thread) {
        osMutexWait (powerMutex,osWaitForever);
        powerFlags | = (Thread << 8);              //Set Thread Pending bit
        osMutexRelease (powerMutex);
}
```

Now sending a message to our application thread looks like this:

```
ThreadIsPending(p_phaseA);
osThreadFlagsSet (tid_phaseA, 0x0001);
```

We must modify the ThreadIsRunning routine to clear this bit when execution of the application thread resumes.

```
void ThreadIsRunning(uint16_t Thread) {
        osMutexWait (powerMutex,osWaitForever);
        powerFlags | = Thread;
        powerFlags & = ~ (Thread ≪ 8);                  //clear the Thread pending flag
        osMutexRelease (powerMutex);
}
```

When the application enters the osRtxIdleThread() thread, we can test the powerFlags. If any of the Thread bits (active or pending) are set then the system is not truly idle. If all the power flags are zero then the system is truly idle and we can safely enter low-power mode. The following code can be placed in the osRtxIdleThread().

```
void enterLowPowerState(void) {
        if (powerFlags == ALL_THREADS_INACTIVE) {
                configureAndEnterDeepSleepMode();       //All Threads are idle so enter
                                                        standby mode
        }
        else {
                configureAndEnterSleepMode();           //A Thread is active but we are
                                                        in a delay so enter sleep mode
}}
```

So now, the osRtxIdleThread() will use the available power modes to minimize the runtime energy usage and also detect when the microcontroller can be placed into a deep low-power mode.

Watchdog Management

We can also use the power management flags to solve another problem. In this case, we want to enable a hardware watchdog on the microcontroller. It would be possible to refill the watchdog timer inside each Thread. However, as we are continually switching between threads, it is likely that the watchdog counter would be refilled by several different threads, and we would be unable to detect if a given thread had failed. To make better use of the watchdog, we need to add a separate system thread that is used to monitor the Thread pending bits. The system monitor runs periodically as a user timer. Each time it runs, it will check the Thread pending bits. If a pending bit is set, then a matching counter variable is incremented. If a threshold is exceeded, then the system is in error, the watchdog will not be refilled, and a hardware reset will be forced.

```
void Monitor (void *param){
uint8_t stalledThreadError = FALSE;
        stalledThreadError = checkForStalledThread();
        if(stalledThreadError == FALSE)
        {
                patWatchdog();
        }
}
uint8_t checkForStalledThread(void)
{
uint8_t index,ThreadIsStalled = 0;
        checkForPendingThreads();
        for(index=0;index<(NUMBER_OF_THREADS-1);index++){
                if(ThreadWatchdogCounters[index]>MAX_THREAD_STALL_COUNT){
                        ThreadIsStalled = TRUE;
                }
        }
        return(ThreadIsStalled);
}
```

To complete the Thread monitor code, we need to reset the Thread watchdog counter each time the Thread runs. This can be done in by adding another line to the ThreadIsRunning() function.

Here, we are using the count leading zeros intrinsic to convert from a bit position to an integer value. This is very efficient and takes the minimum number of CPU cycles.

```
void resetCurrentThreadWatchdogCounter(uint16_t ThreadQuery) {
uint16_t ThreadNumber;
        ThreadNumber = 32- __clz((uint32_t)ThreadQuery);   //intrinsic instruction clz
                                                           = count leading zeros
        if(ThreadWatchdogCounters[ThreadNumber-1]>0){
                ThreadWatchdogCounters[ThreadNumber-1] = 0;
}}
```

Integrating Interrupt Service Routines

In the case of an interrupt that will wake the processor and signal a thread to resume processing, we can use the power flags as a means of providing a timeout that allows the microcontroller to go back to sleep. For example, if the microcontroller will be in a low-power mode waiting for data to be sent to a UART. Then, some random noise may wake up the processor, which will then wait for a full message packet to be sent. In this situation, we can introduce a power flag for the UART ISR that is set in the ISR and only cleared when a full message has been received. In the watchdog routine, we can increment a counter if the threshold value is reached. Then, instead of resetting the microcontroller, we simply need to clear the ISR running flag to allow the processor to fall back to sleep. While the ISR power functions are essentially performing the same task as the thread power functions, we need to provide dedicated functions which do not access the mutex as the mutex cannot be acquired by an ISR routine.

```
void IsrThreadIsPending(uint16_t Thread) {
        powerFlags |= (Thread << 8);
}
void IsrIsInactive(uint16_t Thread) {
        powerFlags &= ~ (Thread);
        resetCurrentThreadWatchdogCounter(Thread);
}
```

Then in the watchdog Thread, we can test the ISR flag and increment the UART counter if it is set.

```
if( ThreadManagementFlags & WIRED_COMMS )
{
        ThreadWatchdogCounters[WIRED_COMMS_COUNTER]++;
}
```

If we exceed the threshold level, we can use the mutex-protected ThreadIsSuspended() routine to clear the ISR flag and also reset the watchdog counter.

```
if(ThreadWatchdogCounters[UART_ISR] > UART_TIMEOUT)
{
        ThreadIsSuspended(UART_ISR);
        ThreadWatchdogCounters[UART_ISR] = 0;
}
```

Now, if all the other threads are in a blocked state, all the power management flags will be clear, and the microcontroller will enter its low-power state.

Exercise 11.3: Power and Watchdog Management

We can see these examples in practice by running a simple program in the MDK-Arm simulator. This is a simple RTX program that uses several Threads to flash a group of LEDs in sequence. The start of each sequence is triggered by a hardware timer interrupt. In this example, we will start with the working program and introduce the power management and watchdog code.

Open the Pack Installer.

Select the Boards::Designers Guide Tutorial.

Select the example tab and Copy "Ex 11.3 RTOS Power Management."

Start the simulator by selecting debug\start::stop debug session.

Open the peripherals\General Purpose IO\GPOIOB debug window.

Start the code running with debug\run.

After an initial few seconds, you should see the upper byte of GPIO B pins being toggled by the application Threads.

Open the View\Analysis window\Performance analyzer.

Here, we can see that in this simple program, most of the run time is being spent in the osRtxIdleThread() (Fig. 11.13).

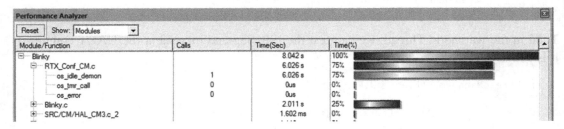

Figure 11.13
In this simple program, the processor is spending all its time in the idle task.

Exit the debugger by selecting debug\start stop.

Open the RTX_Conf_CM.c file and locate the osRtxIdleThread().

Uncomment the __wfi(); call and rebuild the code with project\build.

Restart the simulator with debug\start stop and start the code running.

If we now reexamine the project in the performance analyzer, we can see a dramatic reduction in the number of cycles consumed by the osRtxIdleThread() (Fig. 11.14).

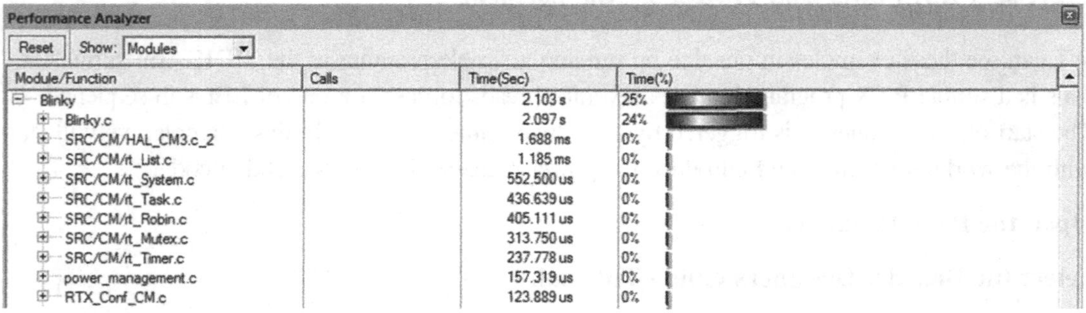

Figure 11.14
Now the processor halts in the idle loop rather than burn energy.

The __wfi() instruction is placing the processor into sleep mode, which is saving power in the Cortex-M processor. However, the peripherals of the microcontroller are still consuming energy. To take advantage of the deeper low-power modes provided by a chip manufacturer, we need to detect when the application is idle and then place the device into a deeper low-power mode until it is ready to restart processing.

Now examine the application Threads.

```
void phaseA (void *argument) {
        for (;;) {
                ThreadIsSuspended(p_phaseA);
                osThreadFlagsWait(0x0001,osFlagsWaitAny, osWaitForever);
                                                /* wait for an event flag 0x0001 */
                ThreadIsRunning(p_phaseA);
                LED_On (LED_A);
                signal_func ();                         /* call common signal function  */
                ThreadIsPending(p_phaseB);
                osThreadFlagsSet (tid_phaseB,0x0001);    /* send event to Thread   */
                osDelay (5);                             /* delay 5 clock ticks */
                LED_Off(LED_A);
}}
```

In each Thread, we identify the Key blocking point and surround it with the ThreadIsSuspended() and ThreadIsRunning() functions. If the Thread signals another Thread, we must first call the ThreadIsPending() function.

Now go back to the osRtxIdleThread()and comment out the __wfi() instruction and uncomment the enterLowPowerState() function.

The enterLowPowerState() function checks the power flags and decides which power mode to enter.

```
void enterLowPowerState(void) {
        if (powerFlags == ALL_THREADS_INACTIVE)
        {
                configureAndEnterDeepSleepMode();
        }else{
                configureAndEnterSleepMode();
}}
```

In each of the low-power configuration functions, we would normally configure the microcontroller power mode and run the __wfi() instruction. However, for demo purposes, we will sit in the function and simply toggle a debug variable.

```
void configureAndEnterDeepSleepMode(void)
{
        uint32_t delay;
//      SCB->SCR |= (1<<SCB_SCR_SLEEPDEEP_Pos);    //enable the deep sleep mode
```

```
//        __wfi();                          //Place the micro in deep sleep
          for(delay = 0;delay<0x100;delay++);  //Simple delay
          debugDeepSleep ^ = 0x01;           //Toggle the debug variable
}
```

Rebuild the code and restart the debugger.

Open the logic analyzer window with view\analysis window\system analyzer.

Start the code running.

After the code has been running for a few seconds, press the Min/Max Auto button and press the zoom out button until you get a view similar to the one below (Fig. 11.15).

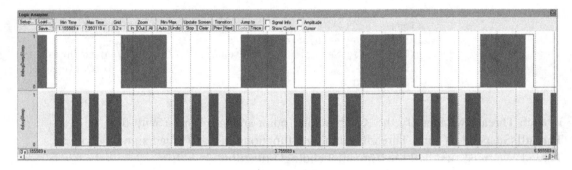

Figure 11.15
Trace of runtime and standby power management.

The upper trace shows the deep sleep debug variable. The solid blocks indicate where we are in the low-power mode. The lower trace shows the sleep debug variable. Again the solid blocks are showing where we are in low-power mode. Here, we can see how the code is saving energy in both its standby mode and also during the runtime execution of the application code.

In blinky.c uncomment the os_tsk_create (Monitor, 10); call on line 171.

This will enable the watchdog monitor Thread which will ensure each of the application Threads is running correctly.

To create an error, uncomment the code in blinky.c 131 to 136, this will let the application run a few times and then delete Thread A.

Build the code and start the simulator.

Open watchdog.c and add the ThreadWatchdogCounters array to the watch window.

Add a breakpoint on the ThreadIsStalled = TRUE; line of code.

Open the debug\os support\RTX Thread and system window.

Start the code running.

After the code has run the led flashing sequence five times, it will delete Thread phase_A() from the system (Fig. 11.16).

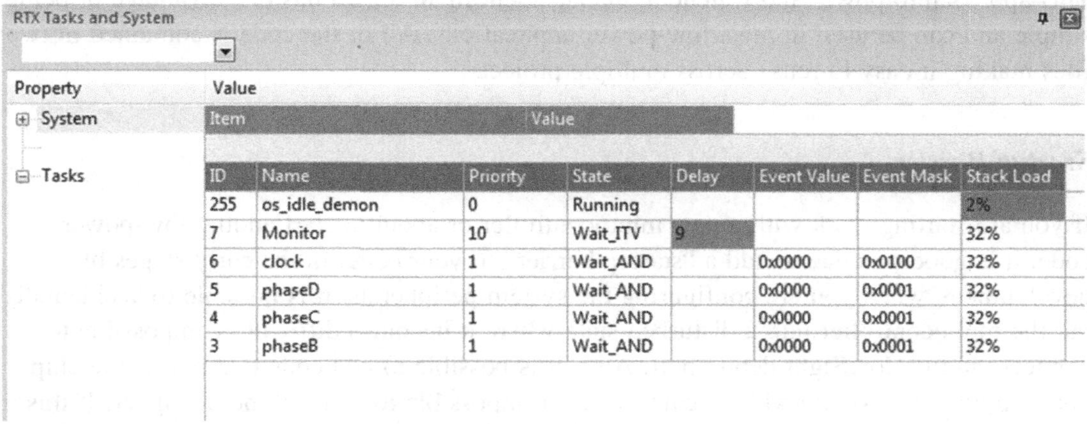

ID	Name	Priority	State	Delay	Event Value	Event Mask	Stack Load
255	os_idle_demon	0	Running				2%
7	Monitor	10	Wait_ITV	9			32%
6	clock	1	Wait_AND		0x0000	0x0100	32%
5	phaseD	1	Wait_AND		0x0000	0x0001	32%
4	phaseC	1	Wait_AND		0x0000	0x0001	32%
3	phaseB	1	Wait_AND		0x0000	0x0001	32%

Figure 11.16
Remaining Threads after Task A has been deleted.

Now the Thread monitor will detect that the Thread A pending bit is set and increment the ThreadWatchdogCounters associated with thread phaseA.

Observe the ThreadWatchdogCounters in the watch window (Fig. 11.17).

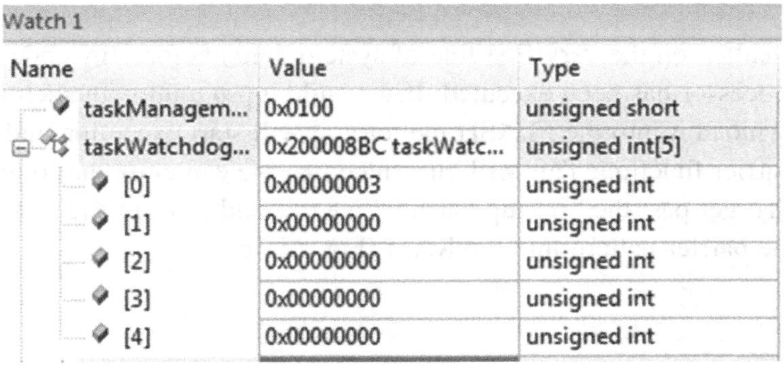

Name	Value	Type
taskManagem...	0x0100	unsigned short
taskWatchdog...	0x200008BC taskWatc...	unsigned int[5]
[0]	0x00000003	unsigned int
[1]	0x00000000	unsigned int
[2]	0x00000000	unsigned int
[3]	0x00000000	unsigned int
[4]	0x00000000	unsigned int

Figure 11.17
Task watchdog counters.

Counter 0 will start to increment. When it hits the MAX_THREAD_STALL_COUNT (five), we will hit the breakpoint.

Start the code running again, the ThreadIsStalled variable will be set. This will stop the monitor Thread from patting the watchdog, and the microcontroller will be reset, causing the application to restart.

This approach to power and watchdog management in an RTOS has the advantage of being simple and can be used in most low-power applications. All of the code is contained in two files making it easy to reuse across multiple projects.

Startup Barrier

If you are starting work with a new microcontroller or about to start adding low-power code, it is good practice to add a "startup barrier" to your code. In the early stages of development, when you are configuring the system peripherals, it is possible to accidentally put the microcontroller into a disturbed state where it becomes difficult or impossible to connect via the CoreSight debug port. Also, it is possible to add code that places the chip onto a power-down state which again makes it impossible to connect the debugger. If this happens, you can no longer erase the flash and program the memory. If you only have a limited supply of development boards, this can be an embarrassing problem! A "startup barrier" is simply a small function that will only allow the code to continue if a specific pattern is in a specific RAM memory location.

```
volatile uint32_t debuggerAttached __attribute__((at(0x2000FFFC)));
void startupBarrier (void)
{
        if(debuggerAttached! = 0x55555555){
                __BKPT(0);
        }
}
```

This function is called on the very first line of system_init() before any code that may upset the Cortex-M processor has been executed. If we build a program with such a "startup barrier" and download it into the FLASH memory, it will start execution and then become trapped in the barrier function. This will guarantee that we can erase and reprogram the Flash memory. To get past the "startup barrier," we can add a script file to the debugger, that programs the barrier pattern into the RAM (Fig. 11.18).

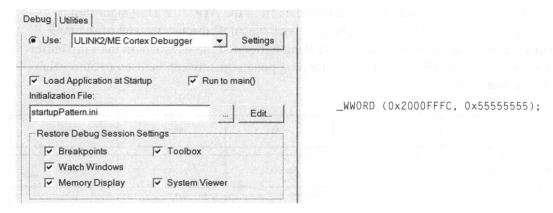

```
_WWORD (0x2000FFFC, 0x55555555);
```

Figure 11.18
Adding the debugger script.

Now when we start the debugger, the barrier pattern will be present in the RAM memory location, and the processor will begin the execution of our application code. If the code does something to latch-up the processor, we can repower the board, and this time it will stop at the barrier allowing the CoreSight debugger to get control of the Cortex-M processor so we can erase the flash memory, modify the source code and try again.

Designing for Real Time

So far, we have looked at the DSP features of the Cortex-M4 and the supporting DSP library. In this next section, we will look at developing a real-time DSP program that can also support event-driven code features like a user interface or communication stack without disturbing the performance of the DSP algorithm.

Buffering Techniques — The Double or Circular Buffer

When developing your first real-time DSP application, the first decision you have to make is how to buffer the incoming blocks of data from the ADC and the outgoing blocks of data to the DAC. A typical solution is to use a form of double buffer as shown below (Fig. 11.19).

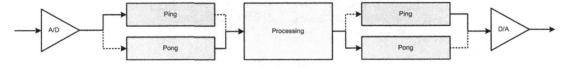

Figure 11.19
A simple DSP system can use a double buffer to capture the stream of data.

While the Ping buffer is being filled with data from the ADC, the DSP algorithm is processing data stored in the Pong buffer. Once the ADC reaches the end of the Ping buffer, it will start to store data into the Pong buffer and the new data in the Ping buffer may be processed (Fig. 11.20). A similar ping pong structure can be used to buffer the output data to the DAC.

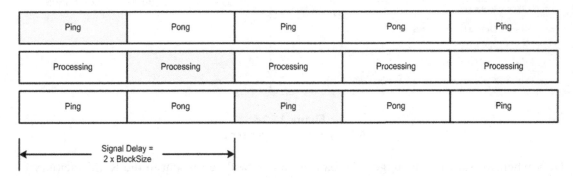

Figure 11.20
The double buffer causes minimum signal latency but requires the DSP algorithm to run frequently. This makes it hard to maintain real-time performance when other software routines need to be executed.

Buffering Techniques FIFO Memory Pool

We can implement an input and output double buffer by using a memory pool object with access controlled by a set of event flags (Fig. 11.21).

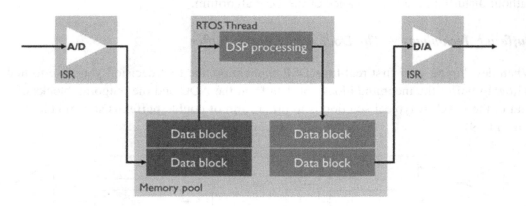

Figure 11.21
Block processing using RTOS memory pool increases the signal latency but provides reliable integration of a DSP thread into a complex application.

Using the RTX RTOS, we can create a Thread that is used to run the DSP algorithm. The inputs and outputs of this Thread are a pair of memory pool data blocks. A timer interrupt running at the desired sample rate is used to trigger an ADC conversion and the results are used to fill a data block. When the block is full, it can be processed by the DSP thread while the second block is being filled with fresh ADC data. The DSP thread will use the CMSIS DSP library to process the ADC data as a block and will generate an output block which will be fed to the DAC one sample at a time on each timer interrupt. The DAC output date is managed using a similar double buffer where one block holds a set of results ready to be sent to the DAC while the second block is used to accept the output from the DSP filter algorithm.

In the first example program, the ADC is sampling at 32 kHz and its results are placed in a memory pool buffer. The memory pool consists of four memory blocks: two for the input buffer and two for the output. They are each declared as arrays of type Q15. DSP_BLOCKSIZE is declared in DSP_IIR.h and has a value of 32 samples.

```
typedef struct _DSP_DataType {
 q15_t Sample[DSP_BLOCKSIZE];
} DSP_DataType;
DSP_MemPool = osMemoryPoolNew (4, sizeof(DSP_DataType), NULL);
```

Once we have created the memory pool buffers, we can allocate an initial input and output block:

```
pDataTimIrqOut = osMemoryPoolAlloc(DSP_MemPool, 0);
pDataSigModOut = osMemoryPoolAlloc(DSP_MemPool, 0);
```

And set the input and output array indexes to zero.

```
dataTimIrqOutIdx = 0;
dataTimIrqInIdx = 0;
```

We must also create an event flag object that will allow signaling between the timer interrupt and the DSP thread.

```
DSP_Event = osEventFlagsNew(NULL);
```

Once the memory pools and event flags are configured, we can start the hardware timer running so that it generates a periodic interrupt at the required sample rate. The timer interrupt can be split into two halves: the first half samples the ADC and fills the active memory pool, the second half writes a filtered sample from the output memory pool block to the DAC.

In the first half, we read the ADC and convert it to Q15 format. The sample is then written to the next free space in the active memory pool block and the index pointer is incremented. When the block is full it is passed to the DSP thread by setting the pDataSigModIn pointer to the full block and setting the EVENT_DATA_TIM_OUT_SIG_IN event flag. Finally, we allocate a new block to start collecting future ADC samples.

A simplified version for the ADC processing section of the timer interrupt is shown below.

```
void TIMER2_IRQHandler(void) {
 uint32_t adGdr; int32_t flags; osStatus_t status; q15_t  tmp;
    adGdr = LPC_ADC->ADGDR;                                  //get ADC sample
    arm_float_to_q15(&adGdr, &tmp, 1);                       // Convert to Q15
   pDataTimIrqOut->Sample[dataTimIrqOutIdx++] = tmp;  //Write to the memory pool block
                                                            //and increment index
    if (dataTimIrqOutIdx >= DSP_BLOCKSIZE){            //if the memory pool block is full
        pDataSigModIn = pDataTimIrqOut;                     //Hand over to the DSP pointer
        flags = osEventFlagsSet(DSP_Event, EVENT_DATA_TIM_OUT_SIG_IN)      //Signal DSP
                                                                            Thread

        pDataTimIrqOut = osMemoryPoolAlloc(DSP_MemPool, 0);   //Allocate a new block
        dataTimIrqOutIdx = 0;                                 //Reset the block index
    }
 }
```

In the main DSP Thread the application code waits for the EVENT_DATA_TIM_OUT_SIG_IN event flag to be set. When the flag is set a full block of ADC data is ready and can be processed by the IIR filter to generate an output block for the DAC. When the filter algorithm has finished the DSP thread will set the EVENT_DATA_TIM_IN_SIG_OUT event flag to notify the timer interrupt that a block of output data is available. Finally, we allocate a fresh memory pool block ready to be filled with the next set of processed samples.

```
void SigMod (void __attribute__((unused)) *arg) {     //The DSP thread
 int32_t flags; osStatus_t status;
 for (;;) {                    //First wait to be signalled that a new block of ADC data is ready
  flags = osEventFlagsWait(DSP_Event, EVENT_DATA_TIM_OUT_SIG_IN, 0, osWaitForever);
  iirExec_q15 (pDataSigModIn->Sample, pDataSigModOut->Sample);    //Process the data
  status = osMemoryPoolFree(DSP_MemPool, pDataSigModIn);    //Free the processed block
  pDataTimIrqIn = pDataSigModOut;                     //Hand over to the timer output pointer
  flags = osEventFlagsSet(DSP_Event, EVENT_DATA_TIM_IN_SIG_OUT);//Signal the DAC output
  pDataSigModOut = osMemoryPoolAlloc(DSP_MemPool, 0);    /Acquire a new block
 }
 }
```

In the second half of the timer interrupt the event flags are tested to see if a block of output data is available. When one is ready a sample is written to the DAC on each timer interrupt.

When all the block data has been written the block is released, the array index is reset and we will check on each interrupt to see if a new bloc is ready to process.

```
/* -- signal Output Section ------------------------------------------- */
if (dataTimIrqInIdx == 0) { // If no data is available check to see if a new block is ready
flags = osEventFlagsWait(DSP_Event, EVENT_DATA_TIM_IN_SIG_OUT, 0, 0);
}
//Check to see if we have data. Eiter an current active block or a fresh block
if ((dataTimIrqInIdx > 0) || (flags == EVENT_DATA_TIM_IN_SIG_OUT)) {
    tmp = pDataTimIrqIn->Sample[dataTimIrqInIdx++];      //read the sample
    arm_q15_to_float(&tmp, &tmpFilterOut, 1);            //Convert to float and write to the DAC
    LPC_DAC->DACR = (((uint32_t)((tmpFilterOut + 1) * (0x03FF / 2))) & 0x03FF) << 6;
    if (dataTimIrqInIdx >= DSP_BLOCKSIZE) {              //If all the data has been read
     status = osMemoryPoolFree(DSP_MemPool, pDataTimIrqIn);       //Free the block
     dataTimIrqInIdx = 0;                               //reset to index
    }
   }
  LPC_TIM2->IR |= (1UL << 0); // clear MR0 Interrupt flag
 }
}
```

Exercise 11.4: RTX Real Time

This exercise implements an IIR filter as an RTX Thread (Fig. 11.22). A periodic timer interrupt takes data from a 12-bit ADC and builds a block of data which is then posted to the DSP Thread to be processed. The resulting data is posted back to the timer ISR. As the timer ISR receives processed data packets, it writes an output value per interrupt to the 10-bit DAC. The sampling frequency is 32 kHz, and the sample block size is 32 samples.

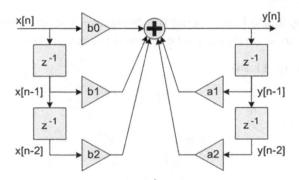

Figure 11.22
An IIR filter is a feedback filter that is much faster than an FIR filter but can be unstable if incorrectly designed.

Open the Pack Installer.

Select the Boards::Designers Guide Tutorial.

Select the example tab and Copy "Ex 11.4 RTOS Real Time."

Figure 11.23
Project file with the DSP_App.c module.

First, examine the code particularly the DSP_App.c module (Fig. 11.23).

The project is actually designed for a Cortex-M3-based microcontroller which has a simulation model that includes an ADC and DAC. The modules ADC.C, Timer.c, and DAC.c contain functions to initialize the peripherals. The module DSP_APP.c contains the initializing Thread, which sets up the memory pool and event flags and creates the DSP Thread called "sigmod()." An additional "clock()" Thread is also created, this Thread periodically flashes GPIO pins to simulate other activity on the system. DSP_App.c also contains the timer ISR. The CMSIS filter functions are in DSP_IIR.c.

Build the project and start the simulator.

When the project starts, it also loads a script file that creates a set of simulation functions. During the simulation, these functions can be accessed via buttons created in the toolbox dialog.

Select view\toolbox (Fig. 11.24).

Figure 11.24
Simulated input to the ADC can be triggered using the toolbox script buttons.

The simulation script generates simulated analog signals linked to the simulator time base and applies them to the simulated ADC input pin.

Open the system analyzer window (Fig. 11.25).

Figure 11.25
The input and output signals can be added to the system analyzer window.

The system analyzer has two signals defined AIN2 and AOUT each with a range of 0−3.3 V. These are not program variables but virtual simulation registers that represent the analog input pin and the DAC output pin.

Start the simulator running and press the "Mixed Signal Sine" button in the toolbox dialog.

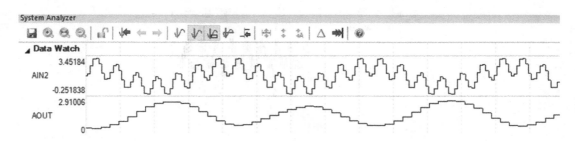

Figure 11.26
The mixed signal script generates a low-frequency signal with distortion. The filter removes the high-frequency component.

This will generate an input signal which consists of a mixed high and low-frequency sine wave (Fig. 11.26). The filter removes the high-frequency component and outputs the processed wave to the DAC.

When we are debugging on real hardware it is possible to see a graphical trace of Thread and interrupt activity (Fig. 11.27) in the system analyzer window. This allows you to visualize the overall loading on the CPU and also to gage the performance of the DSP algorithm. Provided the DSP algorithm completes before the next block of samples is ready, we can guarantee that the output is always going to be continuous. Sadly, this feature is limited to the hardware debug and does not work when using the simulator.

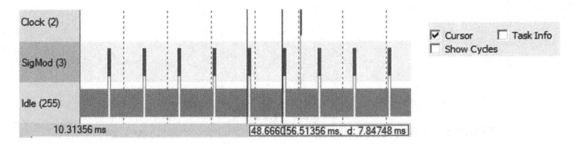

Figure 11.27
System analyzer tThread and interrupt timing diagram (Hardware debug only).

Stop the simulation and wind back the logic analyzer window to the start of the mixed sine wave (Fig. 11.28).

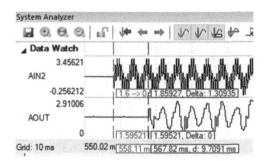

Figure 11.28
The block processing approach introduces a fixed latency between input and output signals.

This shows a latency of around 10 mS between signal data being sampled and processed data being output from the DAC. While part of this latency is due to the DSP processing time, most of the delay is due to the sample block size used for the DSP algorithm.

Exit the simulator and change the DSP block size from 32 to 10.

The DSP block size is defined in IIR.h.

Rebuild and rerun the simulation.

The latency between the input and output signal is reduced to around 4 mS (Fig. 11.29). However, now the SigMod() Thread is running much more frequently and is consuming most of the processor resources and can end up blocking lower priority threads.

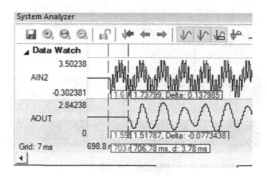

Figure 11.29
Signal Latency with a minimal block size.

Functional Safety

The standard version of the RTX RTOS is suitable for a wide range of embedded applications. However, for use in a safety system, there is a commercial version of RTX, which has been certified to key safety standards. While functional safety is a very big topic in this section, I want to review the additional features within the Functional Safety version of RTX and also introduce another CMSIS specification called CMSIS Zone and the "CMSIS Zone Utility" design tool.

The safety version of RTX called FuSa RTX features a number of extensions that allow you to create a software architecture that supports functional isolation between different software components within the application software. This allows you to create a project that contains both safety-critical components and nonsafety components. The nonsafety components are defined as "interference-free" as follows:

> If a component has been shown to be interference-free, then it has been demonstrated that no failure of that component can cause a failure of any of the safety functions of the product. Therefore, such a component is not subject to be developed with an IEC 61508 compliant development process.

If this is true, we can define our system as a mix of safety-critical and noncritical code and will only need to certify the safety components. This will result in a big saving of both time and cost. To design such a system, we must provide functional Isolation between the safety and nonsafety components (Fig. 11.30).

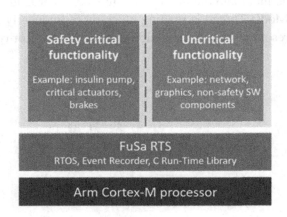

Figure 11.30
Separation of safety and nonsafety related processes.

Functional Isolation within a system may be achieved with the techniques shown in Table 11.2.

Table 11.2: Process isolation techniques within FuSa RTX

Feature	Description
Spatial Isolation	The ability to controls access to the microcontroller resources depending on which software component is running within the system.
Temporal Isolation	Ability to monitor the timing constraints within the system.
Controlled System Recovery	The ability to control system operation in the case of failures. This may involve proceeding to a safe operation state or blocking execution of nonsafety components.

Arm Functional Safety Run-Time System

The Arm Functional Safety Run-Time System (FuSa RTS) is a collection of embedded software components that have been qualified for safety-critical applications in Automotive, Industrial, Rail, and medical systems (Table 11.3).

Table 11.3: FuSa RTS qualifications

Industry	Standard	Safety Level
Automotive	ISO26262	ASIL D
Industrial	IEC61508	SIL 3
Railway	EN50128	SIL 4
Medical	IEC62304	Class C

It is important to note that while the FuSa RTS is provided as a stand-alone package, its certification rests on the use of a safety version of the Arm C compiler. The safety-certified version of the Arm compiler is included in the MDK-Pro edition.

The FuSa RTS consists of five key components (Fig. 11.31).

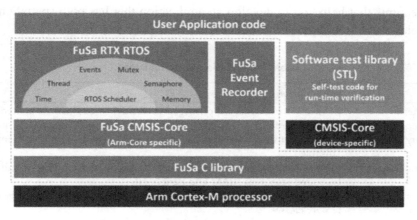

Figure 11.31
The FuSa RTS system with optional software test library.

- FuSa RTX RTOS: deterministic real-time operating system for Arm Cortex-M processors.
- FuSa EventRecorder: implements functionality to easily record events and collect execution statistics in the application code.
- FuSa CMSIS Core: validated vendor-independent software interface to the processor resources.
- FuSa C Library: a subset of the C library functions suitable for developing safety-critical embedded applications.
- Safety Package: documentation set explaining the usage of FuSa RTS in safety context.

Software Test Library

The Software Test Library (STL) is a safety-certified software-based diagnostics suite for detecting random hardware failures within the CPU, RAM, and Flash memory. It is executed within its own Thread and is designed to run periodically alongside the user application. The STL CPU tests aim to check a high percentage of the gate usage within the Cortex-M processor. This requires a deep knowledge of the processor and is surprisingly difficult to achieve. The STL is not part of the FuSa system and will be provided either by the Silicon Vendor or another third party.

RTX Safety Features

In addition to being a safety-certified RTOS, the FuSa version of RTX provides a range of additional features that help you to develop an application that has both spatial and temporal isolation. A summary of the safety features is provided in Table 11.4.

Table 11.4: FuSa safety features

Feature	Description
Kernel executes in Privileged mode	The RTOS code has full privileged access the thread code is limited to unprivileged.
Static memory allocation	The thread stacks are statically allocated. The global Dynamic memory pool is disabled.
MPU protection zone	The MPU is dynamically configured to "sandbox" each Thread and its required resources (RAM and Peripherals).
Safety Class	Communication between threads using RTOS objects is limited by defined safety classes.
Thread watchdogs	The run time execution of each Thread is monitored by thread watchdogs.
Fault Handling	Additional RTOS support is provided to enter safe operating modes.

FuSa RTX Kernel

In the standard version of RTX, the RTOS kernel will always operate in privileged mode. In the rtx_config.h header, it is possible to configure the threads to run in unprivileged mode. This creates a functional divide between the RTOS code, which you can think of as now running at a system level with full access to all features of the CPU and microcontroller. The application code will be running in Thread using unprivileged mode. This means that it cannot access key registers in the CPU SCB. In cases where the user code does need to access the SCB, for example, it may need to enable an interrupt channel in the NVIC, we must provide custom functions that use an SVC instruction SVC to shift execution to privileged mode and allow access to the protected register.

Once the Cortex-M processor has been configured to operate with Privileged\Unprivileged access, we can use the Memory Protection Unit (MPU) to extend this mode of execution over the full range of microcontroller resources.

Spatial Isolation

Within the RTOS, our application code will consist of threads and associated drivers. The FuSa RTX allows us to create Memory Protection Zones that provide spatial isolation between the different software components within the system.

MPU Protection Zones

All of the Cortex-M processors, except Cortex-M0 may be fitted with an MPU when a Silicon Vendor designs the microcontroller. The MPU allows us to define access protection to different regions of memory and peripherals. The FuSa RTX allows us to define a Memory Protection Zone which is a collection of several memory regions (FLASH, RAM, and Peripherals) each with specific access rights. Each RTOS thread is assigned to a Memory Protection Zone (Fig. 11.32) so that it is fully encapsulated within its own device resources. Shared memory regions may also be declared. These will typically contain global RTOS objects such as mail queues that support interthread communication. Any attempt to access a region outside of its Memory Protection Zone will result in a system fault.

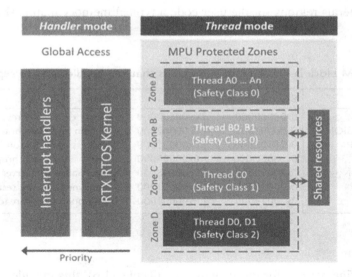

Figure 11.32
MPU Protection Zones.

During the architectural design phase, the microcontroller resources must be logically split into separate regions (RAM and peripherals) and then assigned to different processes (RTOS Threads and drivers) with the same integrity level. The collection of memory resources is called a Zone. This means that within our project, we have to define and manage each of these Zones within the linker script file and statically allocate the RAM requirements for different RTX objects to a suitable Zone within the source code. At runtime, the application code also has to configure the MPU regions to isolate the active Zone resources. As there are a limited number of MPU regions available, the RTOS kernel will also have to reconfigure the MPU regions during a Thread switch to ensure a correctly configured Memory Protection Zone. This means that we have many extra configuration options to maintain throughout the life of the project. While this can be done manually, there would be ample room for errors. Fortunately, the CMSIS Zone specification defines a template markup language and a utility that allows you to design and update the overall memory configuration graphically and then autogenerate the necessary linker script file and MPU configuration code. However, to fully understand what is going on, we will first look at how to set up the configuration manually.

Defining the Memory Map

First, we need to create a project memory map that defines each of the memory regions available to our application code. To make this is easy to manage over time, we can create a set of defines that outline the memory regions in our project. These outline the FLASH, RAM, and Peripherals regions for the user code a typical memory map is shown in Table 11.5.

Table 11.5: A RAM block is divided into subregions that can be allocated to separate processes

Region Name	Start Address	Size	Notes
ER_FLASH	0x08000000	0x00100000	Single executable region for all code
RAM_SHARED REGION	0x20008000	0x00002000	System and Hal library
RAM_NORMAL_OP	0x2000C000	0x00001000	Process control loop
RAM_VERIFY_OP	0x2000D000	0x00001000	Data verification thread
RAM_COM REGION	0x20010000	0x00008000	Non-Safety webserver
RAM_STL REGION	0x20018000	0x00001000	Runtime Software test library
RAM_SAFE_OP	0x20019000	0x00001000	Safe operation Thread

We must also define regions for the system code (Table 11.6), this includes sections for the RTOS kernel, default RTOS threads, C compiler library, and the event recorder debug support.

Table 11.6: The system level code is allocated dedicated regions

Region Name	Start Address	Size	Notes
RAM_PRIVILEGED REGION	0x20000000	0x00007C00	Kernel data
ARM_LIB_STACK	0x20000000	0x00000400	Certified C library data
RAM_EVR REGION_RAM	0x20000000	0x00002000	Certified event recorder
RAM_TIMER REGION	0x20000000	0x00001000	RTX Timer thread data
RAM_IDLE REGION_RAM	0x20000000	0x00001000	RTX Idle Thread data

From these definitions, we can create a linker script file that provides a layout template for each of these regions.

```
LR_FLASH REGION_FLASH_START REGION_FLASH_SIZE {
 ER_FLASH REGION_FLASH_START REGION_FLASH_SIZE {
 }
 ....................
RAM_STL REGION_RAM_STL_START REGION_RAM_STL_SIZE {
 }
 RAM_SAFE_OP REGION_RAM_SAFE_OP_START REGION_RAM_SAFE_OP_SIZE {
 } ....................
 }
```

Placing the Resources

In order to place the application threads and driver code into the correct memory regions, we must segregate our application code into C modules that contain only code for a specific region. A simple example is shown in Table 11.7.

Table 11.7: Application code and controllability class

C Module	Type	Description
NormalOperation	Safety	Thread containing the Application main control loop
Adc_driver	Safety	ADC used as simulated road traffic sensor
Operational verification	Safety	Thread to provide a plausibility verification of the sensor data
STL	Safety	Software test library used to detect hardware errors in the CPU, Flash, and RAM
Communication	Non-Safety	Webserver used to provide remote user interface
Safety Mode Operation	Safety	Thread used to provide a safe operating mode
Watchdog driver	Safety	Driver used to pat the microcontroller watchdog

Then, we can place the data within each C module into the correct region using the linker file template.

```
RAM_NORMAL_OP REGION_RAM_NORMAL_OP_START REGION_RAM_NORMAL_OP_SIZE {
  NormalOperation.o (+RW + ZI)
  adc_mcbstm32f400.o (+RW + ZI)
  }
```

Now when we create a thread, we can statically define the thread stacks within the C module so that they are allocated to the correct memory region.

```
/* Thread stacks located in RAM_NORMAL_OP */
static uint64_t normal_operation_thread_stack[512/8];
```

The stack memory can then be assigned to the Thread when it is created.

```
static const osThreadAttr_t normal_op_thread_attr = {
....................
.stack_mem = normal_op_thread_stack, /* User provided stack */
 .stack_size = sizeof(normal_op_thread_stack),
 .priority = osPriorityNormal,
 }
....................
};
```

This process allows us to layout the application and system code so that each element is placed within a well-defined memory region. The next step is to configure the MPU so that this memory map is enforced during run time.

Configuring the MPU

The CMSIS core specification provides a set of standard functions that are used to configure the different MPU regions. These functions include ARM_MPU_LOAD(), which is used to load an array that defines a set of MPU regions. This allows us to create a two-dimensional array that defines a table of MPU region configurations for each MPU protection zone.

```
static const ARM_MPU_Region_t mpu_table[ZONES_NUM][MPU_REGIONS] = {
{
        /* Zone 'Zone_Normal_OP' */ },
{
        /* Zone 'Zone_Verify_OP' */
},
....................
}
```

A typical Zone will define the FLASH, RAM, and Peripheral resources that the application code assigned to the Zone can access.

To allow the RTX kernel to determine which Zone should be active, each Thread is assigned a protection zone in the tread attribute bits.

```
#define ZONE_NORMAL_OP          ((uint32_t)3U)
const osThreadAttr_t normal_op_thread_attr = {
.name = "Normal_OP",
.attr_bits = osThreadZone(3U)      // assign thread to MPU protected zone 3
};
osThreadNew(Normal_OP, NULL, & normal_op_thread_attr);
```

During a Thread switch, the kernel will trigger a callback function. This callback is provided by the user, and it must configure the MPU to activate the memory protection zone for the Thread that is about to be scheduled. Fortunately, the code in the callback is straightforward.

The kernel will pass the active zone number to the callback, and we then need to activate the MPU Memory Protection Zone using the arm_MPU_Load() function.

```
void osZoneSetup_Callback (uint32_t zone) {
 if (zone >= ZONES_NUM) {
  ZoneError_Handler();
 }
 ARM_MPU_Disable();
 ARM_MPU_Load(mpu_table[zone], MPU_REGIONS);
 ARM_MPU_Enable(MPU_CTRL_PRIVDEFENA_Msk);
}
```

Memory Protection Zone RTOS Functions

The FuSa RTX has a group of additional API calls to create and manage the Memory Protection Zones, as shown in Table 11.8.

Table 11.8: Memory protection zone RTOS functions

Function	Description
osThreadNew	Create a thread and define its memory Zone
osThreadZone	Define a Zone in Thread attribute format
osThreadGetZone	Get Thread Memory protection zone
osThreadTerminateZone	Terminate threads assigned to a specified Memory Protection Zone
osZoneSetup_Callback	RTX Kernel callback to manage the active Memory Protection Zone

CMSIS Zone Utility

The CMSIS Zone specification provides a template markup language and a software utility that allows you to create high-level description of complex memory maps. This description can then be used to autogenerate a coherent set of project files. Within the CMSIS Zone specification, a resource Zone is a flexible concept (Table 11.9). In a multicore project, each Zone is a single processor project that defines how resources are shared and isolated between processors. Within a single project, each Zone can define the execution resources required for each RTOS thread.

Table 11.9: CMSIS zone use cases

Use Case	Description
Multicore System	Allocating memory resources in a multicore project
Security Partitioning	Partitioning secure and nonsecure resources for security projects
Process isolation	Allocating resources to functional code for safety projects

The CMSIS Zone utility is designed as a plugin for eclipse and is available from the link below.

https://arm-software.github.io/CMSIS_5/Zone/html/zTInstall.html

When a project is created, the CMSIS zone utility can import the CMSIS Pack System Viewer Description file and automatically create a default resource map for the target microcontroller.

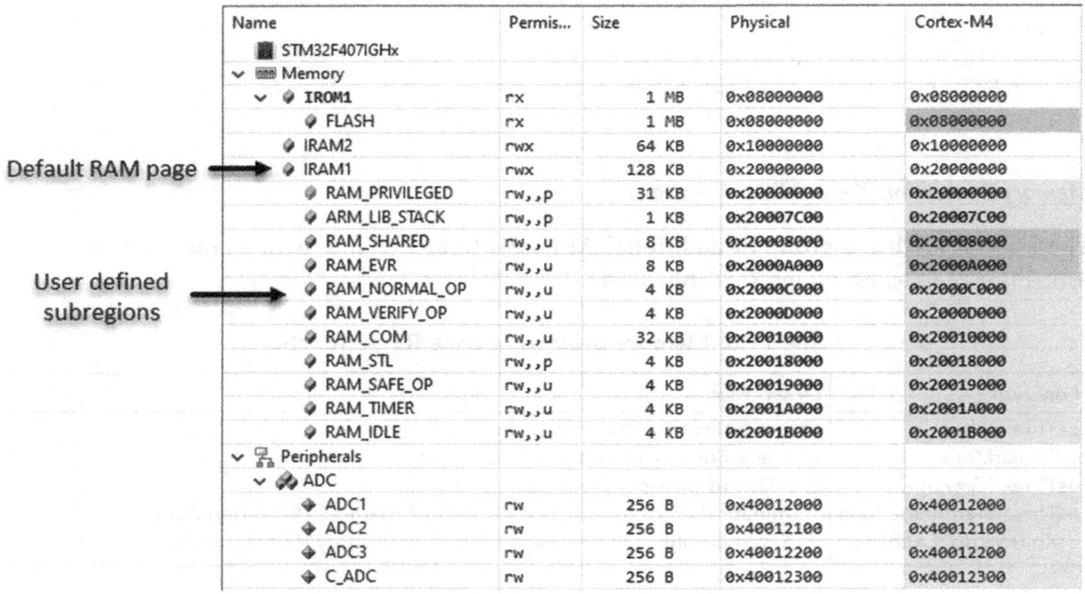

Figure 11.33
Microcontroller RAM and user-defined subregions.

The default memory pages can then be segmented into arbitrary subregions (Fig. 11.33). We can then define Zones and allocate processor resources to each Zone (Fig. 11.34). In this case, each Zone is the execution environment for each Thread and associated driver.

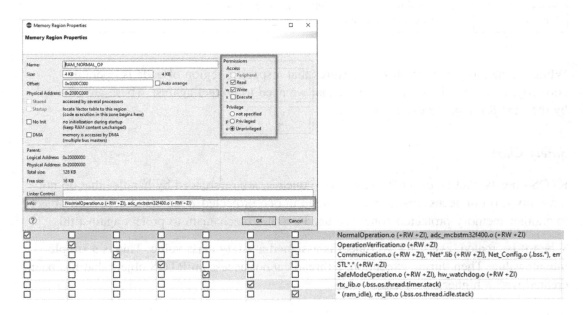

Name	Permis...	Size	Start	End	Zone_Normal_OP	Zone_Verify_OP	Zone_Com	Zone_STL	Zone_Safe_OP	Zone_Timer	Zone_Idle
⊕ RAM_PRIVILEGED	rw,,p	31 KB	0x20000000	0x20007BFF	☐	☐	☐	☐	☐	☐	☑
⊕ ARM_LIB_STACK	rw,,p	1 KB	0x20007C00	0x20007FFF	☐	☐	☐	☐	☐	☐	☑
⊕ RAM_SHARED	rw,,u	8 KB	0x20008000	0x20009FFF	☑	☑	☑	☑	☑	☑	☑
⊕ RAM_EVR	rw,,u	8 KB	0x2000A000	0x2000BFFF	☑	☑	☑	☑	☑	☑	☑
⊕ RAM_NORMAL_OP	rw,,u	4 KB	0x2000C000	0x2000CFFF	☑	☐	☐	☐	☐	☐	☐
⊕ RAM_VERIFY_OP	rw,,u	4 KB	0x2000D000	0x2000DFFF	☐	☑	☐	☐	☐	☐	☐
⊕ RAM_COM	rw,,u	32 KB	0x20010000	0x20017FFF	☐	☐	☑	☐	☐	☐	☐
⊕ RAM_STL	rw,,p	4 KB	0x20018000	0x20018FFF	☐	☐	☐	☑	☐	☐	☐
⊕ RAM_SAFE_OP	rw,,u	4 KB	0x20019000	0x20019FFF	☐	☐	☐	☐	☑	☐	☐
⊕ RAM_TIMER	rw,,u	4 KB	0x2001A000	0x2001AFFF	☐	☐	☐	☐	☐	☑	☐
⊕ RAM_IDLE	rw,,u	4 KB	0x2001B000	0x2001BFFF	☐	☐	☐	☐	☐	☐	☑
⌄ Peripherals											
⌄ ADC											
⊕ ADC1	rw	256 B	0x40012000	0x400120FF	☑	☐	☐	☐	☐	☐	☐
⊕ ADC2	rw	256 B	0x40012100	0x400121FF	☑	☐	☐	☐	☐	☐	☐
⊕ ADC3	rw	256 B	0x40012200	0x400122FF	☑	☐	☐	☐	☐	☐	☐
⊕ C_ADC	rw	256 B	0x40012300	0x400123FF	☑	☐	☐	☐	☐	☐	☐

Figure 11.34
Memory Protection Zones and RAM allocation.

This allows us to very quickly create or update the memory template. Once the template is defined, we can place the application data into a specific region by using the local properties option for each memory region (Fig. 11.35).

Figure 11.35
RAM region local properties are used to allocate code and define access rights.

Once all the resources and application data have been mapped, we can switch to the setup tab and define the CMSIS Zone mode (Fig. 11.36) to either define each Zone as a separate microcontroller project for multiprocessor or security applications or in the case of a safety project, define each Zone as a separate execution zone within a single project.

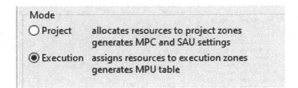

Figure 11.36
CMSIS Zone utility setup dialog defines the code generation mode.

The final step is to generate the project files (Table 11.10) from the underlying template code.

Table 11.10: The resulting autogenerated code defines the project memory map

File	Description
Dump_fzone.txt	Text description of the memory layout
Mem_layout.h	Header file with defines for start and size of each memory region
Scatter.sct	Project Linker scatter file
Zones.c	MPU protection zones as MPU configuration tables
Zones.h	Project Zone definitions

While the memory protection zone ensures that a specific region of code is operating correctly, there are other types of error that we need to protect against. These are covered by the next RTX safety extension.

Safety Class

RTOS objects, that is, event flags, message queues, are referenced by handles that do not have any form of access restriction. A thread in one protected Zone could access an object in another memory-protected zone. We need another mechanism to protect against this.

When it is created, every RTOS object is assigned a safety class value. This is a simple integer value. Then during execution, a thread cannot access an RTOS object that has been created with a higher safety class value.

Safety Class Management

The safety class is managed with a macro, as shown in Table 11.11.

If a thread tries to access an RTOS object with a higher safety class, an error code shown in Table 11.11 will be returned.

Table 11.11: FuSa RTX safety class features

Class Object	Type	Description
osSafety class() osErrorSafetyClass	Macro Return code	Encodes the defined safety class into the RTX object attribute bits Indicated a class error

When an RTOS object is created, its safety class is encoded within its attribute bits

```
const osEventFlagsAttr_t evt_flags_attr = {
.attr_bits = osSafetyClass(SAFETY_CLASS_SAFE_MODE_OPERATION)
};
osEventFlagsId_t evt_flags;
evt_flags = osEventFlagsNew(&evt_flags_attr);
```

when using RTOS object you must check for a safety class violation.

```
status = osEventFlagsSet(evt_flags, 1);
if (status == osErrorSafetyClass)
{
        //handle the safety class error
}
```

Temporal Isolation

In addition to monitoring the memory usage of the application Threads and drivers, we must also ensure that they meet any required timing constraints. The FuSa RTX provides Thread Watchdogs as a kernel service to manage any temporal timing constraints for Thread execution. Each Thread is created with a Thread Watchdog that must be fed using the function shown in Table 11.12.

Table 11.12: FuSa RTX thread watchdog functions

Function	Description
osThreadFeedWatchdog(uint32_t ticks) osWatchdogAlarm_Handler(osThreadId_t thread_id)	Thread watchdog feed function Watchdog alarm callback

The watchdog feed function defines a timeout as a multiple of osTicks. The feed function is normally placed at the start of the Thread for(;;) loop and before any blocking call. If the watchdog times out a watchdog, alarm will trigger a callback function (Table 11.12) that can be used to recover the system or enter a safe operating mode.

We can use the watchdog handler to identify the Thread which triggered the alarm, and depending on its safety class, we can then decide what action to take. In the case of a nonsafety-related Thread, it would be possible to simply suspend or restart the Thread. This can be done with the function osThreadSuspend() or we can suspend an entire safety class with osThreadSuspendClass(). If a safety-related Thread triggers the watchdog alarm, we can decide to suspend the nonsafety threads and then continue monitoring the safety thread or we may choose to place the system into a safe operating mode.

However, you must be careful when suspending threads. If a thread is suspended, its watchdog is still running and will still trigger the watchdog callback alarm function when it expires.

Fault Handling

Alongside the normal Cortex-M processor exceptions and the RTX error handlers, there are a number of additional fault cases that can be detected and managed with the FuSa RTOS (Table 11.13).

Table 11.13: FuSa RTX fault conditions

Fault	Description
Startup fault	Hardware configuration failure at startup
STL Error	Run Time hardware CPU or memory failure
Memory Manager fault	Run time Memory Protection Zone error
Watchdog Alarm handler	Thread Temporal failure
RTX error	RTOS Kernel error
Normal Operation error	The validating Thread discovers a failure in the main control thread

If it is not possible to recover from a failure, then we can trigger a fatal error handler which will place the hardware into a safe state and then reset the microcontroller with NVIC_SystemReset(); or we can suspend execution of the main control loop and place the system into a safe operating mode.

Safe Mode Operation Thread

One way to achieve this is to create a dedicated safe mode thread at startup and then immediately suspend the Thread so that it is waiting for an RTOS signal such as a thread flag. This is the same technique that we used to manage interrupts earlier in this chapter.

If a fault is detected, the error handler can start the safe operating mode by signaling the safe mode thread. To ensure that the safe mode thread remains operational, we can allow it to periodically wake up in order to feed hardware watchdogs and its own thread watchdog. Once this is done, it will return to a blocked to wait for the next timeout or for the SAFE_OPERATION_MODE flag to be set.

```
do {
 flags = osEventFlagsWait(safe_mode_id, SAFE_MODE_OPERATION, osFlagsWaitAny, 250U);
 HwWatchdogFeed();
 (void)osThreadFeedWatchdog(500U);
 } while (flags == osFlagsErrorTimeout);
```

Additional Safety Features

Object Pointer Checking

When a call to an RTOS object is made, the kernel will check that the pointer to the object is valid (non NULL). The kernel will also ensure that the pointer is referencing an object control block that is located in the correct memory section and that the pointer is correctly aligned within the memory section.

Accessing Privileged Resources

Since the user threads are running in unprivileged mode, they are unable to access any privileged memory or registers. In some cases, we may want to enable access through the use of SVCs. Typically, this will be to access registers in the NVIC in order to enable and disable interrupts. We can refine our SVC handler a bit further by checking the Zone ID of the originating Thread. This allows us to further limit the functionality available to each resource zone.

```
void svcNVIC_Handler (uint32_t func_index, IRQn_Type IRQn) {
  /* Only Ethernet IRQ handling from Zone 2 is allowed */
  if (IRQn == ETH_IRQn) {
    if (osThreadGetZone(osThreadGetId()) == ZONE_COM) {
      switch (func_index) {
        case 2U:
          __NVIC_EnableIRQ(IRQn);
          break;
        case 4U:
          __NVIC_DisableIRQ(IRQn);
          break;
  }}
  }
```

SVC Pointer Checking

Most of the RTOS functions are called using an SVC to generate an exception rather than a standard branch instruction. The kernel code places the SVC entry points into a dedicated section of memory. When SVC pointer checking is enabled, the RTOS calling code will check that the calling pointer destination is located in the correct memory region and is also correctly aligned for that region.

Conclusion

One of the chief reasons made by developers for not adopting the use of an RTOS is that it represents too much overhead in terms of processing time for a small microcontroller. This no longer stacks up, just about every Cortex-M-based device is capable of supporting an RTOS. Once you are familiar with an RTOS it should be used for all but the simplest projects. Used correctly the benefits of using an RTOS far outweigh the negatives. Once you have invested the time in learning how to use an RTOS, you will not want to go back to bare metal projects.

CHAPTER 12

CMSIS-Driver

Introduction

The CMSIS-Driver specification defines a generic driver interface for a range of common microcontroller peripherals. The original rationale for the CMSIS-Driver specification was to provide a standardized set of interfaces for middleware libraries. This allows any middleware library that uses CMSIS-Drivers to be easily reused on any microcontroller with a matching CMSIS-Driver. So, for example, we can create a TCP/IP library that uses a CMSIS ethernet driver and then use it on any microcontroller which has the same driver without the need for any porting or other low-level code development (Fig. 12.1). This is great news for anyone developing a reusable software component, as you can instantly support a wide range of microcontrollers while spending all your development effort on your own software.

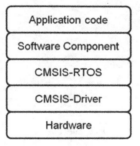

Figure 12.1
CMSIS Driver and CMSIS RTOS allow you to create a software component that can be reused across different hardware platforms.

This has lots of advantages in that once you are familiar with the CMSIS-Driver API, you can reuse that knowledge across many microcontrollers and projects. You can also move your code between different microcontrollers and even different toolchains. The downside is that you are using a generic interface that must be able to work on any microcontroller. This means that the features offered by the CMSIS-Driver API are limited and may not take advantage of more sophisticated peripherals. Also, as the CMSIS-Driver may be implemented as a wrapper over a silicon vendor peripheral library, the code size and performance may not be optimal.

The Designer's Guide to the Cortex-M Processor Family.
DOI: https://doi.org/10.1016/B978-0-323-85494-8.00015-2

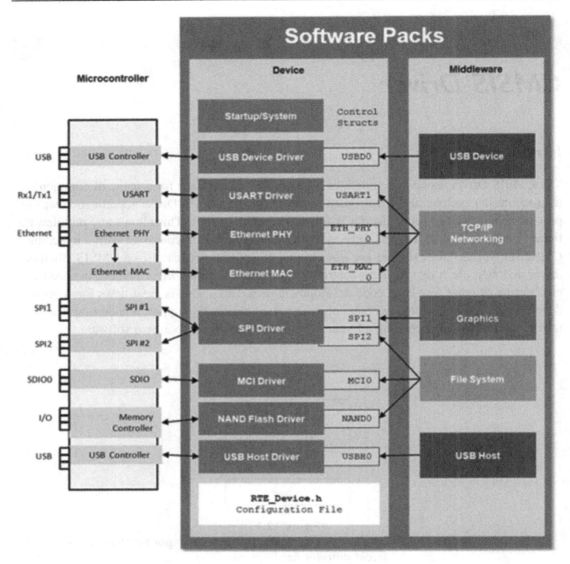

Figure 12.2
CMSIS Driver provides a standard API for a range of peripherals common to many
microcontrollers.

CMSIS-Driver currently supports communication peripherals such as I2C, SPI, USART,
CAN, and SAI, as well as more complex peripherals such as Ethernet, USB Device, and
USB Host, along with a range of WiFi Modules (Fig. 12.2). CMSIS-Driver also has
specifications for Multi Media Card interface and specifies drivers for NAND and NOR

Flash memory to support embedded file systems. Peripherals such as ADC, DAC, and timers are not supported, but later in this chapter, we will have a look at how to extend the CMSIS-Driver specification to create our own custom drivers.

CMSIS-Driver API

Each CMSIS-Driver contains a set of API functions that are used to configure each supported peripheral. The CMSIS-Driver functions have a common structure across each of the supported peripherals as shown in Table 12.1. Once you are familiar with how to use one type of driver, the same logic applies to all the others. Each CMSIS-Driver is capable of supporting multiple instances, so within a project, it is possible to instantiate several drivers to support multiple peripherals of a given type, for example, three SPI peripherals. The common CMSIS-Driver functions are shown below in a generic form. These functions vary a little between different driver types but once you are familiar with how the CMSIS-Driver API works, moving between different peripherals presents few problems.

Table 12.1: CMSIS driver generic API

Function	Description
get Version()	Returns The driver version
get Capabilities()	Returns the supported driver capabilities
getStatus()	Returns the current peripheral status
initilize()	Initial driver setup and registers the driver callback
uninitilize()	Return the peripheral to its reset state
powerControl()	Enable\disable the peripheral power state
control()	Configures the peripherals operational parameters
signalEvent	A user define callback to handle peripheral events
Peripheral Data Transfer Functions	
send() receive()	An additional collection of functions to control data transfer

The CMSIS-Drivers for a given microcontroller are provided as part of the device family pack. Once installed, each of the drivers will be available through the RTE. A typical calling sequence is shown in Fig. 12.3.

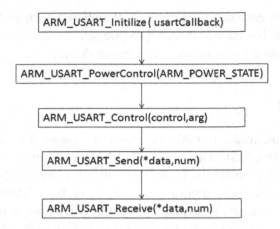

Figure 12.3
The CMSIS driver API is (fairly) orthogonal. Once you have used the CMSIS USART driver you will be able to use any other CMIS Driver.

Exercise 12.1: CMSIS-Driver

In the following example, we are going to look at how to use the CMIS USART driver. The principles learned with this driver can then be applied to any of the other peripheral drivers. This example is a multi-project workspace with two projects, each configured for a different microcontroller. One project is set up for an NXP microcontroller the second is for an ST Microelectronics microcontroller. The UART code is contained in a single module that is common to both projects.

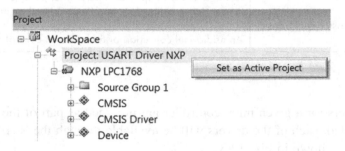

Figure 12.4
Make "USART Driver NXP" the active project.

Select the NXP project. Right-click and select "Set as Active project" (Fig. 12.4).

First, we can add the CMSIS USART driver to our project by selecting it in the RTE, then resolve any required subcomponents, and add it to our project.

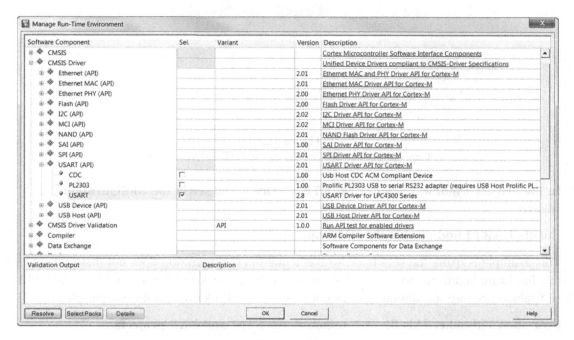

Figure 12.5
Select the CMSIS USART Driver in the Run Time Environment Manager.

Open the RTE and select the CMSIS-Driver: USART::UART (Fig. 12.5).

Press the resolve button and then OK to add the UART driver to the project.

Now, we need to configure the microcontroller pins to switch from GPIO to the USART Tx and Rx pins. The microcontroller pins are configured in the Device::RTE_Device.h file (Fig. 12.6).

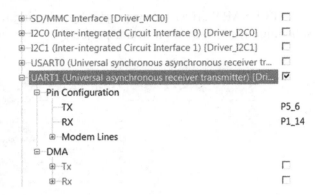

Figure 12.6
Configure the USART pins in the RTE_Device.h file.

In the project window, open device::RTE_Device.h and select the configuration wizard.

Enable UART1 and disable the DMA options.

This example is using the simulator, so the actual pin configuration does not matter. On a real hardware board, we would need to configure the correct pins for Tx and Rx. The DMA is disabled because the simulator does not support the DMA unit within this peripheral.

Now open Thread.c. This file is common to the STM and NXP projects.

To access the CMSIS-Driver API, we can then add the USART driver header file to our source code. The CMSIS-Driver provides support for each peripheral of a given type available on the selected microcontroller. Each driver instance is defined by an access structure that holds a definition of the CMSIS-Driver API for the given peripheral.

```
ARM_DRIVER_USART Driver_USART3 = {
USARTx_GetVersion,
USART3_GetCapabilities,
USART3_Initialize,
USART3_Uninitialize,
USART3_PowerControl,
USART3_Send,
USART3_Receive,
USART3_Transfer,
USART3_GetTxCount,
USART3_GetRxCount,
USART3_Control,
USART3_GetStatus,
USART3_SetModemControl,
USART3_GetModemStatus
};
```

To install the driver support, we need to add the driver include file to our source code and add an "extern" declaration for the UART peripheral we want to access.

```
#include "Driver_USART.h"
extern ARM_DRIVER_USART Driver_USART1;
```

In thread.c, we can make the first API call to initialize the driver.

```
Driver_USART1->Initialize(myUSART_callback);
```

The initializing call starts up the driver and registers a callback function. Once the callback function has been registered with the driver, it will be triggered by hardware events in the USART peripheral, and the driver interrupts service routines. Each time this function is triggered it will be passed a parameter that indicates a specific USART hardware event as shown in Table 12.2.

Table 12.2: USART callback events

USART Hardware Events
ARM_USART_EVENT_SEND_COMPLETE
ARM_USART_EVENT_RECEIVE_COMPLETE
ARM_USART_EVENT_TRANSFER_COMPLETE
ARM_USART_EVENT_TX_COMPLETE
ARM_USART_EVENT_TX_UNDERFLOW
ARM_USART_EVENT_RX_OVERFLOW
ARM_USART_EVENT_RX_TIMEOUT
ARM_USART_EVENT_RX_BREAK
ARM_USART_EVENT_RX_FRAMING_ERROR
ARM_USART_EVENT_RX_PARITY_ERROR

In the callback function, we can now provide code for any of the USART hardware events we want to handle.

```
void myUSART_callback(uint32_t event)
{
switch (event)
{
  case ARM_USART_EVENT_RECEIVE_COMPLETE:
  case ARM_USART_EVENT_SEND_COMPLETE:
    osSignalSet(tid_Thread, 0x01);
  break;
}
}
```

Now when the USART successfully sends or receives some data, the callback will be triggered, and we can send an RTOS signal to one of our application threads to indicate the event.

Once we have initialized the driver and installed the callback, we can switch on the peripheral by calling the power control function.

```
USARTdrv->PowerControl(ARM_POWER_FULL);
```

This function provides the necessary low-level code to place the USART into an operating mode. Typically, this involves configuring the system control unit of the microcontroller to enable the clock tree and release the peripheral from reset. Once the USART has been powered up, we can configure its operating parameters using the control function.

```
USARTdrv->Control(ARM_USART_MODE_ASYNCHRONOUS |
ARM_USART_DATA_BITS_8 |
ARM_USART_PARITY_NONE |
ARM_USART_STOP_BITS_1 |
ARM_USART_FLOW_CONTROL_NONE, 9600);
```

Now that the USART's operating parameters have been configured, we can use the send and receive functions to transfer data.

```
USARTdrv->Send("A message of 26 characters", 26);
USARTdrv->Receive(&cmd, 1);
```

Depending on the driver configuration, the underlying data transfer may be performed by the CPU or by an internal DMA unit provided by the microcontroller.

Examine the CMSIS-Driver code in Thread.c.

Select project "Batch Build" to build both projects.

Start the debugger.

In the NXP project, open the "view\serial window\serial 2."

Open the peripherals\usart\usart1 window.

Run the code.

Check the USART configuration in the peripheral window.

The initial "Hello World" message will be displayed in the serial console window.

Exit the debugger.

Switch to the second project and repeat the above debugger session.

Due to the different naming conventions used by the silicon manufacturers, the STM serial data will be output to "view\serial window\serial 1."

We are now running exactly the same USART code on a different Silicon vendors microcontroller family.

The CMSIS-Driver specification provides us with a way to very quickly bring up key peripherals on a microcontroller without having to spend a lot of time writing the low-level configuration code. Whilst configuring a USART may seem like a trivial matter, many of the higher end Cortex-M-based microcontrollers are becoming quite complex, take a look at the LPC1768 clock tree and fractional baud rate generator if you are in any doubt.

Driver Validation

The CMSIS-Driver specification is an open specification that anyone can download and use to implement their own driver. Ideally, a Silicon Vendor will provide a set of CMSIS-Drivers for their microcontroller as part of a Device Family Pack. Wherever the CMSIS-Driver code comes from, it is still alien third-party code that is being added to your project and as such may contain its own bugs. Before we trust this code and build it into our application, we need to perform some initial tests to check it works correctly. Fortunately, there is an easy way to do this in the form of a CMSIS-Driver Validation Pack. As its name implies, the Validation Pack is designed to test the capabilities of each CMSIS-Driver and output a report on the results.

Exercise 12.2: Driver Validation

In this exercise, we will set up the CMSIS-Driver Validation Pack and create a validation project to test the capabilities of the I2C, SPI, and USART drivers.

Open the pack installer and select the ARM::CMSIS-Driver_Validation pack.

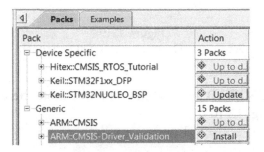

Figure 12.7
Select and install the CMSIS Driver Validation pack.

In the tutorial examples, open the project "Exercise 12.3 CMSIS-Driver Validation" (Fig. 12.7).

This project has the SPI, I2C, and USART drivers selected and configured in the project (Fig. 12.8).

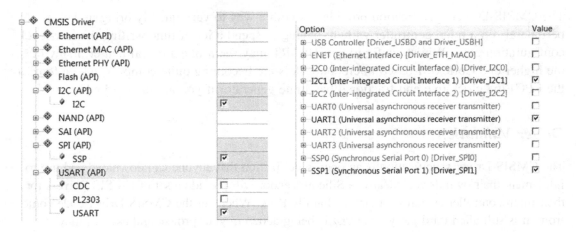

Figure 12.8
Enable the drivers you want to test in the RTE Manager and RTE_Device.h.

The validation report will be sent to the STDIO channel. The ARM::compiler:IO setting in the RTE is configured to redirect the STDIO output to the ITM. All our results will now appear in the view\serial windows\printf viewer (Fig. 12.9).

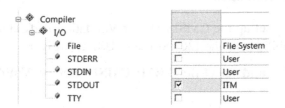

Figure 12.9
Setup the ITM to be the STDIO channel.

In the RTE select the CMSIS-Driver validation and enable the Framework, I2C, SPI, and USART options (Fig. 12.10).

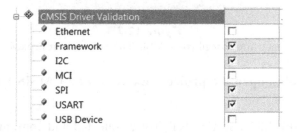

Figure 12.10
In the RTE manager select the test framework and the drivers to test.

Click Ok to add the validation components to the project (Fig. 12.11).

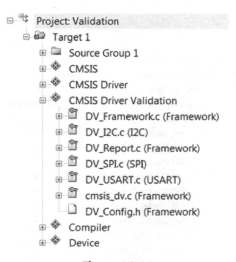

Figure 12.11
The project now has the validation framework added.

Open the CMSIS-Driver Validation::DV_Config.h and select the Configuration Wizard (Fig. 12.12).

For each of the CMSIS-Drivers select Driver instance 1.

Option	Value
⊞ Common Test Settings	
⊟ SPI	
Driver_SPI#	1
⊞ SPI baudrates	
SPI data bits	8
Transfer timeout	1000000
⊞ Test Cases	
⊟ USART	
Driver_USART#	1
⊞ USART baudrates	
USART data bits	8
Transfer timeout	1000000
⊞ Test Cases	
⊞ Ethernet	
⊟ I2C	
Driver_I2C#	1
⊞ Test Cases	

Figure 12.12
DV_Config.h allows you to setup the CMSIS driver test configuration.

Next, open the test cases section for each driver and enable some of the test cases (Fig. 12.13).

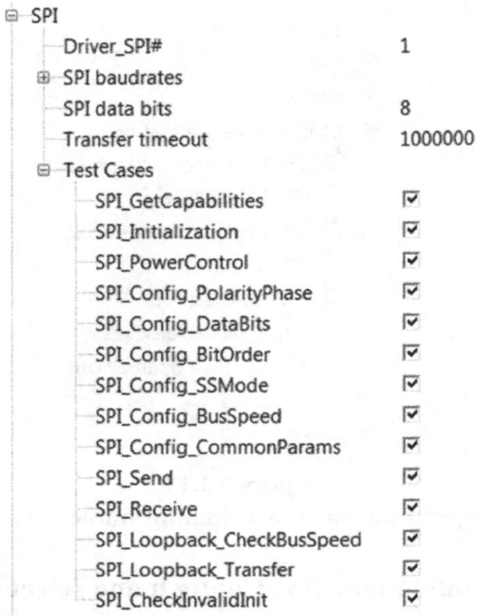

```
⊟ SPI
      Driver_SPI#                        1
   ⊞ SPI baudrates
      SPI data bits                      8
      Transfer timeout                   1000000
   ⊟ Test Cases
         SPI_GetCapabilities              ☑
         SPI_Initialization               ☑
         SPI_PowerControl                 ☑
         SPI_Config_PolarityPhase         ☑
         SPI_Config_DataBits              ☑
         SPI_Config_BitOrder              ☑
         SPI_Config_SSMode                ☑
         SPI_Config_BusSpeed              ☑
         SPI_Config_CommonParams          ☑
         SPI_Send                         ☑
         SPI_Receive                      ☑
         SPI_Loopback_CheckBusSpeed       ☑
         SPI_Loopback_Transfer            ☑
         SPI_CheckInvalidInit             ☑
```

Figure 12.13
Each CMSIS driver has a range of predefined tests.

This configures the validation tests, to run the tests we have to call the test framework.

Open main.c in the project and the cmsis_dv() function to call the validation framework.

```
int main (void) {
    osKernelInitialize();
    cmsis_dv();
    osKernelStart();
}
```

Build the project.

Start the debugger and run the code.

Each of the drivers will be exercised in turn, and the results will be shown in the view \serial windows\debug(printf) window (Fig. 12.14).

Debug (printf) Viewer

```
CMSIS-Driver Test Suite    Nov 14 2015    14:17:12

TEST 01: SPI_GetCapabilities              PASSED
TEST 02: SPI_Initialization               PASSED
TEST 03: SPI_PowerControl
  DV_SPI.c (244) [WARNING] Low power is not supported
TEST 04: SPI_Config_PolarityPhase         PASSED
TEST 05: SPI_Config_DataBits              PASSED
TEST 06: SPI_Config_BitOrder
  DV_SPI.c (315) [WARNING] Bit order LSB_MSB is not supported
```

Figure 12.14
The test results are output to the ITM console window. The local options allow you to save the results to a file.

In the dv_config.h configuration file the "Common Test Settings" allow this report to be generated as plain text or XML (Fig. 12.15).

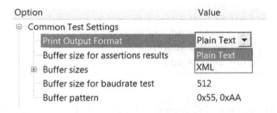

Figure 12.15
The test results can be output as plain text or XML.

CMSIS Virtual IO

CMSIS VIO is a recent addition to the CMSIS-Driver specification. It is intended to allow "high-level" code to be ported directly to new hardware. It is a much lighter approach than the original CMSIS-Drivers in that it does not support interrupts and acts as a wrapper layer over low-level driver library functions. However, it is quick and easy to implement and provides access to peripherals using background thread code. The CMSIS-Driver functions can be used to support a range of device peripherals. The most commonly used are listed in Table 12.3.

Table 12.3: Peripherals supported by CMSIS-VIO

Peripheral	Description
GPIO	Direct GPIO pin control
ADC	Demand a result from an ADC
DAC	Send a value to a DAC

CMSIS VIO API Functions

The CMSIS-VIO driver consists of a C file that provides the following functions (Table 12.4).

Table 12.4: CMSIS-VIO functions

Function	Description
vioInit	Initialize the VIO driver
vioPrint	Print a diagnostic or error message to a buffer
vioSetSignal	Change the state of a bit signal ie GPIO pin
vioGetSignal	Get the state of a bit signal
vioSetValue	Set the value in memory or register
vioGetValue	Get the value in memory or register
vioSetXYZ	Set an array of values X, Y, Z
vioGetXYZ	Set an array of values
vioSetIPv4	Set an IPv4 Address
vioGetIPv4	Get an IPv4 address
vioSetIPv6	Set an IPv6 address
vioGetIPv6	Get an IPv4 address

A supported evaluation board will typically have a CMSIS-VIO driver file that has been ported to match its hardware configuration. This allows you to use a low-cost evaluation board in the early stages of development and then switch to the target hardware by updating the CMSIS-VIO layer. Each of the CMSIS-VIO functions also contains defines that can be used to disconnect the driver from the low-level hardware functions. In this state, the CMSIS-VIO functions will only write to its internal state variables which are held in RAM. This feature allows you to define a range of software tests that can be used to validate software components in the service layer independent of the target hardware.

Exercise 12.3: CMSIS-VIO

In this exercise, we will look at a CMSIS-VIO version of a blinky example to see the driver in action.

In the pack installer select exercise 12.3 and press the copy button.

Open the RTE and the CMSIS-Driver branch (Fig. 12.16).

Figure 12.16
CMSIS-VIO driver and templates.

The VIO driver can be added to the project in three different versions. If a board support package is available this can be added directly to work with existing software components. A second version provides a virtual interface that can be used for testing and finally a blank template is available to develop a new driver. In the case of our example a custom driver template has been added to support the LPC1768 simulation model.

Figure 12.17
Adding a new CMSIS-VIO template.

Close the RTE manager (Fig. 12.17).

Open main.c.

The vioInit() function is called as part of the system initializing code before the RTOS is started.

```
SystemCoreClockUpdate ();     // System Initialization
vioInit();                    // Initialize Virtual I/O
osKernelInitialize ();
```

Open Blinky.c.

The blinky code is a standard file that can be used with any device that provides a CMSIS-VIO driver. It creates an LED thread that is used to flash the boards' LEDs in two different patterns. The vioSetSignal() function is used to control the state of each LED.

```
for (;;) {
if (osThreadFlagsWait(1U, osFlagsWaitAny, 0U) == 1U) {
active_flag ^= 1U;
}
if (active_flag == 1U) {
vioSetSignal(vioLED0, vioLEDoff);      // Switch LED0 off
vioSetSignal(vioLED1, vioLEDon);       // Switch LED1 on
osDelay(100U);                          // Delay 100 ms
vioSetSignal(vioLED0, vioLEDon);       // Switch LED0 on
vioSetSignal(vioLED1, vioLEDoff);      // Switch LED1 off
osDelay(100U);                          // Delay 100 ms
```

```
}
else {
vioSetSignal(vioLED0, vioLEDon);        // Switch LED0 on
osDelay(500U);                          // Delay 500 ms
vioSetSignal(vioLED0, vioLEDoff);       // Switch LED0 off
osDelay(500U);                          // Delay 500 ms
}
}
```

A second thread is used to read the button on the board and set a variable to change the LED flashing pattern. The button state is read by using the vioGetSignal() function.

```
state = (vioGetSignal(vioBUTTON0));     // Get pressed Button state
if (state! = last) {
if (state == 1U) {
osThreadFlagsSet(tid_thrLED, 1U);       // Set flag to thrLED
}
last = state;
```

Build the code and start the debugger.

Open the view\watch window\CMSIS-VIO driver (Fig. 12.18).

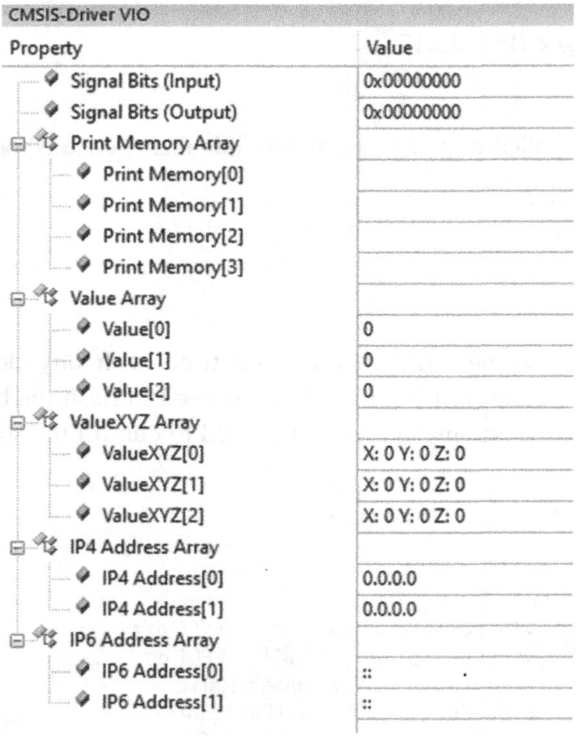

CMSIS-Driver VIO	
Property	Value
🎗 Signal Bits (Input)	0x00000000
🎗 Signal Bits (Output)	0x00000000
⊟ Print Memory Array	
🎗 Print Memory[0]	
🎗 Print Memory[1]	
🎗 Print Memory[2]	
🎗 Print Memory[3]	
⊟ Value Array	
🎗 Value[0]	0
🎗 Value[1]	0
🎗 Value[2]	0
⊟ ValueXYZ Array	
🎗 ValueXYZ[0]	X: 0 Y: 0 Z: 0
🎗 ValueXYZ[1]	X: 0 Y: 0 Z: 0
🎗 ValueXYZ[2]	X: 0 Y: 0 Z: 0
⊟ IP4 Address Array	
🎗 IP4 Address[0]	0.0.0.0
🎗 IP4 Address[1]	0.0.0.0
⊟ IP6 Address Array	
🎗 IP6 Address[0]	::
🎗 IP6 Address[1]	::

Figure 12.18
CMSIS-VIO component viewer.

This provides a component viewer which displays the internal driver state variables.

Open the peripherals\GPIO Fast interface\Port 1 (Fig. 12.19).

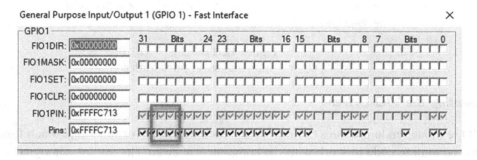

Figure 12.19
GPIO pin view in the debugger.

The LEDs used in this example are located on Ports pins 1.28 and 1.29.

Open the toolbox dialog.

Figure 12.20
Simulator toolbox with GPIO script.

The toolbox provides a script that simulates a button press to a GPIO pin (Fig. 12.20).

Run the project and press the button simulation.

The thread code also writes a diagnostic message to the VIO print buffer.

```
vioPrint(vioLevelHeading, "Test [Heading] = Button Thread");
vioPrint(vioLevelMessage, "Test [Message] = Thread started");
```

As the code runs we can see the messages in the component viewer (Fig. 12.21).

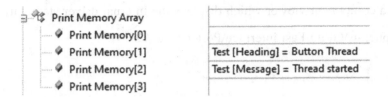

Figure 12.21

CMSIS-VIO component viewer print buffer.

Implementing the VIO Driver

The VIO template code provides definitions for each of the driver functions. Each function contains the outline code to work with the component viewer. The outline set signal function is shown below.

```
void vioSetSignal (uint32_t mask, uint32_t signal) {
#if!defined CMSIS_VOUT
        //Add user variables here:
#endif
vioSignalOut & = ~mask;
vioSignalOut | = mask & signal;
#if!defined CMSIS_VOUT
        //Add user code here:
#endif
}
```

The mask and signal parameters are used to update the internal state variable vioSignalOut. Your user code should be placed in the conditional build regions created by the CMSIS_VOUT define.

The vio_get() and vio_set() signal functions are best used to control the state of GPIO lines. The get and set value functions allow us to read and write data to a range of common peripherals.

The vio_init() function must first be extended to support the additional peripheral. As we may need to support a number of peripherals, the internal state variable is defined as an array. The number of elements in the array is configured by a define at the start of the module.

```
#define VIO_VALUE_NUM 3U // number of values
The vio_Init() template code will zeroise the vioValue array.
memset (vioValue, 0, sizeof(vioValue));
```

Next we can provide code to initialize the peripheral, in this case, we will add support for the internal ADC and GPIO pins used to control the evaluation board LED pins.

```
#if!defined CMSIS_VIN
// Add user code here:
ADC_Initialize();
LED_Initialize ();
#endif
}
```

Now we can extend the vioGetValue function to support the ADC. The vio header file provides a set of predefined ids for common peripherals, including an ADC. In this case, we are using ADC channel 2 so we can use #define vioAIN2 to represent the analog channel.

The vioGetValue function will initial check that the id is valid.

```
int32_t vioGetValue (uint32_t id)
if (index >= VIO_VALUE_NUM) {
return value;
}
```

We can also place the low-level ADC code in the user section of the function. This can register level code it may call or vendor libraries or existing board support functions as shown below. The ADC result is stored in the state array before being returned to the calling function.

```
#if!defined CMSIS_VIN
if(index == vioAIN2 )
{
ADC_StartConversion();
while( ADC_ConversionDone()! = 0);
vioValue[index] = ADC_GetValue();
}
#endif
value = vioValue[index];
return value;
}
```

This structure allows the low-level code to be removed for software testing in a simulation environment.

The vioSetValue() function follows a similar structure and can be used to control peripherals such as a DAC that provides an output value to the hardware.

The CMSIS-VIO driver also supports reading and writing a range of values to support more complex peripherals. The vioGetXYZ and vioSetXYZ functions can be used to read and write a data structure to a set of underlying peripherals. A typedef for the default structure is defined in vio.h and can be modified to meet your needs.

```
typedef struct {
int32_t X;   ///< X coordinate
int32_t Y;   ///< Y coordinate
int32_t Z;   ///< Z coordinate
} vioValueXYZ_t;
```

The vioSetXYZ() and vioGetXYZ() functions follow the same style as the set and get value functions and can be used to support more complex external peripherals such as accelerometers and mems gyroscopes.

```
vioValueXYZ_t vioGetXYZ (uint32_t id) {
uint32_t index = id;
vioValueXYZ_t valueXYZ = {0, 0, 0};
#if !defined CMSIS_VIN
// MEMS variables
float pGyroDataXYZ[3] = {0};
int16_t pDataXYZ[3] = {0};
#endif
if (index >= VIO_VALUEXYZ_NUM) {
return valueXYZ; /* return default in case of out-of-range index */
}
#if !defined CMSIS_VIN
// Get input xyz values from MEMS
if (id == vioMotionGyro) {
BSP_GYRO_GetXYZ(pGyroDataXYZ);
vioValueXYZ[index].X = (uint32_t) pGyroDataXYZ[0];
vioValueXYZ[index].Y = (uint32_t) pGyroDataXYZ[1];
vioValueXYZ[index].Z = (uint32_t) pGyroDataXYZ[2];
}
#endif
valueXYZ = vioValueXYZ[index];
return valueXYZ;
}
```

The CMSIS-VIO driver also supports a print function that can be used to send a formatted message through an IO channel. The vio_print() function provides a configurable number of message buffers. The size and number of print buffers are set by defines at the beginning of the VIO driver.

```
#define VIO_PRINT_MAX_SIZE 64U   // maximum size of print memory
#define VIO_PRINTMEM_NUM 4U      // number of print memories
```

Each of the separate print buffers is accessed using a level variable which is passed as the first parameter of the vio_print() function. The remaining parameters follow the same rules as the standard printf() function.

```
vioPrint(vioLevelMessage, "Test [ADC] = %d", x);
```

Extending the CMSIS-Driver Specification

The CMSIS core specification provides a standardized method accessing the Cortex-M processor registers and the CMSIS-Driver specification provides a standard API to access common communication peripherals. However, we will also need to access additional peripherals such as ADC, DAC, and hardware Timers. One way to achieve this is to extend the CMSIS-Driver specification with our own custom profiles. This provides a coherent driver API across all our projects and promotes the reuse of additional collateral such as documentation and test harnesses. There are two main ways that we can do this within the CMSIS-Driver specification. The first is to provide a layer of basic functionality using the CMSIS-VIO driver. The virtual IO driver can be used to provide a wrapping layer over a low-level driver library to act as a standardized target interface for our software components. However, the functionality of the CMSIS-VIO driver is limited to a few peripherals. The second way is to develop a range of full-blown profiles that extend the twelve standard specifications to unsupported peripherals. In this section, we will take a look at how to define and develop a custom CMSIS-Driver.

Custom CMSIS-Driver

The bulk of additional work is in defining the custom CMSIS-Driver profile. At the start of this chapter, we saw how the CMSIS-Driver has a common API (Table 12.1). Our custom driver must present a similar style of interface for the supported peripheral. Since there may be multiple instances of the same peripheral within the microcontroller, the driver is "code only." This means it must not use any local variables as these would be corrupted in the case of multiple instances.

The control function is the peripheral configuration interface and has the following function definition.

```
<Peripheral>_control(control,arg,arg)
```

The control values are #defines that control the functionality of the driver. This profile will be standard for all devices and hence becomes a target for software component code. In turn, this means we can easily reuse the component code on any device.

A partial profile for the timer control word is shown in Table 12.5. As this is your in-house driver, you can adjust it to reflect the types of features you typically use in a project and leave room to extend it so that additional features may be supported in the future. The standard CMSIS-Drivers are designed to be used without a deep knowledge of the underlying microcontroller peripheral. They allow you to configure the peripheral using a high-level definition, for example, you can configure the CMSIS USART baud rate by simply selecting a standard rate such as 115200. This means that the underlying driver code must calculate the correct values for the baud rate prescaler registers based on the internal device clock settings. When we define our custom driver, it is best to avoid such complexity and pass the register level values directly. If we don't need the full functionality of a driver profile, it can be partially developed to provide all the features required for the current project. Additional functionality can be added in later projects as required.

Table 12.5: Custom CMSIS-timer control parameters

Parameter Control	Bit	Category	Description
ARM_TIMER_STOP	0.0.10	Mode Controls	Disable the timer
ARM_TIMER_START			Start the timer
ARM_TIMER_PRESCALER			Prescaler value held in arg
ARM_TIMER_ENABLE_COMPARE			Enable the active compare channels
ARM_TIMER_COMPARE_CHANNEL(n)	11.0.19	COMPARE Channels	Enable COMPARE channel 0 - n, COMPARE value is held in arg
ARM_TIMER_COMPARE_INTERRUPT			Enable the COMPARE channel interrupt
ARM_TIMER_COMPARE_PRESCALER(n)			
ARM_TIMER_COMPARE_RESET_COUNTER			Enable counter reset on COMPARE
ARM_TIMER_COMPARE_HALT_COUNTER			Halt the counter on COMPARE
ARM_TIMER_COMPARE_PIN_CONTROL(n)			Set the external pin state on COMPARE LOW,High,toggle

Here, each of the control parameters is a flag within the allocated bit field which is defined as shown below.

```
#define ARM_TIMER_START_Pos 1
#define ARM_TIMER_START (0x01UL ≪ ARM_TIMER_START_Pos)
```

Once the driver profile has been defined, it is possible to create the control function in the template code (Table 12.6). Once this is done, there is minimal overhead of creating new drivers.

Table 12.6: Custom CMSIS-driver template modules

Module	Description
<peripheral>_<mcu>.c	Driver source code (private)
<peripheral>_<mcu.h>	Driver resources and defines (private)
Driver_<peripheral>.h	Driver API and API defines (public)
Driver_Common.h	Standard driver definitions. Included by Driver_peripheral.h
RTE_Device.h	GPIO pin definitions

When a driver is required for a project, we would create the minimal viable driver. When the driver is reused in a new project, any additional functionality may be added. The CMSIS-Driver specification provides an unsupported feature return code that should be returned when if an unimplemented feature is called.

If a driver has to be created from scratch for a project, the template code can be used to create a skeleton driver that provides stub functions that simulate basic features of the target peripheral. The application developer can use this version while the full functional driver is created.

The use of a standard driver allows the creation of detailed documentation for the driver API. Once created, this documentation can be reused in all future projects.

Exercise 12.4: CMSIS Timer

In this project, we will examine the structure of a custom CMSIS-Driver based on the hardware timer outlined above.

Go to the exercises in the pack installer and copy Exercise 12.4.

Open timer_LPC17xx.c and go to the end of the file.

Here, we create a structure that defines the driver functions for each supported timer. In this case, we create a typedef structure for the driver API.

```
typedef struct ARM_DRIVER_TIMER {
ARM_DRIVER_VERSION (*GetVersion)(void);
uint32_t (*Initialize)(ARM_TIMER_SignalEvent_t cb_event);
uint32_t (*Control)(uint32_t control,uint32_t arg);
enum RESULT (*isTimer_Running) (void);
int32_t (*PowerControl) (ARM_POWER_STATE state);
int32_t (*Uninitialize) (void);
} const ARM_DRIVER_TIMER;
```

In the driver, we now create matching functions for each peripheral instance we need to support.

```
uint32_t TIMER0_timer_Initialize (ARM_TIMER_SignalEvent_t psignal_event)
```

and then create a structure using the driver typedef.

```
ARM_DRIVER_TIMER Driver_TIMER0 = {
TIMER_GetVersion,
TIMER0_timer_Initialize,
TIMER0_timer_control,
TIMER0_isTimer_Running,
TIMER0_timer_PowerControl,
TIMER0_timer_uninitialize
};
```

In our component code, we can now declare the driver for a particular instance

```
extern ARM_DRIVER_TIMER Driver_TIMER0;
```

and access the functions within it.

```
Driver_TIMER0.Initialize(NULL);
```

Each of the timer instance functions (TIMER0_timer_Initialize() etc.) acts as a wrapper which in turn calls a base driver function (TIMERx_initialize()). When we call the base function, we pass the same parameters plus an additional resources structure.

```
uint32_t TIMER0_timer_Initialize (ARM_TIMER_SignalEvent_t psignal_event) {
return TIMERx_Initialize (psignal_event,&timer0Resources);
}
```

The resources' structure provides the parameters associated with a specific timer. Since this is hidden from the user API, it is possible to modify its contents to match the underlying hardware and any software library provided by the silicon vendor. In the case of the NXP device, we can go with the following definition. This is a minimal definition; you may wish to add additional resources such as support for DMA channels.

```
typedef struct {
TIMER_PINS    pin;
LPC_TIM_TypeDef *reg;
IRQn_Type    irq_num;
TIMER_INFO *info;
} TIMER_RESOURCES;
```

In this case, we are providing the base address and register definitions for timer 0 along with its interrupt channel number and GPIO pin selection. The timer info structure defines a function pointer to the interrupt callback function.

```
typedef struct _TIMER_INFO {
ARM_TIMER_SignalEvent_t cb_event;
uint8_t state;
uint8_t mode;
} TIMER_INFO;
```

The info structure also stores the current state of the driver and will typically move through the following states during its lifetime (Table 12.7).

Table 12.7: CMSIS-driver lifecycle states

CMSIS-Driver Lifecycle State
TIMER_INITIALIZED
TIMER_POWERED
TIMER_CONFIGURED
TIMER_MODE_FAULT

Each of the driver states is a bit flag which is OR ed into the info-> state variable as the driver is used. The info-> mode element is used to reference the peripherals operating mode. In the case of a timer, this may be input capture or output compare PWM generation.

The definition of the PIN structure will vary between different microcontroller vendors and it is best to make use of any vendor library definitions rather than re-invent the wheel. Here it is best to study existing drivers and reuse the GPIO pin configuration code. In the case of our NXP microcontroller the ports are defined as port and pin numbers and a basic GPIO configuration library is provided as part of the project.

```
typedef struct _PIN
{
uint8_t Portnum; // Port Number
uint8_t Pinnum; // Pin Number
} PIN;
```

Now we can develop a generic driver in each of the base functions using the resources' structure. Here, the TIMERx_initilize function provides code to install the interrupt callback function into the timer resources info structure. The resources pin information is also used to configure the GPIO pins as outputs for the timer compare outputs.

```
static int32_t TIMERx_Initialize (ARM_TIMER_SignalEvent_t cb_event, TIMER_RESOURCES*
timer) {
timer->info->cb_event = cb_event;
PIN_Configure (timer->pin.match0->Portnum,
timer->pin.match0->Pinnum,timer->pin.match0_func,
PIN_PINMODE_PULLUP,
PIN_PINMODE_NORMAL);
timer->info->state = TIMER_INITIALIZED;
return ARM_DRIVER_OK;
}
```

Interrupt handling is treated in a similar fashion to the user API calls. The timer interrupt function is used to call the base driver interrupt function, which is passed the timer resources structure. The base interrupt function is used to clean up the timer registers, typically, this will be to clear and interrupt pending bits. It then uses the callback event to trigger the user interrupt service routine. The interrupt status register can be passed to the user function as an event parameter. This allows the service component to identify the cause of the interrupt (compare channel, overflow, etc.).

```
void TIMER0_IRQHandler (void) {TIMERx_IRQHandler(timer0Resources);}
void TIMERx_IRQHandler (TIMER_RESOURCES timer)
{
timer.reg->IR = IR_MR0I_CLEAR;
timer.info->cb_event(0x01);
}
```

Build the project

In this project, our CMSIS timer driver is used to configure two hardware timers to generate separate PWM channels and periodic interrupts.

Start the debugger.

Run the code and follow the behavior of both the application and driver code.

Custom Driver Validation

As we saw earlier in this chapter the CMSIS-Driver is supported by a driver validation framework. This framework is easy to extend, and this allows us to add a group of tests to validate our new driver. Once we have developed a range of tests for our first-timer driver, they can be reused for all future timer drivers. In fact, for future drivers, we can develop a new driver within the framework and get it to pass each test as we add code to the template.

Exercise 12.5: Custom Driver Validation

In this exercise, we will extend the driver validation framework to support our custom CMSIS Timer driver. Since we are going to modify the test framework, the Driver_Validation framework has been added as a set of local files rather than the standard RTE components.

Go to the exercises in the pack installer and copy Exercise 12.5.

This project contains the CMSIS Timer and the CMSIS-Driver validation framework (Fig. 12.22).

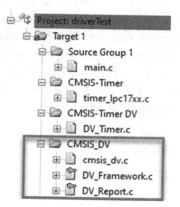

Figure 12.22
CMSIS-Driver validation framework.

Open the RTE Manager and the CMSIS-Driver Validation branch (Fig. 12.23).

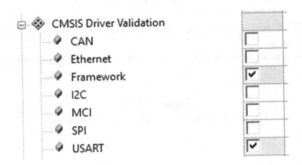

Figure 12.23
Selecting the driver vaildation test cases.

Here, we have enabled the validation framework and enabled the validation tests for a supported peripheral, a USART in this case. This ensures that appropriate paths are set up for the validation framework, include files. The CMSIS-Driver USART template driver is also enabled for the same reason.

Open DV_Timer.c.

We can construct a set of tests using the asserts within the validation framework. These call the driver functions and check the return codes. It is possible to make more complex tests to check the driver's internal code.

```
void TIMER_PowerControl (void) {
int32_t val;
TEST_ASSERT(drv.PowerControl (ARM_POWER_FULL) == ARM_DRIVER_OK);
val = drv.PowerControl (ARM_POWER_LOW);
if (val == ARM_DRIVER_ERROR_UNSUPPORTED) { TEST_MESSAGE("[WARNING] Low power is not
supported"); }
else { TEST_ASSERT(val == ARM_DRIVER_OK); }
}
```

Now open CMSIS_DV.c.

The code contains a set of header files for each test which are conditionally included if the driver test cases are enabled. These are inherited from the RTE definitions. However, since we are adding tests manually, we must place the RTE_CMSIS_DV_TIMER define at the start of CMSIS_DV.c in order to enable our test group (Fig. 12.24).

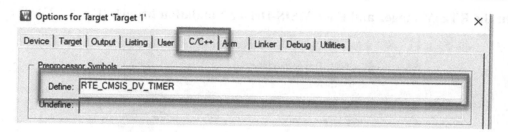

Figure 12.24
Create a define to enable the test group.

```
#ifdef RTE_CMSIS_DV_TIMER
#include "DV_Timer_Conf.h"
#endif
```

The configuration file is used to define the setup parameters to be used for the tests, and a set of defines used to enable each driver test within the test group.

```
#define TIMER_INITIALIZATION_EN 1
#define TIMER_POWERCONTROL_EN 1
```

In CMSIS_DV.c the next section of code provides initialize and uninitialize functions for the test group

```
#ifdef RTE_CMSIS_DV_TIMER
static void TS_Init_TIMER (void) {
TIMER_DV_Initialize ();
}
static void TS_Uninit_TIMER (void) {
TIMER_DV_Uninitialize ();
}
#endif
```

After the init/uninit section we can add an array of tests which will be executed if both the timer test suite and the individual tests are enabled

```
#ifdef RTE_CMSIS_DV_TIMER
static TEST_CASE TC_List_TIMER[] = {
TCD ( TIMER_Initialization, TIMER_INITIALIZATION_EN),
TCD ( TIMER_PowerControl, TIMER_POWERCONTROL_EN)
};
#endif
```

Finally, at the end of the file, we can add our timer tests to the overall test group. Now when the project is built the test framework will be configured to run our CMSIS driver timer test cases to validate the timer.

```
TEST_GROUP ts[] = {
#ifdef RTE_CMSIS_DV_TIMER /* TIMER test group */
{
__FILE__, __DATE__, __TIME__,
"CMSIS-Driver TIMER Test Report",
TS_Init_TIMER,
TS_Uninit_TIMER,
TC_List_TIMER,
ARRAY_SIZE (TC_List_TIMER),
},
#endif
.............
}
```

Build the project and run the test framework in the debugger.

This will execute the tests and provide a report to the debugger console window (Fig. 12.25).

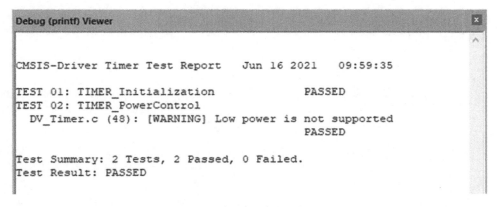

Figure 12.25
Custom CMSIS-Timer validation tests.

The test framework can be extended to provide additional loopback and hardware tests so that it is also useful for board bring up. Once the new test suite has been built it can be used on future projects not just to test the final driver but as a framework that is used to prove the driver code as we write it. This is a form of Test-Driven Development that we will look at again in Chapter 14, Software Components.

Conclusion

In this chapter, we have looked at creating a standard microcontroller independent driver layer using existing CMSIS-Driver profiles and how to extend these by defining our own custom profiles. In Chapter 14 Software Components, we will look at how we can create reusable software components that target our driver layer.

CHAPTER 13

Test-Driven Development

Introduction

In this chapter, we are going to look at how to create software unit tests on a small microcontroller. We will also look at a new technique called "Test-Driven Development" (TDD) for developing code. When I say new, I really mean new to embedded system developers. TDD has become established in other areas of computing since around 2003. Traditionally, most embedded systems developers will tend to write a relatively small amount of code, compile it and then test it on the hardware using a debugger. Any formal testing is then done by a separate team once all of the codebase has been written. Unit testing would then be a long phase of testing, debug, and fix until the final production code is deemed fit to be released. TDD is not a replacement for the formal unit testing phase but introduces testing as a tool that is part of the code development process. Put simply, TDD requires you to add tests for every function that you write, but crucially you write the tests first then write the application code to pass the tests. At first, this seems unnecessary time consuming and complicated; however, TDD does have a number of very positive benefits.

As we develop our application, code tests are written for all the major functions. As the project grows, this provides us with a full recursive test suite. Now, as new code is added or existing code is modified, we can run all the tests and be confident that the new code works and does not have any unexpected side effects.

By writing tests for each incremental addition to our production code, we will catch bugs early. This is the holy grail of project development. Trapping and removing errors as early as possible save a huge amount of debug time. There is nothing worse than trying to catch an elusive bug buried deep in the final production code with a deadline looming (well, possibly there are worse things).

If we develop the discipline of testing code as we go, it increases our overall confidence that the code works. Plus if you are working within a team of developers, a full range of software tests will help speed up any software integration phase.

In order to be able to successfully write unit tests, we need to write the production code so that it is testable. This means that we must be able to decouple our production code functions from the main application. This encourages the writing of well-structured production code. We will look at this in more detail in the next chapter.

The Designer's Guide to the Cortex-M Processor Family.
DOI: https://doi.org/10.1016/B978-0-323-85494-8.00014-0

As our application codebase grows, a traditional compile and debug cycle will begin to take longer and longer to complete. Also, the system complexity grows, it can be harder to debug a portion of the system because it is dependent on another part of the system being in the correct state. The TDD approach only requires us to compile our test code and the module under test. This leads to a very fast compile and test cycle, typically this will be under 30 s. Over the design cycle of the project, this can really mount up. This also means that in order to apply TDD effectively, we must have an automated compile and test process that can be launched by a single action in the IDE.

Software has two aspects, the first and most obvious is that it must fulfill the required functionality. The second is more subtle. Software must also be well crafted and expressive so that it may maintain over the lifetime of the project and ideally reused in future projects. Once we have written the code to be functionally correct, it can then be refactored to become releasable production code. With TDD once the code is functionally correctly, we will have a set of passing tests. We can now refactor the code and go through a very fast compile and test cycle to ensure that we have not introduced any bugs. TDD can save a huge amount of time during this stage of code development.

The written tests are also a very good form of living documentation. You may well write a formal document describing how the code is structured and how it works, but most programmers hate to read it. Instead, we all prefer to look at the source code. In this context, the test cases describe what each function does and how to use it. Examining the test cases alongside source code is an express route to understanding how the application works.

Many microcontroller-based embedded systems are developed by small teams or individual developers. In such an environment, formal testing can really be a luxury and often simply does not happen. In such an environment, TDD is a vast improvement to the development process and also has the benefit of being free to adopt.

Finally, any kind of software development is a marathon rather than a sprint. As you get deeper into the project, it can become hard to see the wood for the trees. If you are a sole developer, you will be holding most of the project details in your head day after day. Using TDD does breed confidence that your codebase is working and gives a feeling of progress and accomplishment through this daily grind.

The TDD Development Cycle

The philosophy of TDD may be summed up in three rules formulated by Robert C. Martin:

1. You are not allowed to write any production code unless it is to make a failing unit test pass.

2. You are not allowed to write any more of a unit test than is sufficient to fail; and compilation failures are failures.

3. You are not allowed to write any more production code than is sufficient to pass the one failing unit test.

While this feels very awkward at first, with a little practice, it becomes second nature. Once you have invested the time and discipline to work with TDD in a project, you will start to reap the benefits.

To use TDD in a real project, we can go through the following cycle:

1. Write the code to add a test.
2. Build the code and run the test to see it fail.
3. Add the minimal amount of application code to pass the test.
4. Run all the test code and check all the tests are passed.
5. Refractor the code to improve its expressiveness.
6. Repeat.

Test Framework

In order to make the TDD approach work in practice, we need to be able to create tests and run them as quickly as possible. Fortunately, there are a number of test frameworks available that can be added to our project to create a suitable test environment. Two of the most popular the frameworks are Unity and CppUtest:

Unity	unity.sofurceforge.net
CppUtest	http://www.cpputest.org

In the example below, we will use the Unity framework. Unity is designed to test "C" based applications. It has a small footprint and is also available as a software pack, so it is very easy to add to a project. CppUtest is a similar framework designed for both the "C" and "C++" languages.

Test Framework Integration

As tests run, the results are sent to the STDIO channel. The Unity framework component requires the Compiler IO options to be configured to define the STDIN and STDOUT channels. In most projects, the test results are best sent via the ITM to the console window in the debugger. A Cortex-M0 does not have the ITM trace, so you would need to use a different IO channel to display the results. Typically, this will be a USART or the Event Recorder.

Test Framework Automation

As well as being able to quickly add tests to our framework we need to be able to automate the build and test cycle. So ideally, within the IDE, we will be able to build the test project, download to the target, run the tests and see the results with one button press. This should then give us a development cycle of under 30 Seconds. In the next two examples, we will look at setting up and automating a TDD framework with a simple project.

This project is configured to use the simulator so we can experiment with the TDD approach without the need for specific hardware, but the same approach will work with a debug adapter and real hardware.

Designing for Testability

When we are writing code for a project, it is normal to focus on getting the functionality correct. However, you must also structure your code so that it is easily testable. The key concept here is to separate the firmware, the low-level code that accesses processor and peripheral registers, from the software, the "pure" C code. The firmware should be collected into its own set of C modules and provide helper functions to the code in the software modules. This allows you to easily separate the two "layers" of code and use the debugger and hardware to prove the driver firmware layer while using repeatable tests for the software layer (Fig. 13.1).

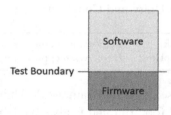

Figure 13.1
Your application must be separated into separate firmware and software modules.

In this chapter, we will look at a testing strategy that can be used to validate the software components in your project. Then in Chapter 15, MCU Software Architecture, we will see how these techniques be incorporated into a software architecture that promotes the reuse of both software components and test frameworks.

We will also need to test the low-level firmware. In Chapter 12 CMSIS-Driver, we saw how to use the CMSIS driver API to access a range of common peripherals. The CMSIS drivers also come with a ready-made validation framework. So by using CMSIS drivers where possible, we can rapidly bring up and validate a hardware platform with minimal effort. The

CMSIS driver API only supports a limited range of peripherals, but as we saw in Chapter 12 CMSIS-Driver, it is possible to develop custom CMSIS-Drivers for any peripheral.

Software Testing with Sub-Projects

As we will see in the next exercise, we need to add a test framework to our project so that we can easily create and execute a large number of tests over our codebase. At the same time, we don't really want to modify or add test hooks into our application source code. The best way to solve this problem is to create a set of subprojects that add the test framework and the software module we want to validate. The test subproject can then be included in a multiproject workspace alongside the main application project. To make this work, we need to structure the software as a set of components that can be tested in isolation and then integrated into the final application project. We will see this in more detail in Chapter 14, Software Components.

Exercise 13.1: Test-Driven Development

In this exercise, we will install the unity framework and add it to a simple project. We can then create and run a set of test cases. Finally, we can automate the test cycle within the Microvision IDE to create an efficient edit built and test process.

The Unity test framework is available as a software pack. For the exercises in this chapter, you must install it from within the pack installer (Fig. 13.2).

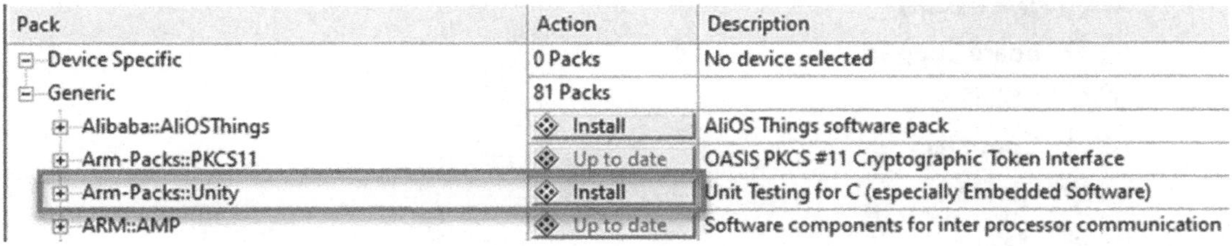

Pack	Action	Description
⊟ Device Specific	0 Packs	No device selected
⊟ Generic	81 Packs	
⊞ Alibaba::AliOSThings	◈ Install	AliOS Things software pack
⊞ Arm-Packs::PKCS11	◈ Up to date	OASIS PKCS #11 Cryptographic Token Interface
⊞ Arm-Packs::Unity	◈ Install	Unit Testing for C (especially Embedded Software)
⊞ ARM::AMP	◈ Up to date	Software components for inter processor communication

Figure 13.2
Pack installer Boards and examples tabs.

You will also need to install the following packs using the pack installer (Table 13.1).

Table 13.1: Required software packs

Pack	Minimum Version Number
Keil::LPC1700_DFP	2.1.0
Keil::ARM_Compiler	1.0.0
ARM::CMSIS	4.8.0

In the pack installer, go back to the book examples and Copy "Exercise 13.1 Unity Developer Test Framework."

The application code is an RTOS-based project that is used to switch on a bank of LED's one at a time. Once all of the LED's have been switched on, they will be simultaneously switched off and the cycle will repeat. We are going to develop it further by adding a function that will introduce different period delays at selected points in the LED cycle. While this is easy enough to do, we will use this example to see how to add a TDD approach to our project.

The main.c module has been removed from the build process using its local options. This allows us to add the test framework which contains its own main() function.

In our project, our new code will be placed in a file called lookup.c (Fig. 13.3). The module Thread.c contains the RTOS thread code, which we will test in the next example.

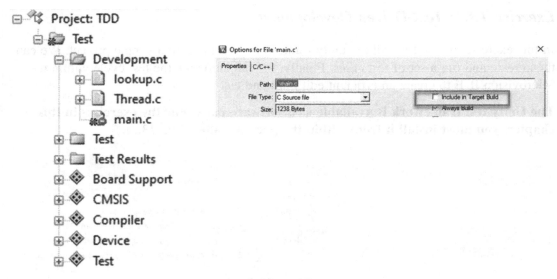

Figure 13.3
Initial project with main.c removed from the build system.

Adding the Unity Test Framework

Lookup.c contains the skeleton of a function called checkTick(). This function is passed an integer value. It must then scan an array to see if this value is in the array. If there is a matching value, it returns the location of that value in the array. If there is no matching value in the array, zero is returned.

In the project window set the test project as the active project.

Open the Run Time environment (RTE).

Figure 13.4
The sel column will be colored orange if an additional component is required.

Select the Utilities::Developer Test:Unity Test Framework (Fig. 13.4).

Next, select the Compiler STDOUT and STDIN and set them to ITM (Fig. 13.5).

The test framework uses STDIO as an output for its results so we must enable this and select an appropriate low-level output channel. Using the ITM will allow us to run on any Cortex-M core which has the ITM fitted. The ITM will give us an STDIO channel that does not use any microcontroller resources that would conflict with the application code. The ITM also relies on the debug adapter and does not need any external hardware resources.

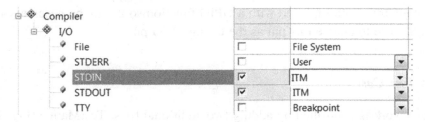

Figure 13.5
Select ITM to be the STDIO channel.

Finally, we just need to make sure that the ITM support in the debugger is correctly configured. This example is using the simulator and does not need any further configuration, but for a project using a hardware debugger, the ITM is configured as follows.

Only change the following debugger settings if you are using a hardware debugger.

Open the options for target\debug menu and press the settings button (Fig. 13.6).

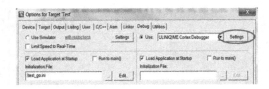
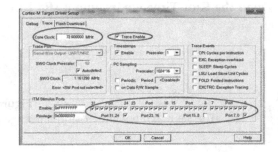

Figure 13.6
For hardware debug you must enable the ITM Trace Unit to get the test results.

Select the trace menu.

Set the core clock speed to match the processor CPU frequency.

Enable the trace and enable ITM channel 31.

It is also advisable to disable any other trace features in order to minimize the amount of data being sent through the Serial Wire Trace channel.

The ITM data is sent to the debugger through the SWO debug pin. On some microcontrollers this is multiplexed with a GPIO function so it may be necessary to use a debugger script file to enable this pin as the debug SWO pin.

Adding the Test Cases

The test framework is controlled by adding two additional files: TestMain.c (Fig. 13.7), which provides the overall test harness, and TestGroup.c, which provides the test cases for a given set of functions. For larger components, you can add multiple test groups to test different areas of functionality. This allows different sets of tests to easily be enabled and disabled within the test framework.

Figure 13.7
The project with the Test harness added.

Open Test Main.c.

This file provides the main() function for the Test Target and calls the Unity test framework. When the tests are finished, execution will reach the stop() function.

Once Unity starts to run, it will start to execute any tests that have been defined within the test framework. Within the framework we can define a set of tests which will exercise a particular function. These are collected together as a test case. Associated sets of test cases can then be collected together as a test group. The Test_Main.c contains the top-level functions that are used to run each defined test group.

In the runAlltest() function rename template default group name "test group" to "lookup."

In the TEST_GROUP_RUNNER() function, rename the "test group" to "lookup" and add the test case "correctTickValues" and "failingTickValues."

```
TEST_GROUP_RUNNER(lookup)
{
        RUN_TEST_CASE (lookup, correctTickValues);
        RUN_TEST_CASE (lookup, failingTickValues);
}
static void RunAllTests(void)
{
  RUN_TEST_GROUP(lookup);
}
```

Open TestGroup.c.

In this file we can define a group of tests for our lookup.c function. This file defines the test group name and provides setup and tear down functions which run before and after the tests. The TEST_SETUP() function runs once before each test in the test group and is used to prepare the application hardware for the test cases. The tear down function is used to return the project to a default state so that any other test groups can run. The final function is our test case which allows us to create an array of tests for our new function.

```
TEST_GROUP(lookup);
TEST_SETUP(lookup)
{
}
TEST_TEAR_DOWN(lookup)
{
}
```

For this simple example, we do not need to provide any setup or teardown code.

Copy and paste the TEST function so we have two test functions.

Rename the first one as follows	**TEST(lookup, correctTickValues)**
And the second as shown below	**TEST (lookup, failingTickValues)**

Now we can create test cases in both of these functions as shown below. Each test is created within a TEST_ASSERT function. In the case below we are calling the function under test and passing a value and then checking it returns the correct value. There are a wide range of TEST_ASSERT functions, which are defined in unity.h.

```
TEST(lookup, correctTickValues)
{
                TEST_ASSERT_EQUAL(1, checkTick(34));
                TEST_ASSERT_EQUAL(2, checkTick(55));
                TEST_ASSERT_EQUAL(3, checkTick(66));
                TEST_ASSERT_EQUAL(4, checkTick(32));
                TEST_ASSERT_EQUAL(5, checkTick(11));
                TEST_ASSERT_EQUAL(6, checkTick(44));
                TEST_ASSERT_EQUAL(7, checkTick(77));
                TEST_ASSERT_EQUAL(8, checkTick(123));
}
```

The first set of test cases will call the function with valid input values and check that the correct value is returned.

```
TEST (lookup,failingTickValues)
{
                TEST_ASSERT_EQUAL(0, checkTick(22));
                TEST_ASSERT_EQUAL(0, checkTick(93));
}
```

The second set of test cases will call the function with invalid values and check that the correct error value is returned.

So far, we have added the test framework and defined some test cases. We can now build the code and execute the test functions. As this is part of our TDD cycle, all the tests should fail because, apart from the function prototypes, we have not written any actual application code.

Build the code.

Start the debugger.

Open the view\serial windows\printf window.

Run the code.

A test fail will be reported in the serial window (Fig. 13.8).

```
Unity test run 1 of 1
.TestGroup.c:18:TEST(lookup, correctTickValues):FAIL: Expected 1 Was -1


.TestGroup.c:30:TEST(lookup, failingTickValues):FAIL: Expected 0 Was -1



-----------------------
2 Tests 2 Failures 0 Ignored
FAIL
```

Figure 13.8
A failing test is reported in the ITM console window.

Automating the TDD Cycle

Before we go and start writing the application code, we can further automate this testing procedure so we can go around the build and test cycle with one click in the IDE.

Set the project target to be the "Test" build.

Open the options for target menu and select the user menu and enable the "Start Debug" option (Fig. 13.9).

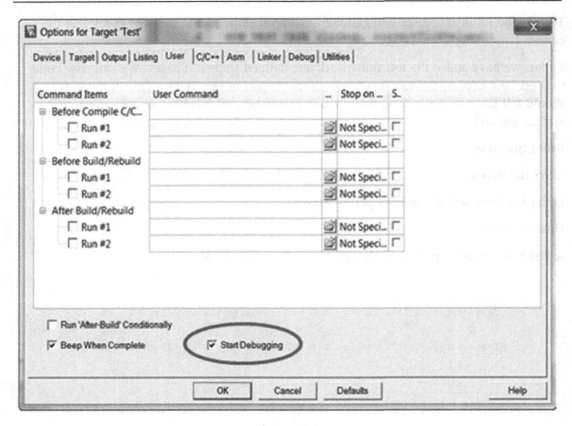

Figure 13.9
The User tab allows you to start the debugger as soon as the build is finished.

Now after a build the code will be downloaded into the target, and the debugger will be started.

Next select the options for target\debug menu and add the test_go.ini **script file which is located in the <project>\RTE\utilities directory as the simulator initialization file** (Fig. 13.10).

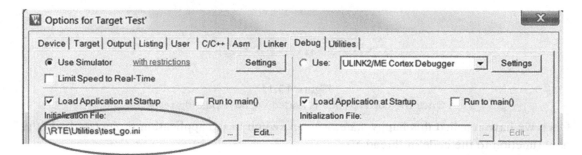

Figure 13.10
Add the automation script as the simulator initialization file.

Next uncheck the "Run to main()" box. Or this will stop the script running correctly.

Now this script will be run as soon as the debugger is started.

```
coverage clear
SLOG≫testResults.txt
g,stop
SLOG OFF
LOG≫testResults.txt
coverage \testgroup DETAILS
LOG OFF
```

The script file starts the code running as soon as the debugger is started. The code will run until it has reached the stop function. All of the test results are printed to the ITM console and also logged to a file on the PC hard disk. At the end of the tests, the code coverage information for the testgroup.c module is also saved to the log file. The code coverage information proves that all the tests have been executed. You could also add the coverage information for the modules under test.

Now if you rebuild the code, it will compiler link start the debugger and execute the tests.

Since we are only building and downloading the test framework and our new code, this is a very fast development cycle, particularly if you are downloading to FLASH. In a large project, this can be a big time-saving.

Now open the lookup.c **file**.

Uncomment the existing code and run the build-test cycle again.

This time the code fails on one of the correct tick value tests.

Correct the failure and re-run the build.

Check all the tests pass. If there is a failure, correct the error and repeat it until the tests pass (Fig. 13.11).

```
Unity test run 1 of 1
..
------------------------
2 Tests 0 Failures 0 Ignored
OK
```

Figure 13.11
Once the correct code is added we have passing tests.

Once we have reached this happy state, we can switch to the application project and add the new function to the code in thread.c.

```
while(1)
{
        ledData = 0;
        for(i = 0;i<8;i++)
        {
                extendedDelay = checkTick(tick);        //call checkTick to calculate
                                                          an extended delay
                totalDelay = 100 + (100*extendedDelay);
                osDelay(totalDelay);
                osMessagePut(Q_LED,ledData,0);
                tick = (tick + 1) & 0xDFF;
                ledData = ledData + 1;
        }
}
```

Switch to the Application project and build the code.

Start the debugger and run the new version of the blinky program.

Testing RTOS Treads

This form of testing also works very well with RTOS threads. In our example, we have four threads, main() plus the three application threads (Fig. 13.12). Each of the threads is a self-contained "object" with very well-defined inputs and outputs. This allows us to test each of the threads in isolation since we can start each thread individually and build a test harness that "mocks" the rest of the system.

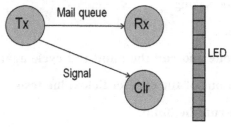

Figure 13.12
An RTOS-based project is composed of thread objects with well-defined inputs and outputs. A thread is an excellent target for testing.

Our application threads consist of a Tx thread that writes values into a message queue, and an Rx thread that reads values from the message queue and writes these to the bank of LEDs. When all the LEDs have been switched on, the Tx thread signals the Clr thread, which switches off all the LEDs.

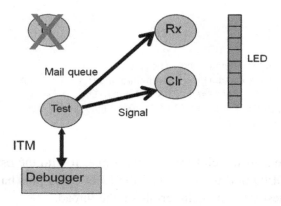

Figure 13.13
A test thread can use the RTOS API to control the application threads. Here we can terminate a thread and apply test cases to the remaining thread using the mail queue and signal flags.

We can test the application at the thread level by using the same testing project. By adding an additional test group called "thread," we can use the TEST_SETUP function to initialize the RTOS objects and create a custom test thread.

```
TEST_SETUP(thread, rxThread)
{
        LED_Initialize ();
        Init_queue();
        Init_rxThread();
        ...........................
```

The test thread can start each thread individually, and then we can apply test data using the RTOS calls such as message queue and signals (Fig. 13.13).

Here we are creating a test group that initializes the LEDs and the application message queue. We can then start the rxThread() running. This thread is now waiting for messages to be posted into the mail queue. This allows us to create test cases that post messages into the mail queue in place of the TxThread(). We can then check the behavior of the rxThread().

```
osMessagePut(Q_LED,(0x00),0);
osDelay(10);
TEST_ASSERT_EQUAL((1<<28),(GPIO_PortRead (1) & 0xB0000000));
```

In our test case, we can post some data into the rxThread() message queue and then read the LED Port settings to check that the correct LEDs are switched on. The osDelay() call ensures the RTOS scheduler will task switch and run the rxThread() code. At the end of the test cases, we can kill the rxThread() and continue to create and test another of the application threads.

```
TEST(thread, clrThread)
{
Init_clrThread ();
osSignalSet        (tid_clrThread,0x01);
osDelay(10);       //block and allow the thread under test to run
TEST_ASSERT_EQUAL(0,(GPIO_PortRead (2) & 0x0000007C));
TEST_ASSERT_EQUAL(0,(GPIO_PortRead (1) & 0xB0000000));
osThreadTerminate(tid_clrThread);
}
```

In the next test case, we start the clrThread() and trigger it with the osSignalSet() function. We can then read the GPIO port to ensure that the LED GPIO pins have been written to zero, at the end of the test we can again terminate the thread.

Exercise 13.2: Testing RTOS Threads

In this example, we will look at testing an RTOS thread using the unity test framework

In the pack installer, select and copy 13.12 TDD and RTOS Threads.

In Microvision, check that the test project is set as the active project.

Add the file testGroupRTX.c.

Open TestMain.c.

Add the thread test group and call it from RunAllTests() as shown below.

```
TEST_GROUP_RUNNER(thread)
{
      RUN_TEST_CASE (thread, rxThread);
      RUN_TEST_CASE (thread, clrThread);
      RUN_TEST_CASE (thread, txThread);
}
static void RunAllTests(void)
{
      printf("\nRunning lookup test case \n");
      RUN_TEST_GROUP(lookup);
      printf("\nRunning Thread test case \n");
      RUN_TEST_GROUP(thread);
}
```

Build the project and let it run through the compile link and test loop (Fig. 13.14).

```
------------------------------Build Time and Date 15:31:39, Nov 19 2015--

Unity test run 1 of 1

Running lookup test case
..
Running Thread test case
...
----------------------
5 Tests 0 Failures 0 Ignored
OK
```

Figure 13.14
Now the original and new RTOS tests run and pass.

This will execute the original set of test cases on the code in lookup.c. Provided these pass ok the new RTOS test group will be executed to test each of the three RTOS threads.

Decoupling Low-Level Functions

Ideally, we do not want to modify the code under test. However at some point we will encounter the problem that the function under test calls a routine that requires a result from a peripheral.

```
int detectPacketHeader (void)
{
    while (!packetHeader)
    {
        cmd = receiveCharacter(USARTdrv);
        ..........
```

In this case the receiveCharacter() routine is waiting to receive a character from the microcontroller USART. This makes automated testing the detectPacketHeader() function difficult to test because it is reliant on the receiveCharacter() routine. To get round this we need to make a "mock" version of receiveCharacter() that can be called during testing in place of the real function. We could comment out the real receivecharacter() function but this means changing the application code or adding conditional compilation using #ifdef preprocessor commands. Both of these approaches are undesirable as they can lead to mistakes when building the full application. One way round this problem is to use the __weak directive provided by the Compiler.

```
__weak char receiveCharacter (ARM_DRIVER_USART *USARTdrv)
```

When the application code is built this function will be compiled as normal. However during testing we can declare a "mock" function with the same function prototype minus the __weak pragma. The linker will then "overload" the original function with the weak declaration and use the "mock" version in its place.

```
int8_t mockSerialData[] = "Hello World";
char receiveCharacter (ARM_DRIVER_USART *USARTdrv)
{
int8_t val,count = 0;
      val = mockSerialData[count++];
      if(count > sizeof(mockSerialdata)) count = 0;
return val;
}
```

Now we can build a test for the detectPacketHeader() function and use the "mock" version of receiveCharacter() to provide appropriate data during the test case.

If you are using a CMSIS driver, it is also possible to mock the driver. In the RTE, each driver has the option to select a template rather than a full driver (Fig. 13.15).

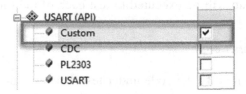

Figure 13.15
Adding a custom CMSIS Driver adds all the necessary build paths to your project.

When this option is selected, the additional paths to supporting header files are added to the project.

We can then add a bare-bones template for the driver to our testing project. These can be found in the following:

```
\Keil\ARM\PACK\ARM\CMSIS\<version>\CMSIS\Driver\DriverTemplates
```

Each template provides a Skellington outline for each of the driver functions. Once added to our project we can populate the functions with suitable code to simulate the activity of the underlying peripheral. In the case of a USART we need to model the send and receive functions. It is also important to note here that if you develop a suitable generic test driver it can be reused on any future project.

Testing Interrupts

An embedded microcontroller project will tend to have code that is intimately interconnected with the microcontroller hardware. Some functions within the application code will be reliant on a microcontroller peripheral. In most projects, we may also have a number of peripheral interrupts active. We can use the same method of overloading functions to test the application functions that are dependent on hardware interrupts. If we have an ISR function we can declare it as __weak and then overload it with a "mock" function that is not hardware dependent.

```
__weak void ADC_IRQHandler(void) {
  volatile uint32_t adstat;
  adstat = LPC_ADC->ADSTAT;              /* Read ADC clears interrupt    */
  AD_last = (LPC_ADC->ADGDR >> 4) & 0xFFF;   /* Store converted value        */
  AD_done = 1;
}
```

So the above ADC interrupt handler which is dependent on the microcontroller ADC peripheral, can be replaced with a simplified "mock" function.

```
void ADC_IRQHandler(void)
{
      AD_last = testValue;
      AD_done = 1;
}
```

The "mock" function mimics the functionality of the original ISR but is in no way dependent on the actual ADC hardware. As we are naming the function using the prototype reserved for the CMSIS ADC interrupt handler, the function will be triggered by an interrupt on the ADC NVIC channel. To use this function during testing, we can also overload the original ADC_StartConversion() function.

```
__weak int32_t ADC_StartConversion (void) {
  LPC_ADC->ADCR &= ~(7 << 24);       /* stop conversion     */
  LPC_ADC->ADCR |= (1 << 24);        /* start conversion    */
  return 0;
}
```

Again the overload "mock" function does not address the hardware ADC peripheral registers but uses the CMSIS NVIC_setPending() function to cause an interrupt on the NVIC ADC channel.

```
int32_t ADC_StartConversion (void)
{
      NVIC_SetPendingIRQ(ADC_IRQn);       //trigger the ADC interrupt channel
}
```

Now we can build test cases around functions that use results from the ADC while using the overload functions to mock the activity of the hardware ADC. It is also important to note that we have made minimal changes (two __weak directives) to the application source code.

Exercise 13.3: Testing with Interrupts

In this example, we will enable the 12-bit ADC and its interrupt. Then in a background function, we will start an ADC conversion. Once the ADC conversion has finished, we will read the result, shift the result 4 bits to the right and then copy the result to a bank of 8 LEDs.

Open the Pack Installer.

Select the Boards:: MCB1700.

Select the example tab and Copy Exercise 13.3 "Unity Framework Interrupt Testing" (Fig. 13.16).

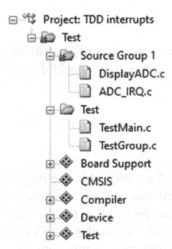

Figure 13.16
The project is setup to work with the test framework.

Open testGroup.c.

In this file we have the overload versions of the ADC_IRQHandler() and the ADC_StartConversion() functions.

The code under test is the function displayADC() in DisplayADC.c.

```
void displayADC (void)
{
uint32_t result;
      ADC_StartConversion();
      while(ADC_ConversionDone()!= 0);
      result = ADC_GetValue();
      result = result>>4;
      LED_SetOut(result);
}
```

In the test case, we can load an ADC result value and then call the displayADC() function.

```
testValue = 0x01<<4;
displayADC();
TEST_ASSERT_EQUAL(1<<28,(GPIO_PortRead (1) & 1<<28));
```

displayADC() will use the value from the ADC and write this value to the LED port pins. We can then check that the correct bits have been set by reading the state of the appropriate port.

Examine the code in the displayADC.c **and** ADC_IRQ.c.

Build the code and execute the test cases (Fig. 13.17).

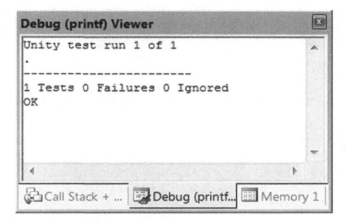

Figure 13.17
The test cases run with the mock overloaded functions.

Conclusion

Hopefully, this has been a useful chapter. If you are new to TDD, it should give you some food for thought. The only real way to get a real sense of how to use TDD in a project is to actually do it. Initially, you will go slower as you have to build experience in how to construct suitable tests. Once you have built some useful testing "patterns," you will start to make more rapid progress. With practice and a degree of discipline, you will start to see some real benefits by adopting a TDD approach to code development.

CHAPTER 14

Software Components

Introduction

In Chapter 12, CMSIS-Driver, we saw how CMSIS-Driver provides a standardized driver layer with a common API that can be used on a very wide range of microcontrollers from multiple silicon vendors. In this chapter, we will look at how to design our application code to use CMSIS-Driver. Here, the aim is to create hardware-independent software components that may be reused across any microcontroller that provides a suitable CMSIS-Driver. Once developed, we can then bundle the component into a software pack so that it can be installed into a toolchain to provide a ready resource for future projects.

Designing a Software Component

A software component is best defined as a collection of functions that provide a "single responsibility" to the application threads and application functions. This is true for the software functionality that the component provides and also for its module layout within the project. In order for the component to be reusable, it must be held in a set of dedicated modules that isolate all the component code from any other code used by the project. A software component (Fig. 14.1) may be a passive collection of functions that are called by the application threads, or it may be active with its own interrupt functions and RTOS thread or threads.

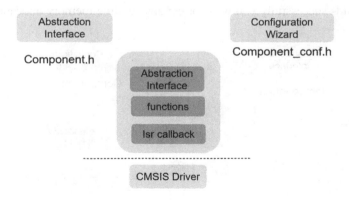

Figure 14.1

Software component structure.

The Designer's Guide to the Cortex-M Processor Family.
DOI: https://doi.org/10.1016/B978-0-323-85494-8.00012-7
521

A component should observe the following principles:

A software component will hide its internal functions and data.
This means that all global data and private functions within a component will be declared static.

All code within the component will comply with the MISRA C coding standard.
Ideally, this will be part of a standard "house style."

Threads that are created by a component must statically allocate their stack space.
This ensures that the necessary memory resources are allocated by the linker when the project is built and prevents runtime allocation errors.

All of the component configuration options shall be stored in a header <component>_conf.h.
This helps standardize the project layout and ease of use through a configuration wizard.

No component data shall be accessed directly from external functions.
Access to all component data will be through helper functions.

The component API shall be provided by a header file <component>.h.
Ideally, the component header file will only contain function definitions.

The component shall only access the microcontroller hardware through CMSIS-Core or CMSIS-Driver functions.
This ensures that the component is pure "C" code independent of a given microcontroller and may be easily ported to a new microcontroller.

The component API shall be closed for modification but open for extension.
Existing function calls must not be changed, or they will break existing code. However, we can add new function calls to extend the component.

In the case of an active component (Fig. 14.2), one that creates its own thread, the thread stack memory should be statically allocated. This ensures that when it is reused, its memory requirement will be automatically allocated without any configuration by the user. Then provided that the project builds successfully you will not get any runtime memory allocation issues.

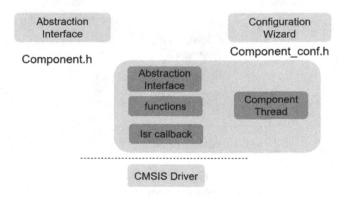

Figure 14.2
An active software component.

Component API

During the initial project definition, we will identify the required components and can then define a user API. This API consists of function calls only and does not expose any data object directly. The API also acts as a "trust boundary" between our component and the rest of the application. This means that all of the component API functions should use secure coding principles, including checking that any passed parameters are within reasonable bounds. Any incorrect parameter values must cause the function to terminate and return a meaningful error code.

Descriptive API comments must be included in the API header file using Doxygen tags so that they are easily available to the application developer. This also allows an automatic documentation generator such as Doxygen to build a separate API document for your application developers.

```
/**
 \fn ARM_USART_MODEM_STATUS USART_GetModemStatus (USART_RESOURCES *usart)
 \brief       Get USART Modem Status lines state.
 \param[in]   usart     Pointer to USART resources
 \return      modem status
\ref ARM_USART_MODEM_STATUS
*/
```

Module Structure

When designing a software component for reuse, it is critical to adopt a standard module layout that ensures that the component is isolated from all other areas of the project code.

A recommended layout of a software component is shown in Table 14.1.

Table 14.1: Software component modules

Module	Description
<component>_<module>.c	Source code of the module functions
<component>_thread.c	The component thread code
<component>.h	Public header file
<component>_conf.c	Source code of user defined functions
<component>_conf>.h	Private header file with configuration options
<component_error.h>	Component error codes for event recorder
<Component>_<template>.c	User template files for the application code

The functional code should be placed in a set of logically named files based on the component's name. In the case of an active component, the thread code should be separated into its own file. Any #defines that are used for compile-time configuration should be placed in <component>_conf.h. The user API functions are declared in a single header file <component.h>. This file will also contain any additional declarations such as typedefs and enums that are required by the user application code. A component may have functions that can be customized, or it may require a callback to be provided by the user. In this case, the component should provide a set of stub functions that and an overall group of application develope can be expanded by the user. These outline functions should be stored in component_conf.c. The component should also provide an example template code in <component>_<template>.c that provides a starting point for the thread application code.

All of the component source files will be stored in a directory <component>\source. This is to separate each of the components while keeping the number of search paths in the project as low as possible. A component may also be compiled as a binary library file, in which case the object file and header will be stored in <component>\library.

Development Workflow

The development of a software component may be split into four stages (Fig. 14.3). This is a loose process. How each stage is conducted will depend on if you are a single developer or part of a team. If you are working on your own, it can be seen as a series of stages. If you are in a team, each stage may be developed in parallel, you may have separate developers for the drivers, different developers for each of the required components and an overall group of application developers.

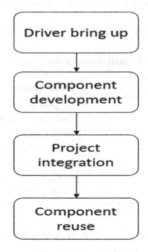

Figure 14.3
Software component lifecycle.

As discussed in Chapter 12, CMSIS-Driver, we can qualify the required CMSIS-Drivers using the driver validation framework. This can also act as a resource for board bring up when the project hardware is available. We can also provide CMSIS-Driver test stubs so that the component can be developed prior to the hardware being available. A stubbed version of the component API can also be provided to the application developers until the full component is available.

Alongside this, we can start to develop the required software components. This can be done within a test framework discussed in Chapter 13 using test-driven development techniques. The Test-Driven Development (TDD) approach can give us a high degree of confidence in the component code as it is incrementally developed within its own test harness. As the component develops, it can be integrated into the main application. If any bugs are found in the main project, we can "regress" back to the subproject to fix them and retest the whole software component to check that we have not introduced further problems.

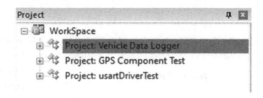

Figure 14.4
Project workspace with test subprojects.

We can use a multiproject workspace (Fig. 14.4) to manage each of these stages. This also isolates each component into its own subproject that becomes the basis for a CMSIS pack that can be added to a local server ready for reuse on a future project.

The test subproject is located in the component directory structure so that it becomes a part of the overall project. When the component is placed in a CMSIS Pack, the test project can be made available as an example project so that the test suite may be recreated when the component is reused on new hardware.

Exercise 14.1 Component GPS Interface

As an example of how to develop, integrate, and reuse a typical software component, we will create a simple GPS software component that is used to read the serial data from a GPS receiver and parse the ASCII data into usable numeric values. The software component uses the CMSIS USART driver to receive data from the GPS Module (Fig. 14.5). This is a low-cost module that is widely available.

Figure 14.5
The UBLOX NEO-6M is a low-cost GPS module that can be connected to a microcontroller through a USART.

The GPS data is sent in the form of a serial protocol called NMEA 0183 (National Marine Equipment Association). The NMEA data is sent in the form of an ASCII serial string. Each NMEA serial sentence starts with "$GP" followed by a set of letters to identify the sentence type. This header is then followed by comma-delimited data. Each sentence ends with a "*" followed by a two-digit checksum. A typical sentence is shown in Fig. 14.6.

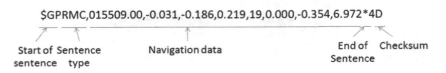

Figure 14.6
A GPS receiver will typically output an ASCII text sentence using the NMEA 0183 protocol.

All of the GPS code is in the module gps_Thread.c and gps_Parser.c. The configuration options are in gps_config.h. The user API functions are declared in gps.h.

This is an active component that creates its own thread that listens on the selected USART. It will detect the GPS header and then read the GPS location data and finally calculate the checksum to validate the sentence. Once the message is received, the ASCII data will be parsed and placed into a structure called gpsData. A set of helper functions are then used to make the location data available to the main application.

As discussed in Chapter 13, Test-Driven Development, the code has been developed using the "TDD" method so we can switch from an application project to a test subproject to further develop or examine how the code works.

In the pack installer, select Ex 14.1 and press the copy button.

The project contains a mythical "Vehicle Data Logger" project that requires location data. A subproject contains the test development framework for the GPS component (Fig. 14.7). A third project contains the driver validation framework for the CMSIS-USART.

Figure 14.7
Software component development subproject.

For testing and initial development, we can switch from the full CMSIS-USART driver to a template that can be populated with code that simulates ASCII sentences from the GPS receiver.

Open the RTE manager.

Uncheck the USART driver and then select the custom driver (Fig. 14.8).

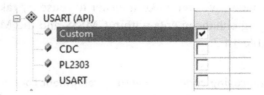

Figure 14.8
The CMSIS Driver custom option adds all the necessary include paths to the project.

This will configure the project paths to the USART_Driver.h header file.

The template is already part of our project, but each CMSIS-Driver template is stored within the pack system as shown below.

```
C:\Keil_v5\ARM\PACK\ARM\CMSIS\<version>\CMSIS\Driver\DriverTemplates
```

Open Driver_USART.c to the project test directory and add it to the project.

This now provides us with the skellington of a USART driver that can be used to provide test input to the component under test.

```
char gpsTestRMC[] = "RMC,123519,A,4807.038,N,01131.000,E,022.4,084.4,230394,003.1,W*6A";
static uint32_t index = 0;
static int32_t USART0_Receive(void *data, uint32_t num)
{
 *(char *)data = gpsTestRMC[index];
 index++;
 if(index >= (sizeof(gpsTestRMC)-1))
 {
   index = 0;
 }
 return ARM_DRIVER_OK;
}
```

Open gpsParser.c.

The gps component uses the CMSIS-USART driver function to receive a single character at a time.

Build the project and run the test cases in the debugger.

During development, the test cases will help detect bugs in the component. Once the component is ready for reuse, the test suite may be used to validate its performance in a future project or new microcontroller.

Adding Custom IDE Support

As we develop the component, we can make it easier to reuse by taking the best advantage of the annotation and debug enhancements within CMSIS and the Microvision IDE. These features are shown in Table 14.2.

Table 14.2: Improving code reuse with development tool enhancements

Feature	Description
Configuration wizard	Annotations within comment blocks to create a guide like configuration menu
Event recorder	Annotations within the code to provide runtime debug messages
Event Statistics	Annotations within the code to provide runtime execution profiling
Component viewer	Custom watch window for component data objects

Configuration Wizard

The CMSIS pack specification defines a set of Configuration Wizard annotations that may be used to convert header files into a GUI-like interface. This provides an intuitive configuration utility that makes setting up a software component easy for a third-party user or even for yourself if you return to the code in 6 months' time.

Component Viewer

As we saw in Chapter 8, Debugging With CoreSight, we can add a custom component view to the debugger. This will allow us to examine the internal data for a software component. Once you are familiar with the component viewer xml structure, a basic window is quick to produce and becomes a very useful resource, particularly during integration testing.

Event Recorder

The event recorder is used to send annotation messages to the debugger, which are expanded into meaningful messages defined in an scvd file. This system is used by the RTX RTOS and the Keil middleware. Since the event recorder uses the minimal serial wire debug features, this technique can be used with any Cortex-M processor and is also supported within the simulation models.

In addition to our software components, we should also create an error message and logging system. Each function in our component will provide a return code of the type int32_t. On successful completion, a function will return zero. Data may be returned using values greater than zero. If an error is encountered, it will return a negative value. This value can be sent using the event recorder and displayed as an expanded text message in the debugger. The structure of the event recorder ID is shown below.

Event Recorder Message ID Format

As we saw in Chapter 8, Debugging With CoreSight, the event recorder provides a message naming standard that allows us to report both error and informational messages from software components within our application. By adopting this system within our software components, we can use the filtering features built into the debugger event recorder viewer and improve the visibility of our component messages within the event viewer window.

Each event recorder message is used to send an ID value alongside application data to the debugger.

```
EventRecord2 (uint32_t id, uint32_t val1, uint32_t val2);
```

The id is a 32-bit value that is divided into three fields that are managed by a macro.

```
#define EventID(level, comp_no, msg_no)   ((level & 0x30000U) | ((comp_no & 0xFFU) << 8) |
(msg_no & 0xFFU))
```

The component number and message number are user defined. The event id level is used to define four types of message that may be selectively filtered in the debugger (Table 14.3).

Table 14.3: Event recorder ID level fields

Event Level	Description
EventLevelError	Reports an error within the component
EventlevelAPI	Reports an API call to the component
EventLevelOP	Details an internal operation within the component
EventLevelDetail	Provides additional details of the component operation

We can define a component number for the GPS component.

```
#define GPS_Evt 0x0B
```

And then, create a component ID that will be used to notify the debugger that the GPS thread has started successfully.

```
#define GPS_Init      EventID (EventLevelAPI,   GPS_Evt, 0x01)
```

Then, within the component code, we can use the encoded Event ID as part of the event recorder message to the debugger.

```
EventRecord2(GPS_Init, 0, 0);
```

In the event section of the SCVD file, we can interpret the Event recorder ID.

```
<events>
<group>
<component name = "GPS_Component" brief = "GPS" no = "0x0B" prefix = "GPS_" info = "GPS
                                                    Component example" />
</group>
<event id = "0xB00" level = "API" property = "Application read" value = "Get Latitude  = %x
                                                                        [val1]" />
<event id = "0xB01" level = "API" property = "Init" value = "Thread and USART Driver
                                                                        started" />
<event id = "0xB02" level = "API" property = "Delete" value = "Thread Deleted" />
<event id = "0xB03" level = "API" property = "Receive NMEA" value = "GGA Sentence" />
<event id = "0xB04" level = "API" property = "Processing" value = "Checksum OK"     />
<event id = "0xB05" level = "API" property = "Idle" value = ""                      />
</events>
```

Then as the component executes, we can see its progress within the event recorder window (Fig. 14.9).

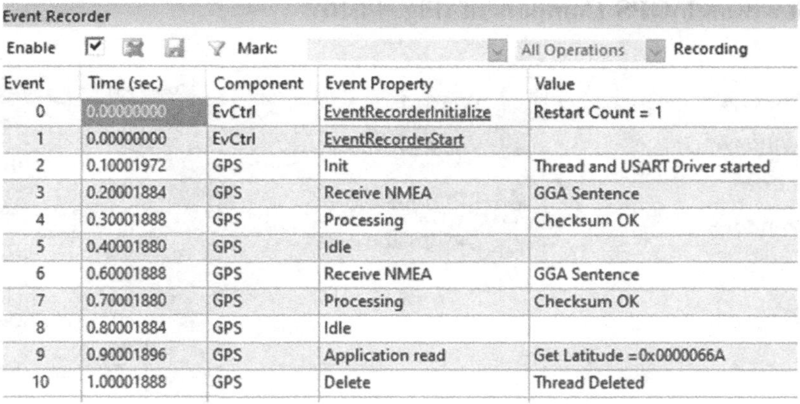

Event	Time (sec)	Component	Event Property	Value
0	0.00000000	EvCtrl	EventRecorderInitialize	Restart Count = 1
1	0.00000000	EvCtrl	EventRecorderStart	
2	0.10001972	GPS	Init	Thread and USART Driver started
3	0.20001884	GPS	Receive NMEA	GGA Sentence
4	0.30001888	GPS	Processing	Checksum OK
5	0.40001880	GPS	Idle	
6	0.60001888	GPS	Receive NMEA	GGA Sentence
7	0.70001880	GPS	Processing	Checksum OK
8	0.80001884	GPS	Idle	
9	0.90001896	GPS	Application read	Get Latitude =0x0000066A
10	1.00001888	GPS	Delete	Thread Deleted

Figure 14.9
Event recorder component API level messages.

This provides a detailed run time trace of component activity and makes execution errors very visible and easy to detect for a minimal amount of extra effort.

Component Characterization

As we develop software components for reuse, it is possible to characterize them in terms of their ROM, RAM, and peripheral usage. As we saw in Chapter 8, Debugging With CoreSight, we can also use the event recorder to obtain key performance timing values.

The event statistics macros can be used to record min, max, and average run time performance. We can also obtain power consumption by using a Ulink Plus debug adapter. While this information is useful during development, it is also a very useful guide to more accurately estimating the performance level of the microcontroller required for the next project when the component is reused.

Exercise 14.2 GPS Component

In this exercise, we will look at the custom debug files that can be added to support a reusable software component.

In the pack installer, select Ex 14.2 and press the copy button.

Switch to the application target.

Build the project.

Start the debugger.

Open the View\watch\GPS Component (Fig. 14.10).

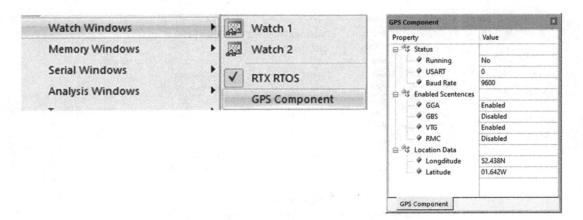

Figure 14.10
Custom component view.

This is a dedicated component viewer window that will show the current configuration, status, and data within the GPS component.

Open the View\Toolbox (Fig. 14.11).

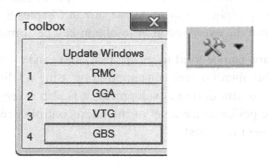

Figure 14.11
Functions in the simulation scripts can be triggered by used define buttons.

Pressing one of these buttons will send a matching NMEA sentence to the simulated microcontroller serial port. Once the NMEA sentence has been received, the parsed data will be updated in the gpsData structure. As the code runs the debug version is instrumented to send event recorder messages to the debugger.

Open the view/analysis/event recorder.

This window will display the event recorder messages sent from the component.

Open the view/analysis/event statistics.

This window displays the timing information sent by the event annotations.

Open the GPS_Characterisation.xls spreadsheet in the project directory.

The spreadsheet provides an overview of the GPS component resource requirements for future projects.

Designing a Configuration Wizard

Once the component files have been added to a project, we will need to set up any configuration options. In a typical component, the configuration options will be grouped together as a set of #defines in the <component>_conf.h file. To make the component more intuitive to use, it is possible to add some annotations within comments that allow the header file to be viewed as a configuration wizard. Any selections made in the configuration wizard will modify the associated #define. Table 14.4 provides an overview of the available tags.

Table 14.4: XML tags are used to build a configuration wizard

Template Tag	Description
<h>	Create header section
<e>	Create header section with enable
<e. i>	Create header section with enable and modify specific bit
<i>	Tooltip text
<q>	Bit values set by a text box
<o>	Text box for numeric value
<o. i>	Modify a single bit in the following #define
<o. x. y>	Text box for numeric value with range
<s>	ASCII string
<s. i>	ASCII string with character limit
<qi>, <oi> <oi. x>, <si>, <si. x>	Skip i #defines and modify the next #define

Once you are familiar with the configuration wizard annotations creating new template files is a relatively quick process.

Exercise Configuration Wizard

The configuration options for our software component are held in the header file gps_config.h. This file contains a set of #defines that are used by the main application code. This file also contains the necessary annotations to be viewed as a configuration wizard. In this example, we will read through this file to see how they work.

In the pack installer, select exercise 14.3 GPS component and press the copy button.

Open the header file gps_config.h.

First, we need to enable this file as a configuration wizard file. This is done by adding the following comment in the first 100 lines of the header file.

```
// <<< Use Configuration Wizard in Context Menu >>>
```

At the end of the file add the following comment to close the configuration wizard.

```
// <<< end of configuration section >>>
```

The closing comment is optional, but it is good practice to add it.

Now we can create some logical sections to group our configuration settings. A section header can be created with the following comment:

```
// <h> section header
```

The end of the section must be closed by

```
// </h>
```

It is also possible to create a section header with an enable option.

```
// <e> section with enable
#define ENABLE 0
```

This tag must be followed by a #define. This define will be modified to a 1 if the selection is ticked. This section must be closed by

```
</e>
```

By adding section headers, we can start to structure the include file template (Fig. 14.12).

```
// <<< Use Configuration Wizard in Context Menu >>>

// <h>NEO-6M GPS_configuration

//<h>Select NMEA Sentences

//</h>

// <h>Select Serial Interface

//</h>

//</h>
```

```
Option
⊟ NEO-6M GPS_configuration
    ⊞ Select NMEA Sentences
    ⊞ Select Serial Interface
```

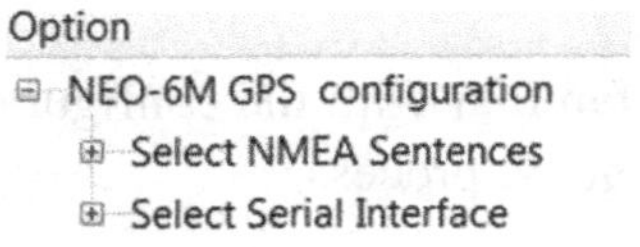

Figure 14.12

Configuration wizard tags are placed in comments. The Editor can then view the source code as a configuration wizard.

Now we can start to populate the sections with configuration options. For the NMEA sentence section, we can specify the GPS messages we want to receive. Using the// < q> tag, it is possible to create tick boxes that modify the #define that immediately follows them.

```
//<q>select option
#define OPTION 0
```

Using this tag, we can expand the sentences section with configuration options to select the NMEA messages that will be processed.

```
//<h>Select NMEA Sentences
//<q> GPGGA
#define GPS_GPGGA 1
//<q> GSA
#define GPS_GSA 0
//<q> GSV
#define GPS_GSV 0
//<q> PRMC
#define GPS_PRMC 0
//</h>
```

Next, we can extend the serial interfaces section. Here, we can create a drop-down selection box by using the following tags:

```
// <o>Selection box
//      <1 =>selection 1
//      <2 =>selection 2
//      <3 =>selection 3
//      <4 =>selection 4
#define SELECTION 1
```

Using the selection box tag we can now create selection options for the USART serial interfaces.

```
// <o>Select USART Peripheral
//      <1 =>USART 1
//      <2 =>USART 2
//      <3 =>USART 3
//      <4 =>USART 4
#define GPS_USART_PERIF 3
// <o>Select USART baud rate
//      <2400 =>2400
//      <4300 =>4300
//      <9600 =>9600
//      <34000 =>34000
#define GPS_USART_BAUD 9600
```

Once we have a functioning configuration wizard, we can provide additional assistance to the user by adding tooltips using the $<$i$>$ tag as shown in Fig. 14.13.

```
//<q>GBS
//<i>Time Position and Fix data
```

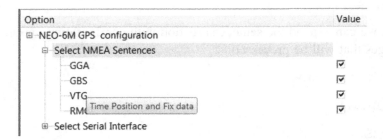

Figure 14.13
Configuration wizard with tooltip.

While the configuration wizard is an easy way to display complex configuration options, it does rely on the user having a basic understanding of the software component. If for example, the GPS unit had two types of serial interfaces, SPI and USART, which could not be used at the same time. The configuration wizard would show you the options for both but would do nothing to stop the user from selecting both. To give the user some warning, it is possible to add error messages that will be emitted during the build process. So, for example, in the case of two serial interfaces:

```
#if (GPS_SPI_EN == 1 && GPS_USART_EN == 1 )
  #error "::GPS Driver:Too many serial interfaces enabled"
#endif
```

If an end-user of the software component selects both serial interfaces, the project will fail to build, and a warning will be issued in the build output window.

Software Component Reuse with CMSIS Pack

Now that we can design a software component independent of the microcontroller hardware, we want to make it easy to use across different microcontroller families. One way to do this is to place all the component software, support, and documentation files in a custom software pack (Fig. 14.14) so that it can be installed directly into the toolchain for reuse in future projects.

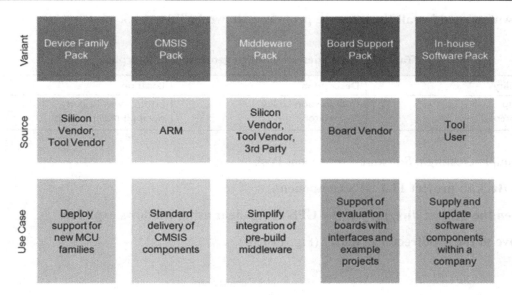

Figure 14.14
The CMSIS pack system can be used to install software components into a toolchain. This can be support for a particular microcontroller family, other use is to install middleware or board support libraries. In house software packs can also be created to reuse code.

To see how this can be done, we will create a custom "in house" CMSIS pack based on the GPS component. This will allow our code to be easily distributed between developers and reused in new projects. Finally, we will look at how to deploy custom CMSIS packs within your organization to support a team of developers.

CMSIS Pack Structure

A CMSIS software pack is simply a collection of software files that you want to add to the Run Time Environment Manager plus additional supporting files such as debugger support, documentation, license information, examples, and templates. To make a software component, you need to generate only one additional file. This is an XML "pack description file" (<pack>. PDSC), which describes the contents of the pack and rules that define any additional support. Once you have created the pack description file, a simple utility is used to check the syntax and generate the pack.

CMSIS Pack Utilities

Before we can begin to create a CMSIS Pack, it is necessary to install a number of programs to help design the pack.

Download and install the software programs shown in Table 14.5.

Table 14.5: Utilities required to generate a CMSIS pack

Utility	Description	Location
7Zip	Compression Utility	http://www.7-zip.org
Notepad++	XML editor	notepad-plus-plus.org

Example creating a Software Component Pack

Go Back to project 14.3 GPS component.

Open the project directory of the GPS component using windows explorer.

Move to the subdirectory "pack" (Fig. 14.15).

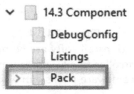

Figure 14.15
From the GPS project move into the Pack directory.

The pack directory contains all the files necessary to create a software pack (Table 14.6).

Table 14.6: CMIS pack build files

File	Description
Gen_pack.bat	Pack generation batch file
PackChk.exe	Pack syntax checker
Pack.xsd	Pack XML schema
Vendor.pack.pdsc	Pack description template

Now, we need to follow a few simple steps to create a pack. There are four steps required (Fig. 14.16). First, you need to decide what the content of the pack is going to be. In our case, we want to make a software pack for the GPS component. The first step will be to remove the code from a project and make it into a standalone component. This will typically be the "C" code modules or library and include header files. Our GPS component consists of the file shown in Table 14.7.

Table 14.7: Software component source files

File	Description
gpsThread.c	Component source code
gps.h	API header file
Gps_config.h	Library configuration file
gpsUserThread.c	Component User template

Next, we need to organize our software component and any associated documentation, examples, and template files into a sensible directory structure. Once we have all the software component files arranged, we need to create an additional XML file that contains a description of the software component contents. Then the final stage is to create the pack by running the gen_pack batch file. This checks the XML pack description against the component files. If there are no errors, the software pack will be generated.

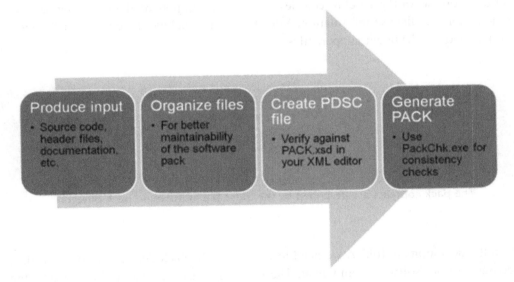

Figure 14.16
A software pack can be created in four steps. Isolate the code you want to reuse. Create a file structure for the software component. Create an XML pack description file. Run the pack generation utility.

The "files" directory contains all of the files that will be included in our pack. We can create any directory structure we want to arrange the pack files (Fig. 14.17).

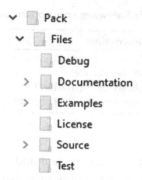

Figure 14.17
Example component directory structure.

This directory structure is completely free form, so here we have decided on a minimal layout that consists of top-level directories for the key component elements such as documentation and license information. You should also include directories for the test subproject and the debugger support files.

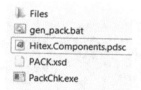

Figure 14.18
The pack contents are described by the .pdsc file in the main pack directory.

Now in the development folder, we need to customize the pack description file to reflect the contents of our software component. The first thing that we need to do is rename the template pack description file to match our vendor name and the name of the software component, in this case, Hitex.Components.pdsc (Fig. 14.18). The final pack name is based on the pack description file name and the pack version number.

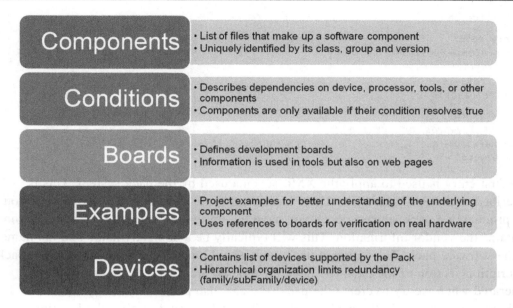

Figure 14.19
The software pack description file is an XML file. It has a number of containers that can be used to describe the contents of a software pack.

The pack description file consists of several XML containers (Fig. 14.19) that describe both the software component contents and how it should be added to the RTE. You can also specify conditions for its dependencies on other components within a project. These conditions can be used as rules that are applied within the RTE that govern how each component is added to a project.

Now open the pack description file, with notepad++ or another XML editor, and we will examine how the description file is structured.

The first section of the file provides an initial description of the file.

```xml
<?xml version = "1.0" encoding = "utf-8"?>

<package schemaVersion = "1.2" xmlns:xs = "http://www.w3.org/2001/XMLSchema-instance"
xs:noNamespaceSchemaLocation = "PACK.xsd">
  <vendor>Hitex</vendor>
  <name>Components</name>
  <description>A collection of device driver libraries</description>
  <url>www.myCompany.co.uk/repository</url>
  <supportContact>developer@myCompany.co.uk</supportContact>
  <!-- optional license file -->
  <license>
  License\License.txt
```

```
   </license>
   <releases>
    <release version = "1.0.0">
     Initial Release
    </release>
   </releases>

   <keywords>
   <!-- keywords for indexing -->
    <keyword>insert_keyword_for_search_engines</keyword>
  </keywords>
```

The first entry is used to apply the XML schema used by the pack system. This is described in the PACK.xsd file. There have been some additions to this file, so if you are producing a pack, always check you have the up-to-date version. The next section contains the vendor information. This will typically be a company name and the name of the software pack. The software pack name should match the name used in the pack description section of the pack file name. If the two are different, an error will be generated when we try to create the pack. We can also provide a description of the pack, that will appear in the pack installer. The URL entry describes the location where the pack can be located. This can be a company file server or an internet webserver. Once the pack is installed this location will be checked for an upgrade pack each time the pack installer is started. This allows you to release a new version of your pack by simply updating the pack file on your webserver, now the new version will be available to any user who has installed the original pack. There is also a section for the release number that is structured as shown in Table 14.8.

Table 14.8: Naming convention for the pack version number

Driver Release Number	Format <MAJOR>. <MINOR>. <Patch >
MAJOR	Major release, may not be backward compatible
MINOR	Minor release, backwards compatible
PATCH	Incremental release for bug fixes

If you want to distribute the pack over the internet, you can also include license information and search keywords. Here we have the path to our license directory and the license document. This will be installed alongside the component and will also be displayed when the component is installed. In the keywords section, you can provide a range of search words to make your pack more visible to search engines.

Now we can start to customize the pack description file to describe our software components. Our software component may be added to any of the pre-defined categories within the run time environment. However, our GPS component does not really fit into any existing category. Within the pack description language, there is a taxonomy section that allows us to extend the range of component categories. Here we are creating a "Components" section within the run time environment (Fig. 14.20).

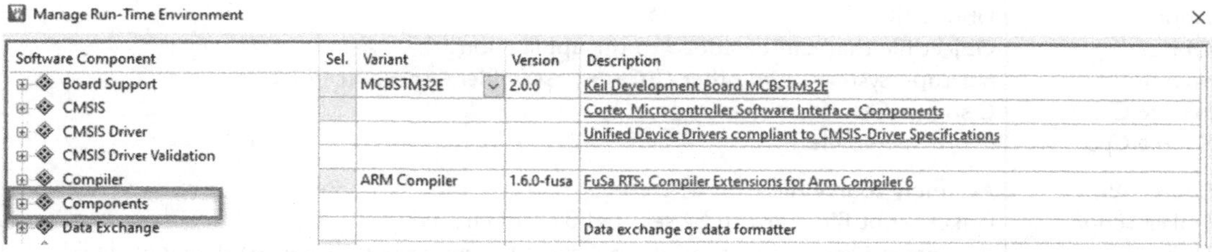

Figure 14.20
The Taxonomy tag creates a new category in the Run Time Environment Manager.

```
<taxonomy>
  <description Cclass = "Components"> Resuable Software components </description>
</taxonomy>
```

Once the pack is installed, the new category will be visible when the Run Time Environment manager is opened.

Now we can start to describe the software components. We can now create a <components> xml container and then define our GPS component within it.

```
<components>
  <component Cclass = "Components" Cgroup = "GPS Module" Csub = "Ublox NEO-6M"
  Cversion = "1.0.0">
    <description>GPS Library for Ublox NEO-6M receiver</description>
      <files>
        <file category = "source" name = "Source/gpsThread.c" />
        <file category = "header" name = "Source/gps.h" />
        <file category = "header" name = "Source/gps_config.h" attr = "config" />
      </files>
  </component>
</components>
```

First, we define its location in the run time environment. In this example, the GPS component will be added to the Components section and a subgroup called GPS. We can then provide a title and description for the component. Next, we can add the files that make up the component. This means defining their type and the path and name of the file (Table 14.9).

Table 14.9: Supported CMSIS pack file types

Category	Description
Doc	Documentation
Header	Header file used in the component. Sets an include file path
Include	Sets an include file path.
Library	Library file
Object	Object file that can be added to the application
Source	Startup-, system-, and other C/C++, assembler, etc. source files
Source C	C source file
Source Cpp	C++ source file
Source ASM	Assembly source file
Linker Script	Linker script file that can be selected by tool-chains
Utility	A command line tool that can be configured for pre- or post-processing during the build process
Image	Files of image type are marked for special processing into a File System Image embedded into the application. This category requires the *attr* being set to *template*.
Other	Other file types not covered in the list above

There is a range of supported file types (Table 14.7). For our component, we need to define our files as "header" and "source." Each file may be given an attribute that affects how the file is added to a project (Table 14.10).

Table 14.10: CMSIS pack source file attributes

Attribute	Description
config	The file is a configuration file of the component. It is expected that only configuration options are modified. The file is managed as part of the component, as a project-specific file typically copied into the component section of the project.
template	The file is used as a source code template file. It is expected to be edited and extended by the software developer. The file can be copied into a user section of the project.

In each of our components, the header file has been given the attribute config. This means that when we use the RTE Manager to add the component to our project, a copy of the header file will be copied to the project directory so that we can edit it. The source file has not been given any attribute, so it will be held as a read-only file in the pack repository within the toolchain. When the component is selected, a path will be set to include the source file. This means that if you are using a version control system, you will need to include all the packs you are using to be able to fully recreate a project. This is the minimum amount of information we need to add to the pack description file to allow a pack to be generated.

Updating and Testing the Software Component

Once you have developed a software component and generated the Pack description file, it is possible to add the unzipped pack files as a local repository. This allows the maintainer to work with the component in the RTE but still be able to edit the component files. This works well with a version control system where we can commit the full pack structure along with incremental changes.

Open the pack installer, and select file/manage local repositories (Fig. 14.21).

Select the component Pack description file and press ok.

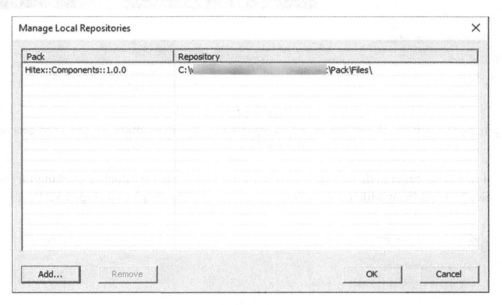

Figure 14.21
Adding a local pack repository.

Now the component will be available in the RTE as if it has been installed, but the component files will not be write-protected and can be updated as required (Fig. 14.22).

Figure 14.22
CMSIS Pack source file attributes.

Generating the Component Pack

An initial pack can be generated by opening a command-line window in the development directory and then running the "gen_pack" batch file (Fig. 14.23). This will parse the pack description file and check against the contents of the software component directories. If there are no errors, then a pack file will be created. This pack file is a compression file of the software component directories and files plus the pack description file.

Figure 14.23
The gen_pack batch file checks the pack files against the pack description file and then generates the final pack file.

Once it has been generated, we can install our GPS component by double-clicking on the pack file. If it has been added, the license agreement will be displayed (Fig. 14.24).

Figure 14.24
The pack file can be installed locally by double clicking on the pack file or by selecting file\import in the pack installer.

Now we can open the RTE manager and see the new modules section with the GPS component. If we select the GPS component, its files will be added to the project (Fig. 14.25).

Figure 14.25
Once the pack is installed the GPS component is available to be added to a new project.

While this is useful at this level, it is just a "pretty" way to add files to your project. To make our component more intelligent, we can start to add some conditions that stipulate if the GPS component requires any other components to work within a project.

```
<conditions>
        <condition id = "GPS_COND">
        <require Cclass = "CMSIS Driver" Cgroup = "USART"/>
        <require Cclass = "CMSIS" Cgroup = "CORE"/>
        <require Cclass = "CMSIS" Cgroup = "RTOS"/>
        </condition>
</conditions>
```

Within the conditions container, we can create a rule that defines the components dependencies. Here we have created a condition called "GPS_COND" that requires a CMSIS RTOS and the CMSIS Core specification to be selected along with a CMSIS USART driver.

We can now add the conditions to the component description.

```
<component Cclass = "Components" Cgroup = "GPS Module" Csub = "Ublox NEO-6M"
Cversion = "1.0.0" condition = "GPS_COND">
```

If the pack is newly regenerated and reinstalled, we can test the rule by selecting the GPS component.

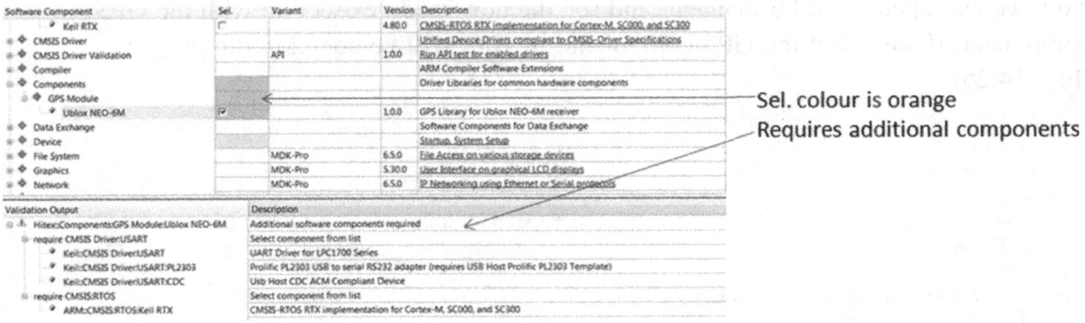

Figure 14.26
Now when the component is selected it will require a CMSIS USART driver.

When selected the GPS component is shown as orange (Fig. 14.26), meaning it requires a subcomponent The verification window tells us that we must include an RTOS and CMSIS Driver::USART in our project. Now press the Resolve button. This will add any component that does not need any further resolution. In the case of the USART there are several Options (Fig. 14.27), so we must manually select a suitable driver.

USART (API)
- CDC
- PL2303
- USART ☑

Figure 14.27
Select an available CMSIS USART Driver.

Once we have selected all the required components, all of the active elements in the select column will be colored green. Now we can click OK, and the finished component selection will be added to our project.

```
<releases>
  <release version="1.0.0">
  Initial Release with GPS driver
  </release>
</releases>
```

As we create new versions of the software pack, we can increment the version number. As we install new versions of the component, the version history may be seen in the pack installer. Once installed, each version is held within the pack repository. This allows you to maintain projects built with different versions of a software component.

```
<files>
        <file category = "source" name = "Source/gpsThread.c" version = "1.0.0" />
        <file category = "header" name = "/Source/gps.h" version = "1.0.0" />
        <file category = "header" name = "Source/gps_config.h" version = "1.0.0"
                                                          attr = "config" />
        <file category = "doc" name = "Documentation/nmea.htm" />
        <file category = "source" name = "Source/gpsUserThread.c" version = "1.0.0"
                            attr = "template" select = "GPS User Thread" />
        <file category = "other" name = "Debug/gps_component.scdv" />
</files>
```

While the component has an overall version number, it is also possible to add version tags for each source code element.

We can expand the component description by adding a documentation link. This uses the doc file category and provides a link to a description of the NMEA protocol.

Figure 14.28
The Description entry is now a hyperlink to the component documentation.

When the pack is regenerated and reinstalled, the documentation will be available through a hyperlink (Fig. 14.28).

It is also possible to add a template file. This is a fragment of code that provides a typical pattern that shows how the software component is used. The file is described as a source file but has a template attribute. We can also provide the selection criteria to be shown within the file manager.

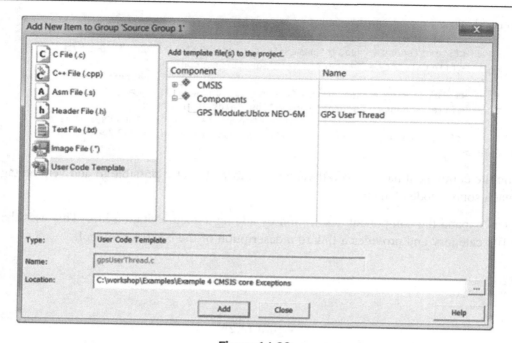

Figure 14.29
Template code is available in the "Add New Item" Dialog.

If we add the GPS component to a project, its template will be available when we add a
new source file to the project (Fig. 14.29).

Finally, by adding the component viewer SCVD file as type "other," we can ensure it is
automatically added to the debugger when the component is selected.

Autogenerated Header Files

In addition to the user-provided source code, the RTE will autogenerate a set of header files
(Table 14.11) which contain a set of #defines that reflect the overall RTE configuration.

Table 14.11: Autogenerated run time environment header files

Include File	Scope
RTE_Components.h	Must be added to user source files
Pre_Include_Global.h	Compiled with all modules in the project
Pre_Include_ < Cclass > _ < component>.h	Compiled with all modules in a given component

Each of these files is placed in the <project> /RTE/_debug directory.

The RTE_Components.h header provides #defines for the active components within the project as well as a standard define for the device header file.

```
#define CMSIS_device_header "stm32f7xx.h"
/* Keil.ARM Compiler::Compiler:Event Recorder:DAP:1.5.0 */
#define RTE_Compiler_EventRecorder
#define RTE_Compiler_EventRecorder_DAP
```

Within our software pack, we can add to this header file by defining an RTE_Components_h section located within the component container.

```
<component Cclass = " Components " Cgroup = "GPS Module">
<RTE_Components_h>

    #define RTE_COMPONENTS_GPS

</RTE_Components_h>
</component>
```

It is also possible to dynamically create additional include preinclude files that are passed to the compiler directly as part of its invocation switches.

We can specify the contents of each header file with additional tags located within the component container.

```
< component Cclass = " Components " Cgroup = "GPS Module" >
        <Pre_Include_Global_h>
        // enabling global pre include
        #define GLOBAL_Component_GPS 0x4
        </Pre_Include_Global_h>
</component>
```

And for the local preinclude file

```
< component Cclass = " Components " Cgroup = "GPS Module">
        <Pre_Include_Local_Component_h>
        // enabling local pre include
        #define Local_Component_GPS 1
        </Pre_Include_Local_Component_h>
</component>
```

While these files can be useful in automatically creating project definitions. However, they are a bit hidden from view, so a bit of caution needs to be applied when using them.

Adding Example Projects

CMSIS Pack also makes it possible to include example projects to demonstrate the software component features. The example project must be placed in a suitable folder within the software component file structure. Then we can add an entry to the pack description file to make it available through the pack installer.

```
<example name = "GPS component configuration Example" folder = "Examples/Configuration"
doc = "Abstract.txt" version = "1.0">
    <description>This example demonstrates how to configure the GPS component</description>
    <board vendor = "Keil" name = "MCB1700"/>
    <project>
     <environment name = "uv" load = "configuration.uvprojx"/>
    </project>
    <attributes>
     <component Cclass = "CMSIS" Cgroup = "CORE"/>
     <component Cclass = "Device" Cgroup = "Startup"/>
    </attributes>
  </example>
```

The pack template contains an <example> container that allows you to add any number of example projects. Each example is described as shown above. The example name and location of the example folder are defined along with a documentation file. Next, a description of the example is added. This will appear in the pack installer example description column. To complete the example description, we provide information about the board vendor, the project development IDE being used, and the project name to load. In the attributes section, we can provide details of any additional components that need to be loaded to allow the project to be rebuilt.

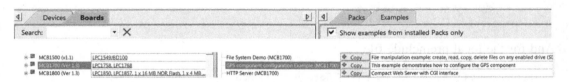

Figure 14.30
Component examples are available through the pack installer.

Now when the pack is installed, our examples will be visible in the pack installer examples tab (Fig. 14.30).

The component test and validation project can also be added as an example so that it can easily be added as a subproject at the start of a design.

Deploying Software Components

Once a software pack has been created, you will need a way to distribute it to your developers. As we have seen, you can copy the pack file to a new PC and then simply double click on the file to install it locally. You can also publish the pack to a team by installing it on a fileserver or intranet. In this case, the location that the pack is installed at should match the URL defined in the pack. Once a user has downloaded and installed your pack, the URL location will be checked when the pack installer is started. If a new version is found, the pack installer will show that a new version is available. This allows you to maintain and update the pack, and any updates are pushed out to your users. If you want the pack to be available to download by anyone, you can simply place it on a webserver to make it publicly available. This is a great way to share software components, and it allows users to add useful code to their projects from a variety of sources.

Currently, the pack installer will automatically check the Keil website for the current list of public and third-party packs. You can view the current listing here:

http://www.keil.com/dd2/pack/

Any new packs added to this listing will automatically be displayed within the pack installer. You can submit your pack to be validated, and then it can be added to this list, making it visible to any tool using the pack system. The pack can be hosted in two ways. Once the pack is validated, it can be listed and stored on the Keil website. This is the easiest way but is harder to maintain. The second method is to list the pack on the Keil website so it can be discovered, but the pack can be hosted on your own server. This way, you can make your pack visible to the entire user base but still be able to maintain and update the pack on your own local server.

Conclusion

In this chapter, we have looked at an important step forward for microcontroller-based embedded systems. Together CMSIS Driver and CMSIS Pack for the first time provide a standardized API for common peripherals and a means of making and distributing software components. At the time of writing, Arm has published a pack plugin for the Eclipse IDE which has been incorporated into a number of commercial and open-source tools.

CHAPTER 15

MCU Software Architecture

Introduction

In this final chapter, I would like to consider an approach to a general RTOS-based software architecture for small microcontroller projects. What is presented here is not intended to be an all-encompassing design system but rather a set of ideas that you can take from and develop your own approach that best suits your needs and programming style.

In any software project, our main concern has always been to deliver the required functionality. However, this is really only part of the equation. Our software design must also consider other important aspects such as future extension, maintenance, testability, and reuse of existing software. Viewed from this angle the key way to provide these less tangible aspects is through a good software architecture. Also, with the huge boost in performance and memory resources from a typical Cortex-M microcontroller, we are no longer fighting for every byte and processor cycle so we can indulge in some code overhead. From experience, it is also common to find that there is no such thing as a simple project. Whatever starts out looking straightforward often gets a lot more complicated as the project moves forward or the goal posts get moved. Therefore we need a design approach that allows us to easily adapt and extend our project code. In the lifetime of our project, we may also need to migrate from one microcontroller to a different device, possibly from a completely different manufacturer. In more recent times this has been because of an acute shortage of semiconductor devices. Other reasons may be to cost reduce the hardware or move to a device with higher performance that is a better fit for our final application.

Our small embedded system is broadly speaking made up of three elements: hardware, firmware, and software. Here, firmware is defined as the code that interacts directly with the underlying hardware. It may seem an odd statement for an embedded systems developer but we really need to minimize the amount of firmware in our projects. Today, many projects make no attempt to do this, and as a result, the whole project consists of firmware that is intimately tied to the underlying hardware. The code in these projects simply rusts, it becomes hard to maintain and even harder to move to a new hardware platform. In short,

The Designer's Guide to the Cortex-M Processor Family.
DOI: https://doi.org/10.1016/B978-0-323-85494-8.00017-6

we need less firmware and more software, software that doesn't even know it is running on an embedded system.

If you are used to writing code as a superloop plus interrupts, then moving to use an RTOS means learning not just the API but also how to use the RTOS objects successfully in an application. This can be daunting at first but after a few projects, you will feel lost without the framework provided by an RTOS. Moving to a Cortex-M microcontroller and adopting an RTOS is also an opportunity to improve the structure of your code and overall workflow.

Through this chapter, I want to bring together many of the topics introduced earlier in this book to outline a general-purpose microcontroller software architecture that can be used for a wide range of applications. This architecture aims to improve both the quality of the final code and your overall productivity, two of the key drivers in commercial software development.

Software Architecture for Microcontrollers

There are a number of software architectures that are currently used on a small microcontroller, below is a review of the most widely used frameworks.

Superloop

The traditional software architecture for a microcontroller is a superloop plus interrupts. While this is still a perfectly valid approach it will start to reach its limitations as the size and complexity of your codebase increases.

Time-Triggered Architecture

Like an RTOS a Time-Triggered Architecture (TTA) divides the application into a set of tasks that are managed by a scheduler. However, unlike an RTOS the scheduler will execute each task in a predetermined order and schedule. In such a system there is a single interrupt that provides a timing tick to the system scheduler. A TTA is often used for safety systems because it has a very deterministic behavior.

Event-Triggered Architecture

In contrast to a TTA an Event-Triggered Architecture (ETA) will schedule software functions in response to real-world events. Such systems are often based on one or more state machines that manage the system events and maintain the system context between different events. While an ETA can be used stand-alone as the overall system scheduler or may be placed within an RTOS thread to act as a coscheduler. The coscheduler approach can be used to provide real-time performance while integrating with existing components, middleware, and drivers.

RTOS

In this book, we have focused on the use of a Real-Time Operating System as this represents the mainstream framework used in place of a superloop architecture. An RTOS is suitable for a very wide range of applications and is often required by middleware and software libraries. When selecting a general-purpose framework an RTOS provides the best solution as it can implement most other approaches. By using software timers an RTOS can implement a TTA. We can also implement an ETA either directly using the RTOS scheduler or by placing a state machine in a thread to act as a coscheduler.

Consequently, in the next section of this chapter, we will look at developing a general-purpose software architecture based on an RTOS framework.

Objectives of our Architecture

When we are considering how to define a software architecture and how to develop applications based on an RTOS we first need to think of the key objectives that need to be realized for our system to be successful.

Requirements Capture

An outline software architecture will act as a template against which we can analyze our project requirements and decompose the overall design problem into the architecture framework.

Modular Design

A modular design approach makes it much easier to navigate a project and update code. In a software team, there will be less need to merge code. Also, modern compilers will take advantage of a multicore PC to build several modules in parallel resulting in a faster build time.

Code Reuse

One of the major benefits of a standard architecture is the ability to reuse code on future projects. This is also an important aspect of modular component design discussed in Chapter 14, Software Components.

Testing

As we saw in Chapter 13, Test-Driven Development, software testing should become an integral part of software development, particularly for software components. When a component is reused, its test project is used to validate the component on new hardware. A structured architecture also allows you to start testing early in the project before the target hardware is available by using CPU simulators. This can also be useful for very resource-constrained devices, in simulation we can provide additional memory not available on the real device which can be used to host the test framework.

Early Software Development

A standard driver layer allows the creation of stub functions that promote development within simulators or on low-cost evaluation boards. This allows a software team to proceed before the hardware is available.

Improved Workflow

A well-structured software architecture will improve collaboration between hardware and software teams and improve workflow within software teams.

Maintenance and Extension

We also need an architecture that is amenable to the addition of new features during and after the main development process. It is fairly common that the full scope of an application isn't known at the outset and there is often a process of "project discovery" which requires new features to be added to the application, to put it politely.

Portability

As mentioned in the introduction, our project code will ideally be easy to port to a new device so we can take advantage of lower-cost hardware and mitigate against component shortages.

Increased Productivity and Quality

The reuse of code has an obvious benefit to productivity. A well-designed software component with unit test harness and documentation increases overall project quality and reliability.

RTOS-Layered Architecture

Our general-purpose architecture needs to make the best use of the existing CMSIS standards, particularly for code portability and reuse. By using CMSIS-RTOS2, we can define a layered model that consists of an application layer, RTOS layer, service layer, and a CMSIS-driver layer (Fig. 15.1). In addition, for most projects, we will need to include a bootloader as a means of validating the integrity of the application code at startup and installing new firmware updates.

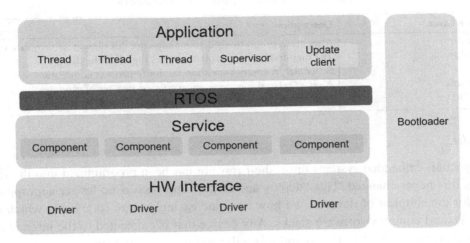

Figure 15.1

The RTOS-Layered Architecture with additional bootloader.

In many projects, although the C code is well written, it is common to find C modules that contain a mixture of functions from different parts of the project mixed with low-level configuration code and interrupt service routines. This makes testing the project difficult and code reuse almost impossible.

In this system, the layout in terms of the C modules is as important as the application's functionality. By separating each layer and siloing each component within a layer into separate modules, we make reuse and testing very much easier.

Our layer structure allows us to develop and test components in each layer in isolation through the use of subprojects. In this approach, each driver and software component are developed in its own self-contained project within a test harness. Once we are happy with the functionality of the component in its development subproject, it can be integrated into the main application. This approach also supports the re-validation of a driver or component on new hardware through the use of its existing test framework.

Each of our components is siloed in its own set of modules which do not contain any functions or data from another component. This allows a developer to work on one area of a project and make commits to the repository without having to merge code with any other team member.

Each of the layers in the system is described in Table 15.1.

Table 15.1: RTOS architecture framework

Software Layer	Description
Bootloader	A separate program used to manage the installation of firmware updates
Application	The high-level threads and "business logic" of the design
RTOS	Scheduler, memory allocation, and interprocess communication
Component	A set of related functions encapsulated in a set of modules
CMSIS Driver	Device-independent low-level peripheral drivers

Bootloader

In many small embedded systems once, the firmware has been programmed into the flash memory it is never changed. This "deploy and forget" approach is no longer appropriate except for the simplest of devices. We now have increasingly large code bases which may contain several complex software stacks. Any device that is connected to the internet must have an appropriate security model and this will need updating as new threats and vulnerabilities are discovered. Many microcontrollers contain a first-stage bootloader that is located in the microcontroller ROM. This provides a way of programming the Flash memory through a local port typically a USART with a simple serial download. In more complex designs, we will need to provide an update client in the application code that is used to download a new image into a temporary storage slot. Then, a second stage bootloader is used to validate the stored image and then copy it to the execution region of the flash before restarting the application using the new firmware version. Creating such a bootloader system is a project in its own right, fortunately, there is an open-source project called MCUboot which provides a sophisticated bootloader that has been designed for use by small microcontrollers. The project is now managed as a community project by Linaro, a nonprofit company that works on Arm-based open-source projects, and has incorporated CMSIS Flash drivers to make it easy to port between different Cortex-M-based microcontrollers.

Firmware Driver Layer

As we saw in Chapter 12, CMSIS-Driver, the CMSIS-driver specification defined a standardized API for a range of peripherals that are common across different microcontroller families. While the CMSIS-driver standard was originally intended to provide a convenient porting layer for third-party middleware libraries, it has become a fast and easy way to bring up supported peripherals in the early stages of a project. CMSIS-drivers provide an ideal abstraction layer between our software component and the microcontroller hardware. For most applications, they provide a sufficient level of performance. The peripheral interrupt is installed as a call-back function which may be part of the software component. This means that any custom interrupt code will remain part of the service layer within the software component. Best of all they are developed and maintained by the Silicon Vendor. If the CMSIS-drivers do not provide the performance or features set you require, there isn't anything to stop you from adding additional functionality to the base driver.

The main drawback of using CMSIS drivers is that only a limited range, of mainly communication peripherals, are supported. In order to adopt CMSIS drivers as our driver layer, we will have to define and develop custom profiles to cover additional peripherals such as an ADC or hardware timer as we saw in Chapter 12, CMSIS-Driver. To develop a custom driver, we have to define a CMSIS-style API, test framework, driver template, and documentation. However, once this work is done, it can be reused on all future projects. This collateral also helps develop drivers faster, since creating a new driver from a template with an existing test framework is faster than creating one from scratch.

Service Layer

The service layer is designed using software components which are constructed as discussed in Chapter 14, Software Components. The objective here is to ensure that the component is in a set of self-contained modules that are fully isolated from both the driver layer and the application layer. This ensures that the software component is "pure" C code and is fully independent of the microcontroller hardware. Since most of our application code will be part of the service layer, we want to make it as easy to test and reuse as possible.

RTOS Layer

The RTOS layer is another reusable component that provides a general-purpose framework within which to develop most types of applications. The use of an RTOS provides scheduling, memory management, and commonly used primitives (delays, signals, message queues, etc.). The RTOS layer is recommended to use the standardized CMSIS-RTOS2 API. This separates our application code from a specific vendor's RTOS. For example, this allows us to select between a commercial-grade RTOS and a safety-certified RTOS depending on the project requirements.

Application Layer

The application layer consists of RTOS threads that are pure software. The application layer threads provide the high-level functionality for the application and will ideally consist of function calls to the service layer and their own business logic. The application threads are defined as the "units of concurrency" within the application specification. This means that each thread represents a function of the application that acts in parallel with other features.

For example, a thread containing a motor control algorithm will operate in parallel with a thread managing a User interface. The thread code must also be organized in its own self-contained modules. Ideally, we should also aim to create some reusable threads which provide common project requirements.

Supervisor Thread

An example of a reusable thread is a supervisor thread. Typically the supervisor will be the first thread to start and is used to create the application threads and other RTOS objects. Once the application is running the supervisor is used to manage system-level features such as error logging and reporting, power management, and providing an update client to manage the download of firmware updates. Once you have a collection of reusable threads, they can be placed in a CMSIS pack for easy reuse.

System Header File

We can also create a system header file. This can be included in every module in the project. As its name implies it contains all of the system-level definitions within the project. The idea is to gather these elements into one file so that they can easily be reviewed and changed as necessary (Table 15.2). This can also be developed as a template file that can be reused across projects.

Table 15.2: System header file

RTOS definitions	Thread priorities
	Thread stack allocation
	RTOS timer
	Message queue fifo size
Interrupt definitions	IRQ priority group
	IRQ priority
Project definitions	System switches
	Enumerated types

Finally, in order to make the code easily testable, the main() entry point must be in a separate module that just contains the minimal code to configure the key system peripherals and start the RTOS.

Design Synthesis

Now that we have an architectural framework it can help us decompose a requirements specification into the elements required for each layer. During the initial design meeting, we can start with the overall requirements and "three pieces of paper" (Fig. 15.2). The first form is used to list the peripherals and associated CMSIS drivers required for the design. Here we can also list the ideal feature set for each peripheral. The second form is used to list the software components required by the project. Finally, the third form lists the application threads and the data flow between them.

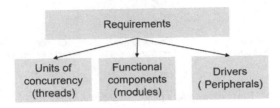

Figure 15.2

Decomposing the requirements into the framework functional layers.

From here we can start to refine the design with a top-down approach focusing on the threads and how they interact. Here the temptation may be to define a large number of threads each providing a small unit of functionality. While this has the advantage of providing good modular design and encapsulation of data, it has the disadvantage of more frequent task switching, higher memory usage more use of RTOS calls. At the initial design stage, you should try to minimize the number of application threads and only add additional threads where absolutely necessary.

A well-defined application layer will guide us to the service layer components. A firm idea of how the threads operate will be a good guide to what functions are necessary for the service layer and how they can be divided into separate software components. Now, we can define a public API for each component and "publish" it to the application layer.

Each component API should be considered a contract between the application threads and the service layer components. This means that in addition to providing the correct functionality the published API will not be modified as this will break existing code. However, it is open for extension to provide new features.

The driver layer is well defined by the CMSIS driver specification except for the functionality of the interrupt service routine which will be part of the components in the service layer.

Implementation

We have looked at implementing both the service as software components in Chapter 14, Software Components, and the driver layer as CMSIS-Drivers in Chapter 12, CMSIS-

Driver. Each of these layers can be developed and tested in its own subproject before integration into the main application project (Fig. 15.3).

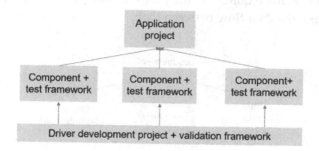

Figure 15.3
Build and test each layer in subprojects before integration into the main application.

Each subproject provides an isolated environment to develop and test the functionality of a given driver or component before it is integrated into the main project. In the case of a software component, the subproject can be used to validate the component on new hardware before it is reused in a later project.

Designing the Application Layer

The application layer will consist solely of the threads identified in the initial design synthesis. The initial application layer design can be drawn as a data flow diagram that identifies each thread and defines the data flow through the system (Fig. 15.4).

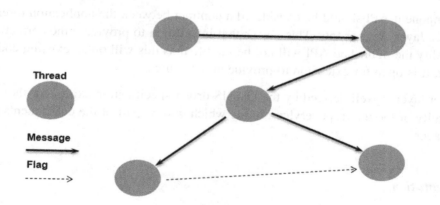

Figure 15.4
Design the application layer as thread objects linked by message queues and event/thread flags.

The code in each thread will consist mainly of calls to software components along with the necessary business logic to meet the functional requirements. Ideally, the code in each thread will read like a high-level description of the project requirements.

When designing the system try to favor the use of message queues, event flags, and thread flags. Message queues in particular act as data pumps between threads that provide both data transfer and buffering. This results in a clean design that is easy to test and reuse on future projects.

Also, try to avoid the use of mutex and semaphores. Or at least don't design a system based on their use. As you work on the design add mutex and semaphores as the need emerges. These objects are accessed from different areas of the project and should be encapsulated in helper functions to minimize the chance of mistakes.

As we saw in Chapter 11, RTOS Techniques, we can manage peripheral interrupts by signaling a high-priority thread that provides the interrupt service code. Try to stick to this approach where possible, ideally, the IRQ thread will be created by a software component.

When you first start designing with an RTOS, it is generally best to create all the RTOS objects at startup and let them run for the lifetime of the code execution. Try to avoid creating and destroying RTOS objects as the code runs, this may possibly lead to unexpected side effects and fragmentation of the RTOS memory pool.

Assigning Thread Priorities

Once we have an outline description of the application threads and how they interwork with each other, we will need to consider how to set the priorities for each thread. In a relatively simple system with a limited number of threads and few interrupts, you can start by setting all of the noninterrupt servicing threads to the same priority and let them run in a round-robin fashion. The interrupt servicing threads can then be assigned higher priorities that reflect the importance of the peripheral interrupt. In a more complex system, we can schedule threads using a system called "Rate Monotonic Analysis" (RMA). In this approach, each noninterrupt thread is assumed to run with a fixed period (T) and a fixed execution time (C) (Fig. 15.5).

Figure 15.5
Synchronous threads are defined with an estimated period and execution time.

Now, we assign thread priorities based on the execution period of each thread. The shorter the period the higher the priority.

In the case of asynchronous threads, we can assume the worst case execution and then assign a matching priority. Asynchronous threads that service an interrupt are a bit different. Here, the thread must have a high priority so that it services the IRQ quickly even though its execution period is quite long compared to the other application threads.

Will it Schedule?

Once we have a prototype set of threads with an estimated set of execution parameters, we can make an initial analysis of their run time behavior and the ability of the system to schedule successfully. RMA is a real-time scheduling theory that can be used to determine if a set of priority-based concurrent threads can be scheduled and meet their deadlines. To start with we will make the following assumptions. All threads are running independently and do not share resources. Each thread runs periodically and must meet its deadline in all cases. Each thread is assigned a priority based on its periodic execution time. The shorter the period the higher the priority of the Thread. We also assume that the context switch time in the RTOS is negligible.

Utilization Bound Theorem

Within RMA the first theorem, we can use is called the Utilization Bound Theorem (UBT). The UBT allows us to make a quick first approximation to determine if the system can be scheduled. As its name suggests the UBT defines an upper bound of the CPU Utilization which depends on the overall number of threads. For a thread that has a period of T and an execution time of C, we can define its utilization of the CPU as C/T. For a given number of threads, we can also calculate the theoretical maximum utilization of the CPU. For a system to schedule the sum of the thread execution utilization factors must not exceed this maximum. The theoretical maximum utilization is given by

$$n\left(2^{1/n} - 1\right) = U_{max}(n)$$

While the utilization for a set of threads is given by

$$\Sigma C_n / T_n = U(n)$$

C = Execution Time
T = period of the thread
N = number of threads
U(n) = Utilization factor

For a system with three threads the maximum CPU utilization is

$$U(3) = n\left(2^{1/n} - 1\right) = 3\left(2^{1/3} - 1\right) = 0.779$$

If we have three threads that have the execution and periods as shown in Table 15.3, we can calculate the sum of their utilization factors. For the system to schedule it must be less than U(3), that is, 0.779.

Table 15.3: Estimating the thread utilization factor provides a quick first approximation guide to scheduling the application threads

Thread	Execution Time C	Period of Thread T	Priority	Execution Utilization Factor
t_1	20	100	High	0.2
t_2	30	150	Medium	0.2
t_3	90	200	Low	0.45
Total CPU Utilization factor must be less than 0.779 to schedule				0.85

In this first pass analysis, we can see that our system may not schedule successfully on the CPU. However, the UBT is fairly pessimistic so if our threads exceed the upper bound, we can take a closer look to see if they can be scheduled using a more exact approach.

Completion Time Theorem

In the example shown in Table 15.3, the three threads have a combined utilization factor of 0.85 which exceeds the maximum CPU utilization of 0.779 so it looks like they will not schedule and meet their deadlines. However, our second RMA Theorem, the completion time theorem gives us an exact test to see if all the periodic threads in our system can be scheduled and meet their deadline. We start by assuming the worst case where all of the threads are scheduled simultaneously and if each thread can meet its deadline, then we can be sure that each thread will meet its deadline for any combination of start times.

A timing diagram of the thread execution is shown in Fig. 15.6. The first two threads will start in order of priority and run for their execution time before starting the lowest priority thread t_3. After 100 ms we reach the first scheduling point P1 where t_1 will again be scheduled. It will preempt t_3 and run for 20 ms before allowing t_3 to run again for another 30 ms. At the second scheduling point P2 thread t_2 is again scheduled and will run for 30 ms before allowing t3 to be rescheduled and run for its remaining 10 ms. By the time we have reached the final scheduling point, all three threads have run and met their deadlines.

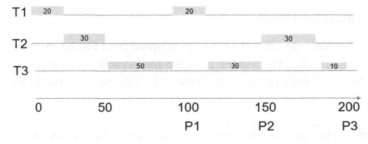

Figure 15.6
The completion time theorem proves that the threads from Table 15.3 can be scheduled.

Asynchronous Threads

In the previous example, we have assumed that each thread is periodically scheduled and has a fixed priority. In most real systems we will also have asynchronous aperiodic threads and threads that are triggered by Interrupt functions. In order to calculate a utilization factor for these threads, we take the minimum period and calculate a worst case utilization factor. We can assign a priority to the thread based on its minimum period and then treat it as a periodic thread for our scheduling calculations. However, IRQ-driven threads need a high priority in order to service the interrupt quickly. In most cases, this will result in them having a higher priority than the priority determined by their period, and we will see how to deal with this in the next section.

Scheduling a Real-Time System

Now, we can look at scheduling a more realistic system.[1] In a real-time design with both periodic and asynchronously scheduled threads, our worst case utilization is given as.

Worst case utilization = Execution Utilization + Preemption Utilization + Blocking Utilization

We have already seen how to calculate the execution utilization. The other utilization factors are calculated as follows.

Preemption Utilization

The preemption utilization is the time lost to the execution of threads with a higher priority. This form of preemption is split into two different cases.

- High-Priority Thread preemption

In the first case, our active thread may be preempted by threads with a higher priority and a shorter period. In this case, we have to add the execution utilization factor for each preemption thread to the active threads utilization factor. We can term this factor P_h.

- High-priority Interrupt Threads

The second case is where an active thread is preempted by a thread that has a higher priority but a longer period. This is often the case for asynchronous interrupt threads. In this case, the preemption will only occur once in the active threads period so we can calculate

---

[1] The examples in this section are taken from *Software Design methods for Concurrent and Real Time Systems* by Hassan Gomaa.

the asynchronous threads utilization as its execution time divided by the period of the active thread. This is then added to the active thread's worse-case utilization factor. We can term this factor P_a.

Blocking Time

Blocking time is calculated as the worst case execution time of the low-priority threads that can block a high-priority thread, this can often happen through accessing a mutual resource such as a semaphore or Mutex. Only one low-priority thread can ever block a high-priority thread because the high-priority thread will always acquire the resource as soon as it is released. To calculate the blocking utilization factor we take the low-priority blocking thread with the longest execution time and calculate its utilization as its execution time divided by the period of the active thread.

To see how this works we can consider an RTOS program with an Interrupt servicing Thread t_a and three threads $t_1 - t_3$ each access a shared resource controlled by a Mutex with the robust attribute bit set. In this case, any thread holding the mutex will be promoted to the highest priority and run until it releases the mutex. This means that a low-priority Thread can block a higher priority Thread while in its critical Mutex protected section. The Thread information is shown in Table 15.4.

Table 15.4: Example threads from a typical real-time system

Thread	Type	Execution Time C_n	Period T_n	Priority
t_1	Periodic	20	100	High
t_2	Aperiodic P_h	15	150	Above Normal
t_a	Interrupt P_a	4	200	Real time
t_3	Periodic	30	300	Normal

The Theoretical maximum Utilization bound for four Threads given by

$$U(4) = n\left(2^{1/n} - 1\right) = 0.69$$

To check if the system can be scheduled, we have to analyze each thread individually starting with the highest priority Thread and ensure that their utilization factor does not exceed the upper bound of 0.69.

The interrupt thread t_a runs at the highest priority and is not blocked by access to resources by any low-priority threads. This means that its Execution Utilization factor accounts for its full run time (Table 15.5).

Table 15.5: Interrupt thread t_a utilization

Utilization	Calculation		Value
$U_E()$	C_a/T_a	4/200	0.02
$U_{Ph}()$	NA	0	0.00
$U_{Pa}()$	NA	0	0.00
$U_B()$	NA	0	0.00
Total			0.02

Thread t_1 is the next highest priority. It is preempted by t_a which has a longer period but a higher priority. Because it also uses the mutex Thread, t_1 can also be blocked you the lower priority Threads. For a worst case calculation, we will assume that we get blocked by Thread t_3 (Table 15.6).

Table 15.6: Interrupt thread t_1 utilization

Utilization	Calculation		Value
$U_E()$	C_1/T_1	0.2	0.20
$U_{Ph}()$	Na	0	0.00
$U_{Pa}()$	C_a/T_1	4/100	0.04
$U_B()$	C_3/T_1	30/100	0.30
Total			0.54

Thread t_2 may be preempted by Thread t_1 that has a higher priority and shorter period. Thread t_2 can also be preempted by Thread t_a and blocked by thread t_3 in the same way as Thread t_1 (Table 15.7).

Table 15.7: Interrupt thread t_2 utilization

Utilization	Calculation		Value
$U_E()$	C_2/T_2	15/150	0.10
$U_{Ph}()$	$U_e(t_1)$	0.2	0.20
$U_{Pa}()$	C_a/T_2	4/150	0.03
$U_B()$	C_3/T_2	30/150	0.20
Total			0.53

Finally, for Thread 3, we can calculate its execution utilization time. As Thread 3 has the lowest priority and the longest period all the other threads will preempt it as it runs (Table 15.8).

Table 15.8: Interrupt thread t_3 utilization

Utilization	Calculation		Value
$U_E()$	C_3/T_3	15/300	0.10
$U_{Ph}()$	$Ue(t_1) + Ue(t_2) + Ue(t_a)$	$0.2 + 0.1 + 0.02$	0.32
$U_{Pa}()$	NA	0	0.00
$U_B()$	NA	0	0.00
Total			0.42

From these calculations, we can see that all four threads each have a Utilization factor below the upper bound of 0.69 and therefore we will be able to schedule all of them successfully.

While there is a certain amount of guesswork about the runtime of each thread doing these calculations at the start of a project does provide a first approximation guide to the performance of the project threads. As we develop the code, we can use the tools within the debugger to measure the execution time of each thread to ensure that it does not exceed its time budget (Fig. 15.7).

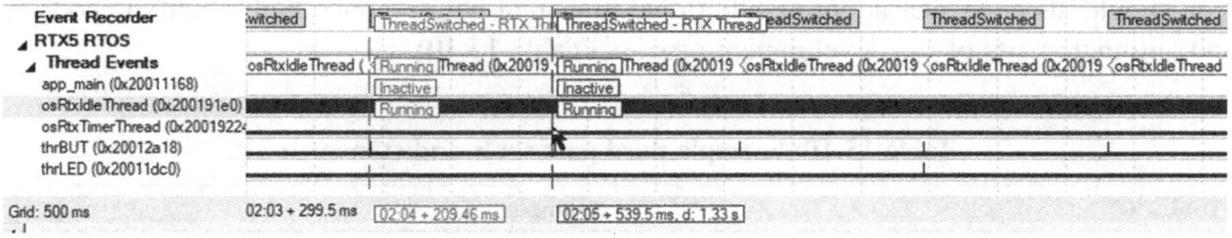

Figure 15.7
The debugger system analyzer allows you to analyze the Thread and interrupt execution times as a timing diagram.

Component Characterization

Once a project has a completed set of software components, we can characterize them to provide some performance benchmarks which can be used as a reference prior to integrating them into a new project. As each component is designed in a subproject it is possible to determine its ROM and RAM usage. We can also list the required range of peripherals and any meaningful timing characteristics. At the start of the project, we can build a table of our components and their characteristics as shown in Table 15.9.

Table 15.9: As we use components in a project they can be characterized, and these values used when planning a new project

Component	ROM	RAM	Execution (ms)	Period (ms)	Utilization	Peripherals	RTOS Thread	Priority
GPS	4078	1452	5	1000	0.005	USART	Yes	osPriorityNormal
Keyboard Scanner	500	50	1	30	0.033	GPIO Timer	Yes	osPriorityHigh
Display	2039	200	2	40	0.05	SPI	Yes	osPriorityNormal
Total	6617	1702			0.08			

We can use this as a means of estimating the minimum requirements for project MCU. Where a component doesn't exist, we can make a best-case estimation. Here, each member of the development team can make an estimate. We can then take an average of the overall values and also make a note of the worst case estimate. Then at the end of the project, we can compare the final real component values with the estimate to help improve the forecasting process.

Additional Tools

You should also consider adopting additional tools that help produce high-quality C code and automate parts of the development process (Table 15.10).

Table 15.10: Example third part tools and standards

Tool	Example
Coding Standard	MISRA-C, BARR Code
Static Analyzer	PC-Lint
Software Metrics	Source Monitor
Documentation Generator	Doxygen
Unit Testing	Tessy
Version Control	Subversion, Git Hub

Coding Standard

As we saw in Chapter 4, Common Microcontroller Software Interface Standard, the CMSIS core standard is written using the MISRA C coding standard. This is a set of rules that try to eliminate common coding errors and clarify gray areas of the C standard. The BARR Code standard also provides a concise style guide.

Static Checker

A static analysis tool should be part of every programmer's toolkit. While the compiler and linker provide errors and warnings a static analysis tool will do a project-wide analysis of both syntax and grammar to provide warnings of suspected bugs. The results of a static analysis tool should also be used during code reviews as it will help spot issues that a human would otherwise miss.

Metrics

Alongside static analysis, you should also use a range of software metrics to manage the code. One of the key metrics to use is a complexity metric such as McCabe's Cyclomatic Complexity. This is an algorithm that calculates the maximum number of paths through a function as a guide to its complexity. This helps identify the most complex regions of your code and acts as a guide to the number of tests required to achieve full coverage of the code.

Documentation Generator

The project is not over till the documentation is done. By using a standard architecture we can write documentation for the system architecture, component design and drivers which like our codebase can be reused over multiple projects with minimal adaptation. As we saw in Chapter 13, Test-Driven Development, the CMSIS Driver validation framework will provide a test report while a similar report can be produced by the component test framework discussed in Chapter 14, Software Components. We can also automate and maintain our API documentation by using a documentation generator such as Doxygen.

Exercise 15.1: Case Study

In this exercise, we will examine a simple project using the architectural model and a set of readymade components.

Open the pack installer and copy Ex 15.1.

This is a multiproject workspace that contains the final application project and test subprojects for each component.

Examine the structure of each project.

Build each project and examine its behavior.

This project acts as a guide to implementing many of the principles discussed in the previous sections.

Continuous Integration

The firmware for many small microcontroller projects is written solely within an IDE with the source code stored within a version control system. If you are working within a team of developers, it is possible for one developer to commit code that introduces a bug or that breaks another part of the project. To help overcome these issues it is possible to add a Continuous Integration (CI) server as part of the commit process. When new code is committed, the CI server will build the current project code base plus the newly committed code. If the project builds successfully, the CI server will then run a set of acceptance tests to ensure the executable works correctly. At the end of this process, an email report is sent to the developer who committed the code. A CI server is also useful to a sole developer as it can be to run periodic builds with more extensive test cases. One approach to the CI server testing process is to use a hardware board farm to execute all the test cases. This can be expensive and the hardware used may not fully reflect the final target hardware. Since our layered design isolates, the bulk of the application software from the details of the low-level hardware it is possible to run most of our tests on an MCU agnostic Cortex-M simulator. This also has the added advantage that we can run many simulations in parallel to achieve a high testing throughput.

Exercise 15.2: Cloud-Based Continuous Integration

The link below provides an example tutorial that creates a cloud-based CI server and repository. GitHub is used as a code repository while the CI server uses an Arm Virtual Hardware simulator running in a cloud computer hosted by Amazon Web Services. An Amazon Machine Image is used to provide an Elastic compute EC2 instance that is preconfigured with all the necessary tools and software (Fig. 15.8).

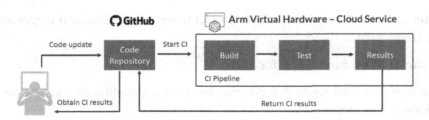

Figure 15.8
Cloud-based Continuous Integration server.

https://arm-software.github.io/AVH/main/examples/html/GetStarted.html

Over the lifetime of this book, this example may change. If you encounter a problem with the link check the examples download page for an update. In fact just as I have finished this book, Arm has announced a partnership with GitHub to integrate its virtual hardware as part of GitHub actions. This will create a CI/CD platform that allows developers to automate their build, test, and deployment pipeline.

CMSIS-Toolbox

To enable any CI server we need to be able to build our Microvision project from a Command Line Interface (CLI) be it Windows or Linux. As we are using the CMSIS-Pack system this can be difficult to migrate from the IDE to a CLI build. The CI server example is driven by a build system called CMSIS-Toolbox that provides the necessary tools to build projects and CMSIS packs from a CLI outside of an IDE (Table 15.11).

Table 15.11: Command line tools provided as part of CMSIS-toolbox

Tool	Description
cpackget	Download add and remove software packs
csolution	Create and manage complex applications. User source files and software packs
cbuild	Command-line build tools
packgen	Create a software pack from a CMake-based repository
packchk	Validate a software pack

An overview of the CMSIS-Toolbox workflow is shown in Fig. 15.9. The cbuild CLI tool uses a standard project format *.cprj that describes the project packs and software layers. The *.cprj format is also capable of supporting multiple projects. This project format can be exported from the Microvision IDE directly. Alternatively, for other IDE's, the project can be described using a set of template files which use YAML (Yet Another Markup Language). These files can then be translated to the *.cprj format using the csolution tool. The cbuild tool is then used to build the project. Cbuild will download any required software packs using cpackget to recreate the project environment and will then use CMake to build the image. Once the project has been built, we can run existing test scripts within a simulation environment such as the Arm Virtual Hardware models or on the real hardware.

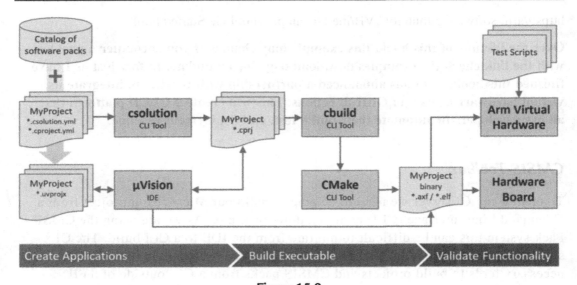

Figure 15.9
Example threads with execution utilization factors that show they cannot be scheduled.

As we can see in Table 15.12 the Microvision project contains a range of tools and components that we need to recreate for the command line build.

Table 15.12: Microvision IDE project items

Project Item	Description
CMSIS Packs	Device family packs with version number
Tool-chain	Compiler and linker with version number
Target Microcontroller	Target device and vendor with device-specific options
Software components	Components with configuration and version information
Project source code	The actual source code

CMSIS-Toolbox Project Format

The CMSIS project format is a text file with the extension *.cprj. We can export the CMSIS Toolbox project format from within Microvision using the project\export option.

In addition to creating a generic definition of the project and its various components, the *.cprj format is designed to support building firmware as a set of software layers.

Layers

The purpose of software layers is to separate different C modules within a project into components to form a hardware-dependent layer and other layers that are purely composed of application software (Fig. 15.10). This way our build system can compiler the higher-level layers and then be able to link these layers to different low-level layers that are hardware dependent. This extends our Layered Software Architecture so that the upper layers of a project can be linked to different hardware layers to target a range of different microcontrollers and hardware boards. This can be run on every commit ensuring that any changes to the upper software layers will work across all your hardware platforms.

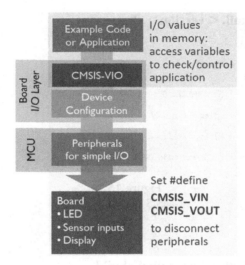

Figure 15.10
CMSIS project format defines a project as a hierarchy of project layers with common interfaces.

Within Microvision, it is possible to define software layers within the project and then assign source code modules and libraries to different layers. Each of the software layers has no further effect within the Microvision IDE but their definitions are exported in the CMSIS project file. Once the software layers are defined across different projects, the CMSIS build system can build projects that combine pure software only layers with different target hardware layers.

Exercise 15.3: Configuring CMSIS-Toolbox

In this example, we will set up the CMSIS Toolbox scripts and environment so that we can build Microvision projects from the command line. The CMSIS Toolbox scripts are designed to run in a Linux environment. On windows, we can add a Linux bash shell by installing Git for windows.

Download and install Git for Windows using the link below.

https://git-scm.com/download/win

We also need to install the CMSIS Toolbox scripts and environment. This is provided as a stand-alone asset on the CMSIS GitHub repository and is located in the releases page.

Open the link below and page down until you find the latest assets section (Fig. 15.11).

https://github.com/ARM-software/CMSIS_5/releases

Download the cbuild_install. < version>.sh.

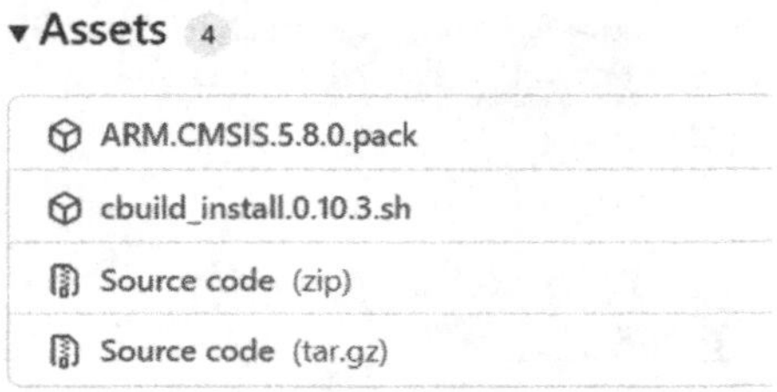

Figure 15.11
CMSIS cbuild assets.

On your c: drive create a directory called cbuild.

Copy cbuild_install. < version.sh into c:/cbuild.

Open a git bash shell and move to the cbuild directory.

Run the cbuild_install < version>.sh script (./cbuild_install. < version>.sh).

You will be prompted for a path to your compiler tools and a pack repository directory.

The scripts use the CMake build utility which can be downloaded and installed from the link below.

https://cmake.org/

Cmake uses a build tool called ninja which we also need to download and install.

https://github.com/ninja-build/ninja/releases

Download ninja-win.zip file (Fig. 15.12).

▾ Assets 5

⬡ ninja-linux.zip

⬡ ninja-mac.zip

⬡ ninja-win.zip

🗎 Source code (zip)

🗎 Source code (tar.gz)

Figure 15.12
Ninja assets.

Unzip the file and copy ninja.exe to C:\cbuild\cbuild\bin directory.

This finishes the tool install section, as we have set up a number of tools and paths close and restart the git bash shell to make sure all the paths are picked up.

When you restarted the bash shell the tool and pack repository environment variables must be recreated using the commands below.

```
cd c:/cbuild
source ./cbuild/etc/setup
```

Our build system is designed to support the pack system used in Microvision so we will need to create a local repository to store any packs that are required by the project.

In the c:/cbuild directory run cp_init.sh c:/cbuild/pack

This will create two subdirectories within the c:/cbuild/pack directory (Fig. 15.13).

Figure 15.13
Cbuild pack directory structure.

The .web directory includes a file index.pidx which provides a list of current software packs and URLs to their repositories.

In the pack installer copy Ex 15.3.

This is a simple blinky RTX project.

Open the project/manage project items dialog and select the Project info layer (Fig. 15.14).

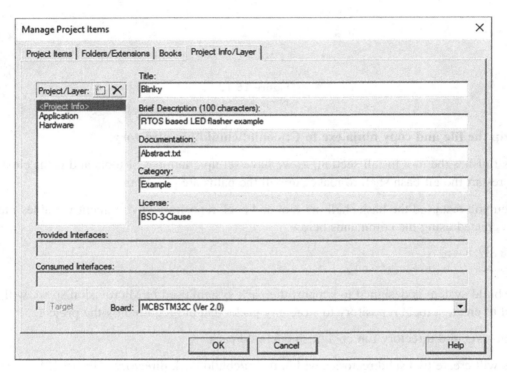

Figure 15.14
CMSIS layer definition within Microvision.

This dialog defines the available layers within the project. Each layer is user-definable along with documentation and license information.

In the Microvision project window, we can now open the local options for each module and assign it to a layer (Fig. 15.15).

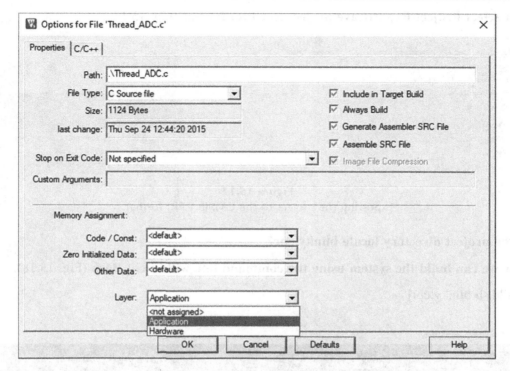

Figure 15.15
Adding a source code module to a layer.

Some RTE components may be available as source code or object libraries. The variant field allows us to select which version we want to use (Fig. 15.16).

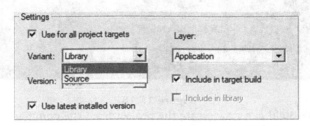

Figure 15.16
Adding a layer code module as source or library.

Now select **Project/Export/save project to CPRJ format** (Fig. 15.17).

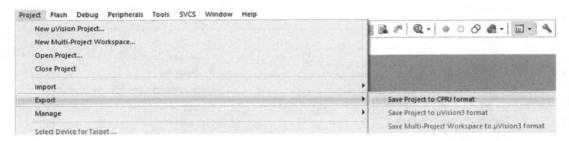

Figure 15.17
Exporting the project to the CMSIS CPRJ format.

In the project directory locate blinky.cprj.

Now we can build the system using the command line within Git Bash (Fig. 15.18).

Cbuild.sh blinky.cprj

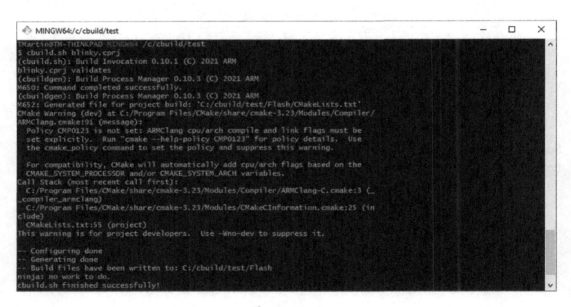

Figure 15.18
CMSIS Build can now create the project from the command line.

The first time the project is built any missing software packs will be downloaded into the local CMSIS-Build repository so that the code can be built successfully.

CHAPTER 16

The Road Ahead

This book is never finished, every time I update a new edition there is a whole truckload of new shiny hardware and software just over the horizon. In this closing section, I would like to briefly cover these new additions and changes which will start to arrive in the near future.

Keil Studio

Keil Studio is a next-generation development tool that will ultimately replace the Microvision IDE. At the time of writing Keil Studio is in an open beta. In its first release, Keil Studio is a cloud-based development environment that provides a browser-based IDE. Keil Studio provides a complete workflow including debugging, Git integration, and support for Continuous Integration testing. Later, iterations of Keil Studio will also include a desktop installation so you can work offline. The Keil Studio debugger can connect to existing evaluation boards and custom hardware but will also support debugging on simulated virtual hardware.

Arm Virtual Hardware

The examples in this book are based on an early simulator developed by Keil which models both the CPU and the microcontroller peripherals. With the explosion of Cortex-M-based devices, it became impossible to develop full simulation models for every family of Cortex-M microcontrollers. Instead, Arm now provides a set of core simulators for each Cortex-M processor. In addition to modeling the processor, a hardware interface layer provides abstracted virtual interfaces and streaming interfaces. Each of the virtual hardware models is available as an Amazon Cloud Image that runs on an Elastic Compute Cloud instance (Fig. 16.1).

The Designer's Guide to the Cortex-M Processor Family.
DOI: https://doi.org/10.1016/B978-0-323-85494-8.00009-7
© 2023 Elsevier Ltd. All rights reserved.

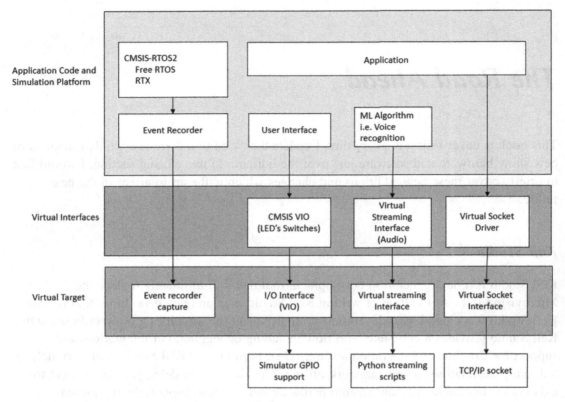

Figure 16.1
The Arm Virtual Hardware provides a CPU simulator with Abstracted Hardware interfaces.

Streaming Interfaces

The simulation model provides up to eight streaming interfaces. Each streaming interface can be configured to model different types of data such as audio or visual data along with data sources such as SPI, CAN, I2C, or USART. The streaming data is managed by a set of python scripts which are used to send and receive data to the model during run time. Each of the Arm virtual hardware models includes a user interface console that provides a simplified user interface including GPIO (switches and LED) and an LCD display. An additional BSD socket interface is used to connect the model to the real internet. This approach allows you to develop a single node for debugging and testing. Multiple instances of the simulation model can be run which allows you to test a design at scale without the need for hardware. Once the application has been developed in the cloud, it can be migrated to evaluation boards and then the final custom hardware.

IoT and Machine Learning

Currently, the big driving forces behind the development of new microcontrollers are support for the Internet of Things (IoT) and the adoption of Machine Learning (ML) algorithms into edge devices. This is reflected in the Armv8.0 and Armv8.1 processors with the addition of the TrustZone security peripheral and the Helium Vector processing extension. So while the hardware has evolved to support these technologies the next step is to simplify both IoT and ML development processes.

Project Centauri for the IoT

Within the IoT space, Arm has created Project Centauri that is intended to define foundational standards to make IoT applications portable across different MCU hardware and Cloud service providers. The Platform Security Architecture (PSA) project is also encompassed by project Centauri (Fig. 16.2). The PSA project defines a best practice security model with reference Trusted Firmware that provides security services (Cryptography, Storage, and Attestation) to the application software.

Figure 16.2
Project Centauri defines standards, best security practice, and an ecosystem for IoT device development.

The Keil MDK is Arm's reference toolchain and will provide the first support for each new standard and each release of new firmware. However, each reference implementation will be provided to the wider ecosystem in the form of an open-source software development kit (Open-IoT-SDK) which can then be incorporated into any third-party toolchain.

CMSIS v6

We have reviewed the current CMSIS specifications through this book. Though each standard will not fundamentally change the next generation, version 6 will see an evolution of how each standard is managed (Fig. 16.3).

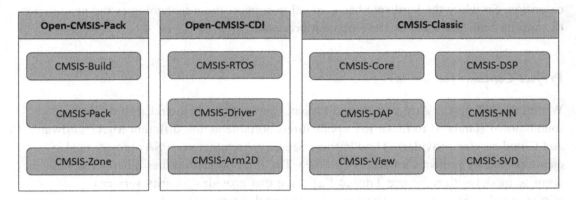

Figure 16.3
CMSIS V6 separates the CMSIS standards into three groups.

CMSIS-Classic

Like fine fizzy drink, the core standards will remain the same with updates to support new processors and hardware. The Event recorder debug agent currently within the Microvision debugger will be incorporated as a CMSIS standard called CMSIS-View.

Open-CMSIS-CDI

Open-CMSIS-CDI groups together standards that provide a Common Device Interface (CDI) for microcontrollers that are used as devices for the IoT. This group includes the existing CMSIS-RTOSv2 and CMSIS-Driver specifications. Over time, the Open-CMSIS-CDI will define a range of additional services that are accessed through a standardized set of APIs. Currently, there are a couple of new additions with more to come. The first of these new additions will include Arm-2D—a graphics driver, that like the CMSIS-Drivers, provides a standard set of API for third-party graphics libraries. Another common component that is increasingly a requirement for most projects and a necessity for IoT projects is a secure firmware update. Based on the MCU-Boot project this is a common code base that is easily adapted to a given microcontroller. Once installed, it is used to manage firmware updates where each new image is cryptographically encrypted and signed. Secure firmware update is also a requirement for cloud-based development where we need to repeatedly update a fleet of IoT devices.

Arm-2D

With the ubiquitous spread of smartphones end users of digital devices have become accustomed to sophisticated and intuitive graphical interfaces. While there are a number of graphics packages that allow a developer to design Graphical User Interfaces, there is a lack of low-level graphics drivers for specific devices. Often only one or two reference devices are supported by a given graphics package. This means it is often necessary to develop a complex graphics driver before the main design process can get underway. This is time-consuming and leads to lots of wasted effort "reinventing the wheel." The Arm-2D project is intended to standardize the low-level hardware driver interface in a similar fashion to the existing CMSIS-Driver specifications (Fig. 16.4).

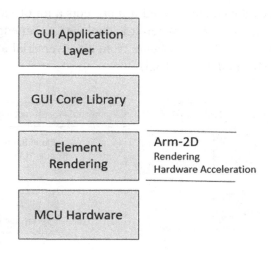

Figure 16.4
The Arm-2D driver standardizes the low-level hardware and rendering layer.

The Arm-2D project creates a new ecosystem between the hardware silicon vendors and the software GUI providers which will ultimately allow everyone to concentrate on developing their own core expertise.

Open-CMSIS-Pack

As we saw in the last few chapters of this book a component-based software architecture improves productivity and reliability of software designs. The Open-CMSIS-Pack project aims to create a standard for software component packaging and create a broader ecosystem to support pack-based workflows across multiple toolchains. Over time there will be additional support for software layers and extensions to the pack description format to improve component usability across the complete design workflow.

Machine Learning

The introduction of ML algorithms for use on a small microcontroller is a very broad topic and I could only briefly touch on it in this book. There is a considerable amount of research and development work currently underway to make the deployment of common ML algorithms as easy as possible for an Embedded Systems Developer. The CMSIS-NN standard provides optimized primitives that can be used with any ML framework however the current focus is on the Tensor flow lite micro framework and microTVM deployment framework.

Confidential AI

When an ML model is developed and trained it represents a lot of value in terms of time, effort, expertise, and confidential data. When it is deployed on an edge device, we face a new set of challenges to ensure that it is secure from an adversarial attack that may attempt to steal or "poison" the ML model (Fig. 16.5).

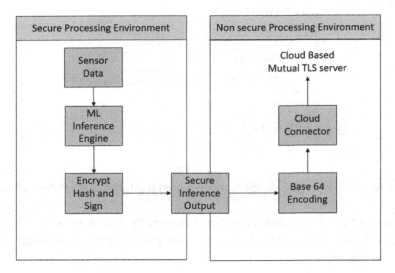

Figure 16.5
The ML inference can be located within the PSA secure processing environment. The inference data can be exported to the cloud service as a securely encrypted token.

To protect the ML model, we can place it within the PSA secure processing environment so that the ML inference becomes a service to the nonsecure application code. If the result of the inference is to be delivered directly to a cloud service it can be delivered as an encrypted, hashed, and signed token so the sensitive data is never exposed to the nonsecure application code.

Conclusion

By working through this book, you should now have a firm grasp of the fundamental concepts and techniques required to develop software for Cortex-M microcontrollers. However, this is just a starting point as the world of Cortex-M devices is constantly in flux with the addition of new processors, software stacks, and new technological drivers. There are many ways to keep up to date from in person seminars and trade shows, blogposts, webinars, and online training. Enjoy the journey!

Appendix A

This appendix lists some further resources that are worth investigating once you have worked through this book.

Chapter 1

Accompanying videos and webinars

There are a set of companion videos which are available by registering using the link below.

https://www2.hitex.com/cortexm

A YouTube channel of past webinar recordings is also available

https://www.hitex.com/company/news/webinars

https://www.youtube.com/user/GermanyHitex

Keil device database

The Keil website maintains a list of supported devices. If you are selecting a device always check here to see the range of software support.

https://www.keil.com/dd2/

Cortex-M Wikipedia page

A good overview of processor features and silicon manufacturers

https://en.wikipedia.org/wiki/ARM_Cortex-M

PSA certified

Introduction the Platform Security Architecture. The website also has a gallery of certified devices so is a good starting point for device selection.

https://www.psacertified.org/

Training companies

Training for Cortex-M processors and associated software is available from these companies.

https://hitex.co.uk/training/

https://www.feabhas.com/

https://www.doulos.com/

Chapter 2

Books

A couple of useful books for programming in C and code development in general

Kernigan and Richie

Clean Code A Handbook of Agile Software Craftsmanship ISBN-10 0132350882

Robert C. Martin

Keil website resources

Useful resources within the Keil website

MDK online documentation

https://www.keil.com/support/man_arm.htm

A collection of Microvision IDE and Cortex-M Application Notes

https://www.keil.com/appnotes/list/arm.htm

STM32 Cube MX Download

Download link for the STM32 Cube MX configuration tool.

https://www.st.com/en/development-tools/stm32cubemx.html

Cortex-M development tools

A range of commercial development tools. You can usually find free tools on the Silicon vendor website.

http://www.iar.com

http://www.keil.com

http://www.tasking.com

Online development tools

There are a couple of online tools that are designed to support Cortex-M processors.

The Mbed OS project allows you to rapidly prototype code for a supported evaluation board.

https://os.mbed.com/

Keil Studio is the next-generation development tool and uses a browser based IDE. Currently in Open Beta.

https://www.keil.arm.com/

Chapter 3

Books

The Definitive Guide to the Cortex-M3, Joseph Yui

A deep guide to the Cortex-M3 processor written by a member of the Cortex-M processor team. Joseph has also written several other books covering different Cortex-M processors.

Insiders Guide to the STM32

You can download a free ebook that provides an introduction to a real microcontroller family. A bit dated now but still a valid introduction.

https://www.hitex.com/fileadmin/documents/tools/dev_tools/dt_protected/insiders-guides/stm32/isg-stm32-v18d-scr.pdf

Arm documentation

The Arm documentation for each processor is provided in two manuals. An architectural reference manual and a processor technical reference manual. You can access the documentation for each processor from the link below.

Architecture reference manual

Armv6-M

https://developer.arm.com/documentation/ddi0419/latest/

Armv7-M

https://developer.arm.com/documentation/ddi0403/latest/

Technical reference manual

Cortex-M3

https://www.arm.com/products/silicon-ip-cpu/cortex-m/cortex-m3

Cortex-M0 +

https://www.arm.com/products/silicon-ip-cpu/cortex-m/cortex-m0-plus

Cortex-M0

https://www.arm.com/products/silicon-ip-cpu/cortex-m/cortex-m0

Chapter 4

Each of the CMSIS specifications are available online and the current version of CMSIS can be downloaded as a repository.

CMSIS online specification

https://www.keil.com/pack/doc/CMSIS/General/html/index.html

CMSIS Github repository

https://github.com/ARM-software/CMSIS_5/

MISRA-C

You can purchase the latest MISRA-C specification in book or pdf format directly from MISRA.

http://www.misra.org.uk

Chapter 5

Armv7 Memory protection unit

https://developer.arm.com/documentation/100699/0100/Introduction?lang = en

Changes for the Armv8 Memory protection Unit

https://developer.arm.com/documentation/100699/0100/Introduction/MPU-programmers−model-changes-for-the-ARMv8-M-architecture?lang = en

CMSIS Core MPU support

Cortex-M0 + , M3,M4,M7

https://www.keil.com/pack/doc/CMSIS/Core/html/group__mpu__functions.html

Cortex-M23,M33,M55,M85

https://www.keil.com/pack/doc/CMSIS/Core/html/group__mpu8__functions.html

Chapter 6

Technical reference manual

Cortex-M7

https://www.arm.com/products/silicon-ip-cpu/cortex-m/cortex-m7

Chapter 7

Architecture reference manual

Armv8-M and Armv8.1

https://developer.arm.com/documentation/100688/latest/

Technical reference manuals

Cortex-M23

https://www.arm.com/products/silicon-ip-cpu/cortex-m/cortex-m23

Cortex-M33

https://www.arm.com/products/silicon-ip-cpu/cortex-m/cortex-m33

Cortex-M55

https://www.arm.com/products/silicon-ip-cpu/cortex-m/cortex-m55

Cortex-M85

https://www.arm.com/products/silicon-ip-cpu/cortex-m/cortex-m85

Chapter 8

Microvision debugger manual

https://developer.arm.com/documentation/101407/0537/Debugging

Ulink debug adapters

https://www2.keil.com/mdk5/ulink

Ulink support notes

https://www.keil.com/support/index/ulink.htm

Chapter 9

Books

Digital Signal Processing: A Practical Guide for Engineers and Scientists

Understanding Digital Signal Processing

Technical Reference Manual

Cortex-M4

https://www.arm.com/products/silicon-ip-cpu/cortex-m/cortex-m4

Tools

ASN filter designer

An intuitive graphical filter designer that can be used to generate C code based on the CMSIS-DSP libraries.

https://www.advsolned.com/asn_filter_designer_digital_filter_software/

Chapter 10

Books

Real Time Concepts for Embedded Systems

Allen B. Downey

RTOS

There are a large number of RTOS available. These are (probably) the most widely used.

Chapter 11

CMSIS Zone specification

https://www.keil.com/pack/doc/CMSIS/Zone/html/index.html

CMSIS Zone utility repository

https://github.com/ARM-software/CMSIS-Zone

Functional Safety

In addition, the RTX5 RTOS Keil provides a fully certified Functional Safety Run Time System.

Chapter 12

CMSIS Driver specification

https://www.keil.com/pack/doc/CMSIS/Driver/html/index.html

Driver template repository

https://github.com/ARM-software/CMSIS-Driver

Driver validation repository

https://github.com/ARM-software/CMSIS-Driver_Validation

Chapter 13

Books

James W. Grenning

Test Frameworks

There are a number of C and C++ test frameworks. The most popular are listed below.

Unit test tools

Tessy

Tessy is a commercial unit test tool that has been developed for use with constrainedmbedded devices.

https://www.hitex.com/tools-components/test-tools/dynamic-module/unit-test

Chapter 14

The CMSIS Build and CMSIS Pack specifications are now managed as a Linaro project under the open-CMSIS-Pack project.

https://Linaro.org

Open-CMSIS-Pack

https://www.open-cmsis-pack.org/

CMSIS-Pack specification

https://open-cmsis-pack.github.io/Open-CMSIS-Pack-Spec/main/html/index.html

Chapter 15

Books

Clean Architecture

Software Design Methods for Concurrent and Real Time Systems

This may be out of print now. A good section on RMA which provides the examples used in this book.

Rate Monotonic Analysis for Real-Time Systems

https://resources.sei.cmu.edu/asset_files/TechnicalReport/1991_005_001_15923.pdf#

Tools

CMSIS-Toolbox

The CMSIS build and CMSIS Pack specifications are now managed as a Linaro project under the open-CMSIS-Pack project.

https://Linaro.org

CMSIS-Toolbox specification

https://www.keil.com/pack/doc/CMSIS/Build/html/index.html

CMSIS-Toolbox repository

https://github.com/Open-CMSIS-Pack/devtools/tree/main/tools

Chapter 16

Keil Studio

https://www.keil.arm.com/

Project Centuari

https://www.arm.com/solutions/iot/project-centauri

Open-CMSIS-Pack

https://www.open-cmsis-pack.org/

Open-CMSIS-CDI

https://www.open-cmsis-cdi.org/

Open-IoT-SDK

https://github.com/ARM-software/open-iot-sdk

Arm-2D

https://github.com/ARM-software/EndpointAI/tree/master/Kernels/Research/Arm-2D

Confidential AI

https://static.linaro.org/assets/ConfidentialAI-LinaroWhitePaper.pdf

Arm Virtual Hardware

https://avh.arm.com/

Index

Note: Page numbers followed by "*f*" and "*t*" refer to figures and tables, respectively.